AF522485

Seed Science and Technology

Seed Science and Technology

D. Yashwanth kumar

RANDOM PUBLICATIONS
NEW DELHI (INDIA)

Seed Science and Technology

ISBN 978-93-5111-321-8

Published in 2014 in India by

RANDOM PUBLICATIONS

4376-A/4B, Gali Murari Lal, Ansari Road
New Delhi-110 002
Phone : +91-11-43580356, +91-11-23289044
e-mail: randomexports@gmail.com, sales@randompublications.com, info@randompublications.com

Reprinted 2023

Type Setting by : Keystoneprintads, Delhi-110051
Digitally Printed at : Replika Press Pvt. Ltd.

Preface

A seed is a small embryonic plant enclosed in a covering called the seed coat, usually with some stored food. It is the product of the ripened ovule of gymnosperm and angiosperm plants which occurs after fertilization and some growth within the mother plant.

The formation of the seed completes the process of reproduction in seed plants (started with the development of flowers and pollination), with the embryo developed from the zygote and the seed coat from the integuments of the ovule. All seeds are different size, shape and colour.

Seed is a mature integumented megasporangium or mature ovule consisting of embryonic plants together whit store food material covered by a protective coat (mega sporangium i.e. female gametophyte –the pistils are the female reproductive organs called megasporophylly in the flowering plants) the ovary of the carpel contains ovules (megasprorangia).

Seeds are the delivery systems for agricultural biotechnology. High quality seed leads to excellent seedling performance in the field. It is the ultimative basis of successful companies that breed crop plants for seed production. Seed quality is a complex trait that is determined by interactions between multiple genetic factors and environmental conditions.

Modern approches to improve seed quality therefore combine classical genetics, plant molecular biology and a variety of seed technologies. These "seed biotechnologies" enhance physiological quality, vigor and synchronity to establish a crop in the field under diverse environmental conditions. Seed technology is the science dealing with the methods of improving physical and genetical characteristics of seed.

It involves such activities as variety development, evolution and release of varieties, seed production, seed processing, seed certification and storage. Thus seed technology is interdisciplinary science which includes broad range of subjects. It also includes seed quality control, seed marketing, seed distribution and research on above aspects of seed.

With a compendious presentation of the subject, readers associated with agricultural science would find this book highly useful.

I would like to thank my team for standing beside me throughout my career and writing this book. My special thanks go to "Random Publications" who have published the book.

— D. Yashwanth kumar

Contents

1

Seed Quality

INTRODUCTION

No single test is sufficient to distinguish all the attributes that contribute to the physiological quality of a given seed lot. From a growers view point, a high quality lot is defined by results. A high quality lot is one that gives fast, uniform, high percentage emergence, resulting in healthy, near 100 per cent stand under the conditions in the growers field or greenhouse. Unfortunately, defining quality by historical results is not enough a predictive test of field performance is needed to provide growers a reasonable assurance of obtaining given good cultural practices and typical (or at least not greatly abnormal) weather, insect or disease conditions. The stand potential needs to be assessed both before and after coating. The goal for a commercial seed or coating company is to develop a fast, reliable series of tests that will predict accurately the stand establishment of a seed lot under a range of possible seed conditions, including stress.

The Federal Seed Act and various state laws stipulate that seed germination be labelled according to standardised tests established by the Association of Official Seed Analysts (Copeland 1981). For example, lettuce must be labelled with the percentage of germinated seeds counted after 7 days at a constant 2°C in the light. Celery is tested under the same conditions but the percentage is counted after 21 days.

The official germination figure is extremely useful, but additional tests are needed to determine if the seed lot is suitable for coating. This single laboratory test does not provide sufficient information to determine if the seed lot will retain its germination pattern (percentage, speed, uniformity, reaction to high or low temperatures and shelf life) after coating or ho it may perform under actual conditions in growers fields or greenhouses. Additional tests can include variations on the standard lab test, thermogradient table tests and greenhouse tests.

LABORATORY TESTS

Variations on the standard germination test increase the stress level, usually using temperature and/or darkness. Lettuce might be tested at 15C or 25C or alternating at 20/30C, perhaps in the absence of light. The lettuce variety Etna was rejected for coating based on such a test, while Domingos 43 was accepted (Table). Etna failed to meet the coating companies standard of > 93 per cent germination with a < 3 per cent abnormal seedlings at 20C and was not coated.

Time	**Percentage**			
	Raw Seed at 20C	**Raw Seed @ 20/30C Coated Seed**	**Coated Seed at 20C**	**Abnormal Seeds**
Etna lot 1292				
1 day	89	90	–	* % abnormal seeds
7 days	92 + 7*	92 + 2*	–	* % abnormal seeds
Domingos 43 (lot 5980)				
1 day	100	68	99	* % abnormal seeds
7 days	99 + 1*	99	98 + 1*	* % abnormal seeds

This standard is empirical, not theoretical, and was chosen based on years of experience with lots that give satisfactory or unsatisfactory field results. The standard is conservative and if no other lot is available,, the seed might still be coated provided that the grower knows, and explicitly accepts the limitations of the seed and the risk of a weak stand.

In another test, the onion variety Sweet Perfection tested as an excellent raw seed lot (Table). Germination patterns for onions differ from lettuce; onion seed lots with total potential 80 per cent to 85 per cent frequently are encountered and accepted by growers. Lettuce seed lot germination < 90 per cent are strongly resisted by commercial growers, who prefer seed with a germination of > 95 per cent.

Table. Germination per cent of Raw and Coated "Sweet Perfection" Onion Seed. Lot 57412, after Two Time Periods in the Lab at 20C and Four Time Periods in the Greenhouse with no Temperature Control.

Time	**Percentage**		
	Raw Seed at 20C	**Coated Seed**	**Abnormal Seeds**
	Laboratory at 20C		
3 days	74	83	* % abnormal seeds
7 days	94 + 2*	95 + 2*	* % abnormal seeds
	Greenhouse, no Temperature Control		
6 days	–	18	* % abnormal seeds
8 days	–	82	* % abnormal seeds
10 days	–	95	* % abnormal seeds
14 days	–	95 + 2*	* % abnormal seeds

Thermogradient Table Tests

A Thermogradient table is a uniform sheet of metal with a cold water bath along one side and a hot water bath along the other. This creates a continuous temperature gradient, typically with a range of 20C. *e.g.*, 12C to 32C or 16C to 36C. ten individual tests can be run on blotters across the width of the table, at 2C increments.

This type of test determines the maximum and minimum temperatures at which an individual lot will germinate, as well as its ideal, or preferred temperature range for fastest germination and growth The different tests are used in combination. For example the lettuce variety Domingos 43 showed heat dormancy at 20/30C, with reduced germination after 24h. The severity of the heat dormancy was tested on the thermogradient table. The test indicated a potential germination of >87 per cent in 24h and 96 per cent in 48 h in temperatures up to 23C. This is acceptable for this variety at the usual soil temperatures when it is sown in Southern California and Arizona.

Greenhouse Tests

Greenhouse tests give a stand establishment estimate under real conditions. These tests also can be prolonged to give a subjective estimate of seedling quality. The greenhouse test can, for example, identify seedlings with necrotic tissue on the cotyledons that might not be apparent in other tests. The percentage of seedlings with significant necrosis is deducted from the stand establishment potential of the lot. Seed quality testing is an integral part of seed coating.

Without it the performance of coated seed would be erratic and unreliable for commercial growers. Precision sowing without dependable, precise emergence wastes the advantages of coated seed. For example, the Etna probably would have a field or greenhouse stand well below 90 per cent. Depending on circumstances, this might be acceptable, but it would fail to meet the expectations of many growers.

No single test is adequate to accept a raw seed lot for coating, but several in combination can identify acceptable and unacceptable lots with a high level of accuracy. Prudent coating companies repeat all or part of their battery of tests on the final coated product to ensure that the commercial product meets the expectations of the pre-coating tests. Testing methods, which differ among companies, have been developed to evaluate coatability of a raw seed lot and to check the coated lot's quality.

This intensified testing for subtle quality distinctions has improved understanding of quality and how to measure it. This helps both seed and coating companies to improve quality with techniques employed in seed growing, harvesting, conditioning storing and coating. When it is possible to measure the quality differences it is also possible to develop and choose superior techniques. The result has been a gradual increase in quality standards throughout the seed industry. Thus, "The Stimulus to Seed Quality" in the title of this article. For example 15 years ago, lettuce could be sold at 85 per cent germination and 90 per cent was considered good. Today, 90 per cent is considered a weak stand and growers seek 95 per cent or better. Commercial growers have higher expectations of physiological performance, mirroring their higher expectations of genetic performance by new varieties of each crop species.

This developing ability to measure and deliver higher seed quality reinforced by the markets higher expectations, is the major long term benefit of seed coating. The particular economic benefits of precise placement, reduced thinning costs, ease of handling etc., motivate growers to change from raw to coated seed and pay the associated research and production costs. Meanwhile, the on going pressure to maintain and exceed existing quality standards benefits all participants.

QUALITY OF SEED COLOUR TESTING

Tests of the quality and other characteristics of seed need to be made at several stages during its progress from the parent tree to the seed bed. Tests for maturity and soundness of seed before and during collection in the forest have been described on pp. 34–36 and are usually associated with surveys of the abundance of the seed crop; they are intended to ensure that both the quantity and the quality of seed justify the effort and cost of collection. Several tests may be required at the seed processing depot, after extraction and cleaning and before the seeds are despatched to the nurseries or placed in storage.

Testing of moisture content is essential for many species before they are stored for a long time, but is usually unnecessary if the seed is to be sown immediately in the nursery; on the other hand, partial drying may be needed before despatch of seed to a distant forest station, to reduce the risk of deterioration in transit, and any deliberate drying operation must be accompanied by the testing of moisture content. Testing of germination or viability should be repeated at the end of the storage period if this is more than a few months and, in the case of long-term storage for conservation of genetic resources, testing should be done at intervals throughout the storage period.

Efficiency and success in raising plants in the nursery and in their subsequent establishment in forest plantations depend to a great extent on the quality of the seeds used. It follows that foresters, nurserymen, seed dealers and others need accurate estimates of the quality of the seeds in which they deal or which form the basis of their afforestation projects (Turnbull 1975 d).

This is particularly important when seeds are bought and sold, or move between countries in international trade. In practice, the number varies considerably from year to year and from seed lot to seed lot, so that it is necessary for the forester to have an accurate estimate of germination for every individual seed lot which he receives, if he is to fulfil planting targets while at the same time avoiding the waste of valuable seed caused by sowing too much. Similarly the nurseryman must have a good estimate of germination if he is to adjust sowing rates to produce the optimum spacing of seedlings in the seedbed.

Justice has defined the following objectives for developing rules for seed testing:

- To provide methods by which the quality of seed samples can be determined accurately;
- To prescribe methods by which seed analysts working in different laboratories in different countries throughout the world can obtain uniform results;
- To relate the laboratory results, insofar as possible, to planting value,
- To complete the tests within the shortest period of time possible, commensurate with the above-mentioned objectives; and
- To perform the tests in the most economical manner.

For the practising forester, the most important object of seed testing is the provision of an accurate estimate of the capacity of a given seedlot to produce healthy, vigourous plants suitable for field planting. In the present context, seed "quality" refers to the physiological vigour of the seed rather than its genetic quality. The essence of good seed testing is the application of reliable standard methods of examination to ensure that uniform and reproducible results are obtained (Turnbull 1975 d).

Standardisation has been greatly facilitated through the adoption by a number of countries of the International Rules for Seed Testing formulated by the International Seed Testing Association (ISTA). ISTA was founded in 1921, drew up its first set of Rules in 1931 and made major revisions of the Rules in 1953, 1966 and 1976. In the early stages ISTA's primary concern was for agricultural seeds, but trees and shrubs have gradually assumed greater importance and the 1976 Rules (ISTA 1976) give (firm) "prescriptions" or (more tentative) "suggestions" as to testing methods appropriate for 61 different genera of these, compared with 26 in the 1953 Rules. The Rules are published in English, French and German.

Although tree species and genera from the tropics and southern hemisphere (notably eucalypts) have begun to figure in the ISTA lists, the balance is still overwhelmingly weighted towards north temperate species. Thus important species such as Tectona grandis, Pinus patula, P. oocarpa and P. kesiya are listed, but only in the list of "suggested" (non-prescriptive) seed testing methods, while a recent survey has shown that 20 other important tropical species are not included at all; they include Cupressus lusitanica, Gmelina arborea, Cordia alliodoraand the genera Albizzia, Araucaria, Casuarina, Swietenia, Terminalia and Triplochiton. Their omission simply reflects the lack of reliable information from research on the best methods of testing these species and genera. ISTA testing methods involve the maintenance of controlled laboratory conditions and the use of some items of equipment which are relatively expensive.

They are therefore well suited to large, well-equipped seed laboratories but are impracticable for small forest stations which have no laboratory and must carry out germination tests in the nursery or the office. Lack of equipment is no reason to omit seed testing altogether. For example simple germination tests carried out in the nursery can produce results which are perfectly satisfactory for local use. But precise details of method should always accompany results whenever this is not standardised, so that the seed user can interpret the results in relation to his own nursery conditions.

An up-to-date guide to the range of equipment available for seed testing is contained in the "Survey of Equipment and Supplies" published by ISTA (ISTA 1982). It lists equipment by type, use, model specification and supplier. The earlier directory, "Equipment and supplies for collecting, processing, storing and testing forest tree seed" (Bonner 1977) includes seed testing equipment and gives the addresses of user laboratories as well as suppliers. Van der Burg *et al.* (1983) give a description of how a seed testing station in tropical or subtropical areas could be established.

Two alternatives are described: Seedlab 2000, that can test about 2000 samples per year and Seedlab 5000 that can test at least 5000 samples per year. Directives and general considerations concerning the staffing, the organisation of the work, the lay-out of the building, and the equipment needed are given.

Equipment which in the experience of the authors has been found suitable for the work is recommended and detailed descriptions and company addresses are mentioned. Some equipment that is not commercially available is described and plans for construction are included. A list of books and journals for a basic seed testing library is appended. The tests which may be required are purity, authenticity, seed weight, germination, indirect testing of viability, moisture content, and seed health and damage.

SEED QUALITY AND SEED TREATMENTS

A number of factors affect the quality of seed. Seed-borne fungi, insect feeding damage, poor seed storage (temperature and humidity too high), and damaged seed coats can reduce the quality of seed.

Seed quality may be reduced due to infection by fungi when harvest of the seed crop is delayed several weeks or when seed diseases occur due to wet conditions. Infected seed can be detected at harvest as discoloured, shriveled, or moldy in appearance.

Insect feeding that may occur in the field prior to harvest or in storage may affect seed quality. Germination of seed lots may be reduced by the insect partially eating the seed or opening the seed coat to fungal infections occurring at a later date.

The seed coat is the primary protective layer surrounding the seed. Seed coat damage can result from rough handling during harvest, and cleaning or planting operations. Small cracks in the seed coat increase the chances of seed-rot by permitting water-soluble nutrients to escape into the soil which activate soilborne fungi that quickly enter the damaged seed.

Grain intended for seed should be handled as little as possible, dried using low temperature drying, and stored at moisture levels below 13 percent. High humidity and temperature during storage promotes growth of fungi and insect damage. Seed lots should be closely monitored for mold growth and insect activity during storage.

Seed treatments can play an important role in achieving uniform seedling emergence under certain conditions. The selective use of seed treatments can protect seeds or seedlings from early season disease and insect pests affecting crop emergence and growth. A few seed treatments may also be used to enhance crop performance during the growing season. Treatments containing Rhizobium inoculant have commonly been used to enhance the nitrogen fixing capability of legume crops.

Increased germination can be observed when seed treatments containing fungicides are used when poor germination results from a fungal infection. Fungicide are available to protect seed and seedlings from many seed-borne or soil-borne pathogens. Seed treatments are most beneficial for seeds infected prior to planting or when cool, wet soil conditions exist at planting which results in delayed germination.

Seed treatments containing insecticides can serve to protect seed and seedlings from certain insect pests found in the soil. Cool, wet, seed beds resulting in delayed germination or plowing down green manure crops may warrant including an insecticide seed treatment.

Seed treatments should not be considered a cure-all for the selection of poor seed lots. Seed treatments will not increase poor germination due to excessive mechanical damage, poor storage conditions, genetic differences in variety, or other damage.

The effect of a seed treatment on plant stands will depend upon seed quality as well as field conditions at planting. Extensive research has shown seed treatments can increase stands from poor quality seed or when planting conditions are less than optimum. However, seed treatments do not always increase yield. Generally, seed treatments result in increased yield by preventing death of seedlings causing frequent skips within crop rows.

Grain producers should realise that any one seed treatment will not control all diseases or insects that may attack seeds and seedlings. Protection is also short lived, generally only lasting as long as it takes for the crop to emerge.

Before selecting a seed treatment it is essential for the grower to consider potential problems associated with the seed and the history of problems associated with each field. Each planting situation is different requiring consideration of seed quality; planting rate; tillage and seed bed preparation; soil moisture and temperature; time of planting; and the likelihood of rapid emergence.

GERMINATION TESTING AND SEED QUALITY

The purpose of laboratory testing of seed germination is to assess seed quality or viability and to predict performance of the seed and seedling in the field. Seed processed for sale must be tested by a qualified laboratory under the Association of Official Seed Analysts Rules for testing seeds. Several different kinds of testing are available depending on the type of seed to be tested, the conditions of the test, and the potential uses of the seed.

The most common tests are the warm germination test, cold germination test, accelerated aging test, and the tetrazolium test. Each test is designed to evaluate various qualities of the seed. Factors that can affect the performance of seed in germination tests include; diseased seed, old seed, mechanically damaged seed, seed stored under high moisture, and excessive heating of seed during storage or drying.

In most cases a seed treatment will improve germination of seed only if the poor quality is due to seed-borne disease.

The most common test is a warm germination test because it is required by seed laws to appear on the label. The percentage of germinating seed in a warm germination test must be printed on the label of the seed if it is to be sold as seed.

Germination is defined as: "the emergence and development from the seed embryo of those essential structures which are indicative of the ability to produce a normal plant under favorable conditions."

The warm germination test reflects the stand producing potential of a seed lot under ideal planting conditions. Usually 400 seed from each seed lot are placed under moist conditions on blotters, rolled towels, or sand and maintained at 77 degrees F for about seven days. At the end of this period the seedlings are categorized as normal, abnormal, or diseased, and dead or hard seeds. The percentage germination is calculated from the number of normal seedlings from the total number of seeds evaluated.

The cold germination test is designed to measure the ability of seeds to germinate under high soil-moisture content and low soil temperature. This vigor test simulates early season adverse field conditions and usually represents the lowest germination that would be expected from a seed lot planted under such conditions.

Actual field germination would normally fall between the cold test result and that of the warm germination test. Seeds are planted in a sand-soil mix at high moisture content and maintained at 50 degrees F for seven days. The test is then placed at 77 degrees F for four days. The percentage of healthy seedlings that emerge at least one inch above the soil is reported.

The accelerated aging test estimates the carryover potential of a seed lot in warehouse storage. The seeds are exposed to high temperatures and high relative humidity for short periods of time that cause seed deterioration. Seed are suspended over water in a chamber for 72 hours (wheat and soybeans) or 96 hours (corn) then tested in a standard warm germination test. This test only would be used on seed whose longevity was in question.

The tetrazolium test is a "quick test" for seed lot viability. It is useful when an approximate germination percentage is needed immediately. Seeds are soaked overnight in water then treated with tetrazolium to give an indication of viable, abnormal, and dead seeds in the seed lot. This test will not detect seed-borne disease, thus is limited in its ability to estimate seed quality. This test is highly reliable for determining viable seed of corn, wheat, oats, barley, and other grasses.

If growers wish to use bin-run seed and has not had a germination test conducted by a competent laboratory, they can get an approximate germination test using the following procedure in their own home.

Place two paper towels in the bottom of a dish or tray, one on top of the other; wet the towels thoroughly and tilt the tray up on one end so that excess water runs off the tray. Select a random sample of 100 seeds from the seed lot and place them in between the moist paper towels. Put the tray in a plastic bag and tie the end shut to prevent the towels from drying out.

Place the tray in a location of diffuse, not direct light, such as a north window. The location should be warm enough for good plant growth. An ideal location is with well growing house plants. After five to seven days, open the plastic bag and count the number of germinated seed with intact tap roots and shoots. Do not count moldy seed or diseased seedlings. Testing 400 seeds in this way will give a good indication of the germination percentage.

Choose the Right Fungicide for the Problem

The basic component of most seed-treatment products is a fungicide. Insecticides are contained in some products to give insect protection when such control is desirable. The grower should select a seed treatment product that will control the most likely pest to be encountered within a given field. It is important to note that not all fungi are alike. One fungicide seed treatment may be very effective against one fungus but will have no effect on another. Rates of active ingredient used also affect a seed treatment performance on different disease problems.

Fungicides have fungicidal activity, that actually kills the fungus, or fungistatic activity, that slows the growth of the fungus to prevent extensive damage. Fungicides used in seed treatment are of three types: protectants that generally prevent infection of seeds and seedlings by soilborne fungi, surface disinfestants that kill fungi and fungal spores on the surface of the seed coat (seed-borne fungi) and "systemic" disinfectants that kill fungi already established within the seed (seed-borne fungi). Insecticides used in seed treatments act as surface protectants killing insects as they feed on the seed or emerging seedling.

The marketing of seed treatments by manufacturers may result in the same chemical being sold under several brand names and by several different companies. All chemicals have three names on the packaging; chemical name, common name, and trade or brand name. The chemical name represents the molecular structure of the active ingredient and is often long and complex. The common name is used by the manufacturer in lieu of using the longer chemical name. The trade or brand name is used by the manufacturer in product marketing.

The chemical name (R,S)-2-[(2,6-dimethlyphenly)-methoxyacetylamino]-proprionic acid methyl ester has the common name mefenoxam and is marketed under the trade or brand name Apron XL LS.

Using mixtures of different seed treatment fungicides or insecticides is a common practice. Mixing two or more fungicides or insecticides results in a product that will control a broader spectrum of seed pests.

Applying Seed Treatments

Seed treatments come in a variety of formulations: dry flowables (DF), liquid flowables (LF), true liquids (TL), emulsifiable concentrates (EC), dusts (D), wettable powders (WP) and others. These materials may be used in slurry and mist type seed treaters upon mixing with water as specified in instructions on the product label.

Special adhesives are incorporated into the dust formulations to adhere pesticide particles to the seed surface.

Commercial Treatment

Commercial seed treatment is often desirable due to the specialized equipment required to properly apply treatments or to treat large volumes of seed. The primary concern of the commercial treater is equipment calibration to ensure delivering the proper amount of active ingredient on the seed. This has become especially important with more modern fungicides that require only very small amounts of material (less than 0.5 oz) per hundred weight of seed. Detailed information on the process of treater equipment calibration can be obtained from the equipment manufacturer. Seed treatment manufacturer labels provide information related to mixing formulations and mixing procedures.

Information needed to properly calibrate a seed treater includes:

1. The correct labeled rate of the seed treatment material that will be used,
2. How many pounds of seed is dumped each time the weigh pan arm dumps,
3. The size of the cup being used in the treater.

This information will allow you to calculate the correct application rate for the seed and treatment material used. A new generation of seed treater is becoming available that offers precision distribution of treatment chemicals on individual seed through atomization. These advanced controlled droplet application (CDA) machines offer complete seed coverage with lower water volume, shorter drying time, and continuous seed flow.

These machines also reduce chemical exposure to employees and are capable of applying current and future products that may include both chemicals and biologicals.

On-farm Application

Many seed treatment materials are available for on-farm use. These are known as hopper-box or planter-box treatments. Most hopper-box treatments are dry treatments formulated with talc or graphite which adheres the treatment chemical to the seed. Liquid hopper-box treatments are available as a fast-drying formulation. Good seed coverage is required for maximum benefit from any seed treatment formulations.

Obtaining thorough seed coverage can be difficult when attempting to treat seed in the field at planting time. Most seed treatment companies recommend that the planter box be filled only half full with seed, and half the required amount of treatment product added. Mix the product into the seed volume with a stick. Then add the remainder of seed to the planter box and add the rest of the required amount of product, again mixing thoroughly. Good results are also obtained by premixing seed with the product in a suitable container, then pouring treated seed into the planter box.

On-the-farm treaters are available that apply liquid or dry formulations to seed as it passes through an auger from the transport bin or truck to the planter boxes. These treaters are a very convenient way to apply seed treatment onto bulk seed. For additional information on on-the-farm treaters, contact a representative of the company selling planter-box treatments.

Labeling Treated Seed

Once seed is treated it can no longer be used for food feed or other purposes. Under the federal and state seed laws, the following information is required to be shown on the label: a word or statement in type no smaller than eight points indicating that the seeds have been treated; the commonly accepted, chemical (generic) name of the applied substance including the application rate, and a caution statement if the substance used in such a treatment has an amount remaining with the seed that is harmful to humans or other vertebrate animals. Seed treated with highly-toxic substances shall be labeled in red ink to show a statement such as "Poison Treated." In addition, the label should show a representation of skull and crossbones. Seed treated with less toxic substances shall be labeled to show a caution statement in no smaller than eight point type such as "Do not use for food, feed, or oil."

Colouring Seed Treated with Poisonous Substances

All treated seed must be coloured an unnatural colour to distinguish it from untreated grain and prevent inadvertent use as food for man, feed for animals, or for oil purposes. All treated seed must be coloured with a dye, colour coat, or colour film which must cause no apparent injury to seed germination or danger to personnel processing or using the seed.

Dyes are designed to stain the seed with colour. Colour coat pigments cover the seed with colour, and colour films are made from a polymer that creates a coloured film around the seed. Seed treatment companies currently offer a wide variety of colour and surface texture options to seed processors.

SAFETY PRECAUTIONS

Seed treatment chemicals are pesticides and should be handled as such. Product labels provide information on safe handling and application. Be sure to follow all instructions as listed.

Hazards associated with Handling, Sorting, Mixing and Storage of Treated Seed

Read the label and follow instructions carefully. Over treatment may injure the seed and under treatment will not control the pests in question. The label will also provide information on the toxicity of the active ingredients in the form of code words.

The label will provide basic information on first aid, environmental hazards, directions for use, proper storage and disposal of containers, and pesticide and general product information. Use the label and be familiar with it. Material safety data sheets (MSDS) are available from the company for each of the products sold. Consult the MSDS for more detailed information before applying the material.

When handling toxic substances, operating personnel must be provided with protective clothing such as coveralls, cap, protective glasses, rubber apron, rubber boots, rubber gloves, and a respirator designed for use with the substance. Personnel should not inhale the dust or vapour nor permit the material to contact the skin or eyes.

The operator should wash thoroughly with soap and water before eating and smoking. Bathe immediately after work and change all clothing and wash clothing thoroughly with soap and hot water before reuse. In case of contact, immediately remove contaminated clothing and wash skin thoroughly with soap and water. A safety shower should be installed in the immediate vicinity of the treater.

An exhaust system should be installed to remove vapours and dust from the operating area. The exhaust air should discharge into a cyclone or bag-type dust collector. The treater should also be vented and tied into the exhaust system. Seed treaters should be isolated and not operated in the vicinity of other personnel or farm commodities that are to be used for food, feed, or oil. Special multiwall paper bags or tightly woven bags are recommended for seed that has been treated with toxic substances. Seed should be thoroughly dry before going into the bag, as excessive moisture can cause rapid deterioration of the seed. Store treated seed in a cool, dry place away from food or feed products.

Seed treaters should be thoroughly cleaned after use as some of the pesticides are corrosive and others settle out and cause clogging of the equipment. Do not run contaminated water into a stream or public sewer, but discharge into a shallow ground pit.

Hazards associated with the use of treated seed and the disposal of unused treated and Empty Pesticide and Treated Seed Containers

The user of treated seed should read and follow the label carefully. The planter hoppers should be filled outdoors. Do not breath dust or fumes from treated seed or get in eyes or on skin. Wash thoroughly with soap and water after handling treated seed. Use treated seed as seed and do not introduce them into the food or feed channels because serious injury to livestock and humans may occur. Follow instructions on the label for disposal of seed treatment containers and bags containing treated seed.

Do not reuse pesticide containers. Triple rinse and, if possible, recycle containers. Containers can be punctured and disposed of in a sanitary landfill or incineration, or, if allowed by state and local authorities, by burning. If burning, stay out of the smoke. Old seed with low germination should be destroyed by burying at least 18 inches deep in an isolated area away from the water supplies. Treated seed exposed on soil surfaces will be hazardous to birds and other wildlife.

Biological Seed Treatments

Recent research has focused on biological control of certain seed and seedling diseases of crops. The most promising biological control agents tested have been either fungi or bacteria. These biological seed treatments control seed pests by parasitizing the pest organisms, competing for food on the root system, or producing toxic compounds that inhibit pathogen growth. Several biological seed treatments are currently on the market.

Insecticide Seed Treatments

A limited number of insecticides are labeled for use as seed treatments to protect seedlings of various field crops from soil inhabiting insects.

ENHANCEMENT OF SEED QUALITY

Oilseeds, the raw material for vegetable oil, occupy a significant place in India's economy. Next to food grains, oilseeds account for 10 per cent of the cultivated area and value of all agricultural produce. Nearly 85 per cent of the oil and fat needs of the country is primarily met by vegetable oils. India is the third largest producer of oilseeds in the world. No other country has its range of perennial and annual oilseeds. In terms of area, India ranks first in groundnut, sesame, linseed, safflower, niger and castor. Although India has 20.8 per cent of the world's area under oilseeds, it accounts for less than 10 per cent of world's production.

In terms of vegetable oils, India is the fourth largest oil economy in the world after the U.S, China and Brazil. Favourable agro-ecological conditions in the country have supported commercial cultivation of seven annual edible and two non edible oilseed crops, besides a number of minor oilseeds of horticultural and forest origin, including in particular coconut and oil palm. With the growing population along with raising per capita income, the demand for vegetable oil is likely to grow unabated in the coming years.

The impact of green revolution has made our country self sufficient in the production of cereals. The situation so far as oilseeds are concerned is not very bright. Although India is one of the world's largest producer of oil seeds, the quantity of edible fat available falls far short of the country's requirement.

Seed being the basic input in agriculture, production and supply of quality seeds to the farmers will go a long way to achieve the goal of self sufficiency in oilseed crops. Very often the seed may retain the required germination standard but its vigour, which is yet to be quantified and productivity. Seed pelleting or coating is the most applicable technique in direct sown crops which need initial vigour for sustained crop growth and development.

Coating provides an opportunity to package effective quantities of materials such that they can influence micro environment of each seed which supplies not only micro and macro nutrients but also protects the crop from pests and diseases from the earlier stages due to the inclusion of Pesticides. Because of increased awareness of organic farming, use of plant products in agricultural research is gaining importance. Seed coating is an important area in which plant materials could be widely used.

To increase the productivity, various presowing seed treatment methods are followed but such treatments have not come into practice among farmers because of the non-availability of the chemicals to the marginal and small farmers and they are also expensive. Hence, an attempt was made to explore the use of easily available materials for pelleting to increase germination and seedling growth in early phase.

Researchers are recommending several pre-sowing seed management techniques with the benefit of invigouration, protection and production Among the various seed treatments, the physical seed treatments *viz.*, pesticide seed treatment, seed colouring, seed coating and seed pelleting are widely under the usage by the seed producers and are normally included as a continuous treatment in sequencing the various post harvest seed handling techniques.

Basu revealed that the seed treatment implies an improvement in seed performance resulting in better field performance than the corresponding untreated seed and the establishment of a seedling in the soil is an important and foremost need for the better crop production.

This depends largely on the germination and vigour potential of the seeds used for sowing. Seed treatment provides an opportunity to package effective quantities of materials such that they can influence micro environment of each seed which supplies not only micro and macro nutrients, but also protects the crop from pests and diseases from the earlier stages due to the inclusion of pesticides.

To increase the productivity of the oilseeds, various physical seed enhancement techniques were followed but such treatments have not come into practice among farmers because of the non-availability of the chemicals to the marginal and small farmers and they are also expensive. With this background, experiments have been conducted to study the influence of physical seed enhancement techniques such as halogenation, fungicidal and pelleting on productivity of sesame cv VRI 1, sunflower cv Morden and groundnut cv VRI 2 kernels.

MATERIALS AND METHODS

Genetically pure seeds of sesame (Sesamum indicum) cv. VRI 1, Groundnut cv. VRI 2 and Sunflower cv. Morden obtained from the Oilseed Research Station, Virudhachalam and Agricultural Research Station, Bhavanisagar formed the base material for the study.

The bulk seeds were graded for uniformity using appropriate round perforated metal sieves of sizes *viz.*, 5/64" for Sesame, 9/64" for Sunflower and 18/64" for Groundnut kernels. The field experiments were conducted at the Department of Agricultural Botany, Annamalai University, Annamalai Nagar (11°24′N latitude and 79°44′E longitude with an altitude of +5.79 mts above mean sea level).

The seeds of above oilseed crops were imposed with the following physical seed enhancement techniques:

- T0 - Control.
- T1 - Dry dressing with Thiram @ 2 g/kg of seed.
- T2 - Slurry treatment with Thiram @ 2 g/kg of seed.
- T3 - Dry dressing with Halogen mixture @ 3 g/ kg of seed for sesame and sunflower and 4 g/kg for groundnut.
- T4 - Slurry treatment with Halogen mixture @ 3 g/kg of seed for sesame and sunflower and 4 g/kg for groundnut as slurry.
- T5 - Pelleting with Arappu leaf powder @ 200 g/kg of seed.
- T6 - Pelleting with Neem leaf powder @ 200 g/kg of seed.
- T7 - Pelleting with Pungam leaf powder @ 200 g/kg of seed.
- T8 - Pelleting with Vasambu rhizome powder @ 200 g/kg of seed.

The treatments were evaluated for field stand, plant height, number of branches, pod yield, seed yield, seed weight, seedling length, dry matter production, for all the three crops. In addition to the above characters, capsule morphological characters for sesame, head morphological characters for sunflower and pod morphological characters for groundnut were also observed.

The experiments were conducted for two consecutive years, their average values were worked out and subjected to statistical analysis. The physiological parameters such as chlorophyll content, leaf area index, relative growth rate, net assimilation rate, germination per cent and vigour index were also evaluated using following methods. Chlorophyll content.

A quantity of 250 mg of leaf tissue (third leaf from top) was homogenised with 80 per cent acetone and centrifuged at 3000 rpm for 10 minutes. The supernatant was collected and made upto 250 ml with 80 per cent acetone. The optical density of the extract was read in a spectronic 20 using 652 nm for total chlorophyll using 80 per cent acetone as blank. The total chlorophyll content was computed using the formula and expressed as Mg/g of fresh leaf.

COATINGS

Coatings first were developed for cereal seeds in the 1930's by Germains, a British seed company. Large scale commercial use of coating began in the 1960's with precision sowing for the European greenhouse transplant industry. When California outlawed the short handled hoe in the mid 1970's, the use of coated seed for precision field seeders increased.

Precision sowing greatly reduced the number of skips and doubles in soil blocks or cell trays for transplants. Field precision sowing spaced seeds and thus, individual seedlings sufficiently to permit accurate thinning with a long handled hoe, while reducing damage to the root systems of remaining plants.

The combination of U.S., field precision seeding and greenhouse transplant production created a demand for high quality coatings to achieve accurate sowing, satisfactory seedling emergence and stand establishment. University and commercial research programmes responded to this demand, producing coatings now widely used for seeds of vegetables, flowers and some field crops. (Kurosawa, 1976; Mayberry and Robinson, 1982; Markey,1990; Robinson et a, 1983; Valdes and Bradford 1987; Bradford *et al*, 1985)

In the United States, the major high-volume vegetable crop being using seed coatings is lettuce. Roughly 95,000 Ha (235,000 acres) are sown with coated seed. Brassicas, carrot, celery, endive, escarole, onion, pepper and tomato also are coated to a significant extent, varying with growing season, individual grower preference, the use of direct sowing or transplants, the economics of seed and coating costs etc. Begonia is the flower crop most frequently sown in coated form. Impatiens, marigolds and petunias are also coated commercially and the market for coating these and other species is growing strongly. Alfalfa and tobacco are two agronomic crops that are coated.

CONCEPT

Seed is coated when growers need a precision-sown crop and the non-coated ("raw") seed is too small, light or variable in size or shape to be sown accurately with existing equipment. Precision sowing is desirable when growers need singulation, *e.g.*, for cell-tray plant production in a greenhouse or strict control of spacing or depth of placement (*e.g.*, onion spacing is critical to achieve desired bulb size at harvest) Singulation and controlled spacing are also vital for crops that are direct sown and then thinned back to the desired population. The field thinning operation is faster, cheaper and more accurate when coated seeds are used.

One Florida grower who used raw seed before 1984 informed me in 1985 that reduced thinning costs paid nearly all of the additional cost of coated seed. He stated that the stands were superior, and that this improvement was essentially "free" Superior stands meant that the incidence of skips or doubles was reduced and the plants were spaced closer to the ideal of 11.5 to 12 inches on center in the row.

The objective of coating is to deliver the seed in a form that is larger, rounder, smoother, heavier and more uniform than the original seed. The coated seed can then be sown with a belt, plate, cup, vacuum or other type of seed. The coated seed or "pills" can be placed individually, with improved spacing and depth control. The pills also flow better through the seeding mechanism, because their surface is smoother than that of non-coated seed.

Coating can be a carrier. Fungicides and beneficial microbials that protect the seed and emerging seedling are carried in the coating. For example, alfalfa seed coating with incorporated rhizobacteria is used to inoculate the field with beneficial microbial.

PROCESS

Seed coating relies on technology developed by the pharmaceutical industry to make medicinal pills. Commercial seed coating operations put seed in a rotating pa, mist with water or other liquid and gradually add a fine inert powder, *e.g.*, Diatomaceous earth, to the coating pan. Each misted seed becomes the center of an agglomeration of powder that gradually increases in size. The pills are rounded and smoothed by the tumbling action in the pan, similar to pebbles on the beach. The coating powder is compacted compression from the weight of material in the pan.

Binders often are incorporated near the end of the coating process to harden the outer layer of the pill. Binders can also reduce the amount of dust produced by the finished product in handling, shipping and sowing. Care must be taken with binders to avoid delaying or reducing the germination percentage. Specific details of the materials used as binders are closely held as proprietary information by the coating companies.

We are unaware of any public information on the classes of materials used as binders. Blanks and doubles are eliminated by intensive screening and other techniques. Uniform size and uniform rate of increase in size are evaluated throughout the process with frequent hand screening. At intervals during coating, and at the end, all of the pills are removed and mechanically sized on a set of vibrating screens.

Smaller pills are returned to the pan and built up to the size of the remainder of the lot. After drying, usually with a forced air system at controlled, moderate temperatures, the pills are screened a final time before packaging. Undersized pills may be built up or discarded. The recovery rate (number of pills divided by the original number of seeds) has been 97 per cent +/–2 per cent for commercial seed lots at one commercial company for the past 10 years.

Species	Seeds / lb (Raw)	Pills / lb (Coated)	Weight Increase (%)
Begonia	70,000,000	250,000	28,000
Lettuce	430,000	12,500 - 40,000	3,400 - 1,100
Onion	120,000	15,000 - 30,000	800 - 400
Tobacco	5,000,000	165,000	3,000

Size uniformity after coating is a major criterion of coating quality. The usual tolerance for coated seed is +/–1/64th inch (0.4mm). This is the US seed trade standard for sizing, established long before coatings were introduced. For example, coated lettuce seed is sown most frequently with a belt planter through a 13/64 inch diameter round holes in the belt.

This hole size requires that the lettuce pills be sized over a 7.5/64 inch screen and through an 8.5/64 inch screen. These tolerances result in levels of singulation well above 95 per cent in the field with placement in the row controlled within < 1/2 inch.

Accuracy of seed placement can vary with weight of the pill as well as the size tolerances. Sowing accuracy also depends on the skill of the equipment operators, the adjustment and the wear of the seeder, and the speed of the tractor through the field.

The same constraints are true for greenhouse seeding: experience, attention to details and appropriate equipment are necessary to obtain the full benefit of coated seed.

TYPES

Two basic types of pill produced with inert coating powders are dissolving or "melt" coats and "split" coats. The melt coats dissolve when wet and gradually wash away from around the seed. Split coats initially retain their shape when wet and, by capillary action, pass moisture through the pill to be imbibed by the seed. The seed swells and cracks the pill by internal turgor pressure. The split coat often permit germination with less water, as they split, allow uniform, rapid oxygen access to the surface of the seed. The melt coats often require more water to wash the coating material away from the seed, and more time for the oxygen to reach the seed through the saturated coating material. Melt coats may offer advantages when soils are saturated, but oxygen availability always influences the speed, uniformity and total percentage of germination.

Powder coatings, both split and melt, multiply raw seed rate and depending on the coating, the number of seeds per pound may decrease dramatically. Generally, the heavier pills are easier to because they bounce an roll less than the lighter pills. However, the sowing equipment must be able to handle the pill weight. This can be critical, especially for vacuum type seeders.

In addition too the types of coating products described above, there is recent and increasing use of "film-coating". A thin polymer film smoothes the surface of the seed for better flow ability. The polymer also influences water uptake and the adherence of chemical fungicide treatments. Film coating only increases the raw weight of the seed 1 per cent to 5 per cent, far less than the powder coatings.

Seed coating aims to influence the external physical properties of the seed, affecting the sowing characteristics only. By itself an ideal coating would be neutral in its influence on the speed, uniformity and percentage of germination when compared to the original raw seed lot.

The ideal seed coating would perform in the same manner as the raw seed under a wide range of environmental conditions: light, moisture, temperature, pH, soil type etc. Also the stress of the coating process should not influence the germination pattern or longevity (shelf life) of the seed lot adversely, nor induce secondary dormancy — *i.e.,* affect seed quality.

SEED EVALUATION

A seed is considered to have germinated after the emergence and development from the seed embryo of those essential structures which are indicative of the seed's capacity to produce a normal seedling under favourable conditions. Abnormal seedlings are not included in the germination count because they rarely survive to produce plants.

The ISTA Rules (ISTA 1976) recognize four groups of abnormal seedlings:

a. Damaged seedlings,
b. Deformed seedlings,
c. Decayed seedlings,
d. Seedlings with unusual hypocotyl development.

These groups and their characteristics are defined in detail in the ISTA Rules. In a laboratory test the majority of the normal seedlings are usually removed at the interim counts, but the assessment of many of the doubtful and abnormal seedlings must be left until the end of the test, to ensure that slower growing but otherwise normal seedlings are not incorrectly classified.

For many species the initial count is carried out one week after starting the test and assessments may be carried out at weekly intervals until the test is ended. More frequent assessment is needed if a more precise picture is needed of the speed of germination.

At the end of the test period, all remaining ungerminated seeds should be cut and examined, and the number of fresh, firm and possibly viable seeds recorded (Bonner 1974). At the same time observations may be made on the condition of unsound ungerminated seeds e.g. an abnormally high incidence of insect - damaged or mechanically damaged seeds which would indicate the need for improvements in seed hygiene or seed processing methods.

The record of the germination test commonly shows the percentage of germinated and the percentage of ungerminated but apparently sound seeds separately, e.g.

Germination %	=	82 %
Sound ungerminated seeds	=	6 %
Viability %	=	82 % + 6 % (= 88 %)

It can be seen from Tables that the duration of a test commonly lasts from two to five weeks, but this does not allow any period for pretreatment to overcome dormancy.

Many species require no pretreatment and some types of seedcoat dormancy can be successfully treated in a matter of hours. But the prescribed pretreatment of Tectona takes 18 days and some prescribed prechilling treatments 3 – 9 months. When planning a testing programme it is necessary to take account of pretreatment as well as testing time. In extreme cases it may be necessary to replace the germination test by one of the indirect tests of viability described below.

Fig. Acorns of *Quercus Alba* Germinating on Kimpak in USA. Note spacing between seeds.)

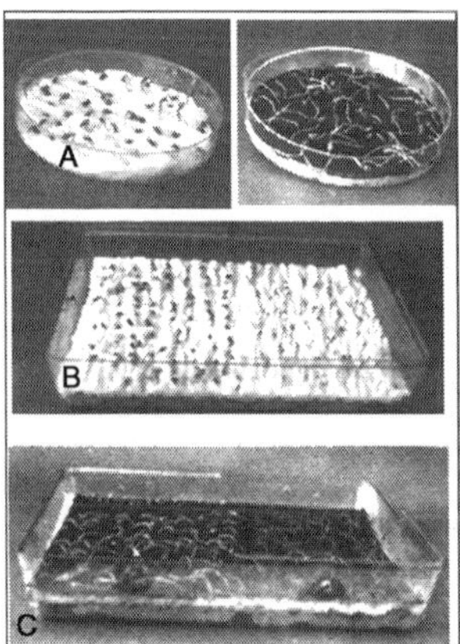

Fig. Germination of Douglas-fir and Lodgepole Pine seeds on Kimpak (A left) and blotter/Kimpak (A right) in Petridishes (A) and Canadian clear Germination Boxes (B-C).

GERMINATION ENERGY

Germination Energy has been defined in more than one way: (1) The percent, by number, of seeds in a given sample which germinate within a given period (defined as the energy period) e.g. in 7 or 14 days, under optimum or stated conditions or (2) The percent, by number, of seeds in a given sample which germinate up to the time of peak germination, generally taken as the highest number of germinations in a 24 hour period.

In both the above definitions the length of the energy period is considerably shorter than the full test period prescribed by ISTA. Germinative energy is a measure of the speed of germination and hence, it is assumed, of the vigour of the seed and of the seedling which it produces. The interest in germinative energy is based on a theory that only those seeds which germinate rapidly and vigorously under the favourable conditions of the laboratory are likely to be capable of producing vigorous seedlings in field conditions, where weak or delayed germination is often fatal. Little experimental evidence has been published on this theory, but excessively delayed germinants must get eliminated automatically from the nursery, either because they are suppressed by older and more vigorous competitors or because, if transplanting has already been completed, they are not worth the trouble of a special supplementary transplanting. An example of calculating germination energy is given on pp. 234–236.

Another method of comparing germination energy of different seed lots is to record "Rate of Germination" i.e. the number of days required to attain 50% of germination capacity. The shorter the period, the greater the germination energy. Yet another method is to evaluate the stage of development of germinated seeds and classify them into a number of development or vigour classes. For example Wang used seven classes for germinated normal seeds of Picea glauca, in addition to ungerminated and abnormally germinated seeds; they ranged from seedlings with healthy root, fully developed hypocotyl and seedcoat completely shed to seeds with seedcoat burst open but as yet no radicle emergence.

GERMINATION VALUE

The concept of Germination Value, as defined by Czabator, aims to combine in a single figure an expression of total germination at the end of the test period with an expression of germination energy or speed of germination. Total germination is expressed as (final) Mean Daily Germination (MDG), calculated as the cumulative percentage of full seed germination at the end of the test, divided by the number of days from sowing to the end of the test. Speed of germination is expressed as Peak Value, which is the maximum mean daily germination (cumulative percentage of full seed germination divided by number of days elapsed since sowing date) reached at any time during the period of the test.Germination Value (GV) can then be calculated from the formula.

$$GV = \text{(final) } MDG \times PV$$

Germination Value, as an integrated measure of seed quality, has been used by several tropical seed workers e.g. for Terminalia ivorensis and for Pinus kesiya.

An alternative method of calculating Germination Value has been proposed by Djavanshir and Pourbeik, who found that it was more closely related to survival of plants in field nurseries than was Czabator's method, in the case of Pinus ponderosa and P. eldarica in Iran. The formula they proposed was:

$$GV = (\Sigma DGS / N) \times \frac{GP}{10}$$

where:

GV	=	Germination value
GP	=	Germination percent at the end of the test
DGS	=	Daily germination speed, obtained by dividing the cumulative germination percent by the number of days since sowing
DGS	=	The total obtained by adding every DGS figure obtained from the daily counts
N	=	The number of daily counts, starting from the date of first germination.

The calculations required are somewhat lengthier than those of the Czabator method and, for many patterns of germination, the simpler method is likely to give sufficiently accurate comparisons between seed lots.

QUALITY OF SEED DEVELOPMENT

Imprinted genes are expressed predominantly from one allele in a parent-of-origin–specific manner. The endosperm, a seed tissue that mediates the transfer of nutrients from the maternal parent to the embryo, is an important site of imprinting in flowering plants. Imprinted genes have been identified in maize and *Arabidopsis thaliana*, but crosses in a variety of species suggest that the effect of imprinting on seed development is widespread throughout the angiosperms. The mechanism of imprinting in plants, as yet only partially understood, likely involves differences in DNA methylation and chromatin structure between differentially expressed maternal and paternal alleles.

Seeds develop from the ovules in the female plant, after they have been fertilised by the pollen from the male parent plant. This is termed 'sexual reproduction', as seeds contain the genes of both the male and female parent, and usually both male and female flowers are required to produce seeds. Sometimes, the male and female flowers are from separate plants, sometimes they are from the same plant, and sometimes a flower may be fertilised by its own pollen. The ovules are the embryos from which the seed will develop. At the time of fertilisation, they are very small compared to the mature seed. Without being fertilised (in most cases), the ovules will not develop into seeds, any more than pollen can. So in order to develop into seeds, the ovules must first be fertilised. We can often judge how close the seeds are to being ripe from the size or colour of the fruit or seedpod.

This is a Passionflower. In the photos, the petals are brown and shrivelled at the top, and the dried stigma is at the bottom. The ovary is a small oval underneath the stigma, which in this species is divided into three. Underneath the ovary are the five stamens. At this stage, the seeds are too small to be visible.

As the seeds develop, the ovary gets bigger, and at this stage we can recognise it as a seedpod, or, in this case, a fruit. Inside the developing seedpod, the seeds are still barely visible around the edge of the fruit, and most of the seedpod is empty space.

Each seed is attached to the wall of the ovary by a small stalk, through which it receives the food it needs to grow. It is therefore vital that the seeds remain attached to the ovary while they are developing.

The seeds need to form a tiny root and shoot, and to collect a store of food for the plant to use when it germinates before it has developed sufficiently to gather its own food supply. The seedpod grows larger, but even when it has almost reached its full size, the seeds are still soft and green, most of the seedpod is still empty space, and the seedpod is still hard and green. The seeds still need to develop a hard covering that will protect them until they are able to germinate.

The mature seedpod has reached its full size and has begun to change colour. The inside of the seedpod is full of seeds and the juicy flesh that will attract animals to eat the fruit and disperse the seeds. The seeds are large, fully formed, have a hard protective coat and may no longer be connected to the wall of the fruit. If we left the fruit on the plant, the flesh and juice would dry up, and the pod would wither and dry, because it is no longer being supplied with food from the parent plant. Eventually the pod would split open to release the seeds. In the example, the seedpod is a Passionfruit, but the development of all seeds follows the same pattern. Seeds cannot be produced unless the flower has been fertilised, and seeds are not viable unless they are mature and ripe. Unripe seeds are soft, white or green, and enclosed in a hard green fleshy fruit (pod). When seeds are ripe and mature, they change to brown or green and develop a hard covering, and the seedpod dries out and changes to white or brown and eventually splits to release them.

Producing seeds uses a lot of the plant's energy. If the plant is dead or dying, it will not have enough energy to complete the process. This is one of the reasons you are unlikely to be able to get seeds from cut flowers. The plant needs to be able to gather food while it is developing seeds, and the seedpod needs to remain attached to the parent plant until the seeds are fully developed. Development of the seeds, from fertilisation to shedding of the ripe seeds, can take several weeks or even months, but seeds gathered before the process has been completed will not be viable.

FERTILIZATION

Seeds consist of three genetically distinct components: embryo, endosperm, and seed coat. In plants, mitosis follows meiosis to produce the haploid phase of the plant life cycle, the male and female gametophytes. The angiosperm female gametophyte, the site of fertilization, is completely embedded within the maternal sporophytic tissues of the ovule. The most prevalent type of mature female gametophyte is a seven-celled organism consisting of three antipodal cells, two synergid cells, an egg cell, and a diploid central cell.

The function of the antipodal cells is unknown. The micropylar synergid cells help attract the pollen tube to the female gametophyte (Higashiyama *et al.*, 2001). The egg cell lies adjacent to the synergid cells and is the progenitor of the embryo. The large central cell, the progenitor of the endosperm, contains two polar nuclei. These fuse before, or at the time of, fertilization to form a diploid central cell nucleus. The male gametophyte, or pollen, develops in the anther from microspores. A mature male gametophyte consists of two haploid sperm cells encased by a haploid vegetative cell.

Seed development begins upon double fertilization. The pollen tube, formed from the vegetative cell of the male gametophyte, enters the female gametophyte through the micropylar end and releases two sperm into a synergid cell. One sperm fertilizes the egg cell and the other fertilizes the central cell. The resulting embryo and endosperm are genetically identical except for ploidy level: the embryo is diploid and the endosperm is triploid. Fertilization also initiates changes in maternal tissues. The ovary develops into a fruit and the ovule integuments differentiate to form the protective seed coat.

ENDOSPERM DEVELOPMENT

The endosperm, a terminally differentiated tissue that nourishes the embryo during seed development, is the only known site of imprinting in plants. It has been known that the endosperm is a product of double fertilization for >100 years, but its evolutionary origin is still a puzzle. The endosperm might be derived from a supernumerary embryo that took on an embryo-nourishing function or from a female gametophyte that was later sexualized.

Arabidopsis endosperm development begins with the division of the triploid primary endosperm nucleus, which precedes the division of the zygote by several hours.

Endosperm nuclei continue to divide without cytokinesis to create a syncytium of nuclei. Each nucleus is surrounded by dense cytoplasm and organelles; these compose nuclear cytoplasmic domains. The syncytium, as analysed by the expression of a histone H2B fusion protein, is organized along the anterior–posterior axis into three distinct mitotic domains: the micropylar endosperm (MCE), the peripheral endosperm (PEN), and the chalazal endosperm (CZE).

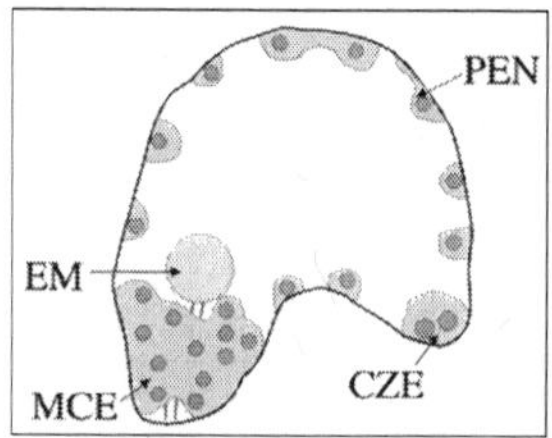

The nuclei of each domain divide synchronously with each other and asynchronously with the nuclei of the other domains. The three domains also are distinct morphologically and cytologically. Dense nuclei surround the suspensor and the base of the developing embryo in the micropylar endosperm. The endosperm nuclei of the peripheral endosperm lie in a thin, evenly spaced layer around the large central vacuole. The CZE is organized into individual nodules and, at the most posterior pole, a large multinucleate chalazal cyst.

The chalazal cyst has larger nuclei than the rest of the syncytial endosperm, perhaps as a result of DNA replication without nuclear division (endoreduplication). The morphology and position of the chalazal cyst are suggestive of a role in the transfer of resources from mother to seed. The cyst sits atop the maternal nucellar proliferating tissue and underlying vasculature. Projections extend from the basal portion of the chalazal cyst into the maternal tissues.

Imprinting may be particularly important in the CZE because of its direct connection with the female parent. The syncytial endosperm begins to cellularize at the heart stage of embryogenesis, in a wave that starts in the micropylar chamber and proceeds through the peripheral endosperm. The CZE is not affected, remaining syncytial until later during seed maturation. The embryo begins to absorb the endosperm after cellularization. As the embryo reaches maturity, the cotyledons function as storage organs and the endosperm disappears almost entirely, leaving only one or two cell layers in the mature seed.

IMPRINTING OF SPECIFIC GENES

Work with entire genomes or with parts of chromosomes indicates that the parental source of genetic information is important in determining its function. Imprinted genes whose expression varies based on the parental mode of inheritance have been identified in maize and Arabidopsis. All are imprinted in the endosperm, and some have effects on endosperm and seed size, as predicted by the parental conflict theory. Two types of imprinting have been described, allelic imprinting, in which only alleles from a certain background are subject to parent-of-origin–specific gene expression, and locus imprinting, in which all known alleles from different backgrounds are under parent-of-origin control.

ALLELIC IMPRINTING

For many years, the only example of an imprinted angiosperm gene was in alleles of the maize *R* gene. The *R* gene conditions anthocyanin accumulation in the aleurone (the outer cell layer of the endosperm) of maize kernels. When an *RR* female (red) is mated to a *rr* male (colourless), all of the kernels have a fully coloured aleurone.

However, the reciprocal cross gives rise to kernels with mottled aleurone pigmentation, indicative of irregular anthocyanin distribution. This phenomenon is specific to the endosperm, and no reciprocal differences are observed in embryos or seedlings demonstrated that the *R*-mottled phenotype is not a dosage effect (*i.e., RR/r* endosperm versus *rr/R* endosperm) but is attributable to the mode of inheritance of the *R* allele. Kernels are mottled regardless of the number of *R* alleles inherited paternally and are always solidly coloured if an *R* allele is inherited maternally. However, this phenomenon is observed only with certain *R* alleles; others (*i.e.,* R^{st}) respond in a dosage-dependent, sex-independent manner.

Alleles of other maize genes, *dzr1* and α-zein, also are imprinted in the endosperm. The *dzr1* locus post-transcriptionally regulates the accumulation of 10-kD α-zeins in the endosperm.

Imprinting of specific α-zein alleles also has been found in the maize endosperm. Using RNase protection assays, it was demonstrated that members of the SF2 subfamily (α-zeins are divided into four subfamilies based on their sequence homology) are expressed maternally but not paternally. This is specific to the W64A inbred background.

LOCUS IMPRINTING

The imprinting of the Arabidopsis *MEDEA* (*MEA*) gene has been the subject of intense study. Mutations in *MEA* were isolated in screens based on two different phenotypes: silique elongation (reproductive development) without fertilization. In the absence of fertilization, the diploid central cell nucleus of the *mea* female gametophyte divides to form a multinucleate central cell, reminiscent of syncytial endosperm, that develops to the point of cellularization.

There is no evidence that the proliferating central cell constitutes an endosperm tissue capable of nourishing an embryo (Friedman, 2001). Sporophytic fertilization programs also are activated in the absence of fertilization: the maternal seed coat develops and the silique elongates. The seed-like structures eventually atrophy. This phenotype is only partially penetrant. Thus, one function of MEA is to prevent replication of the central cell nucleus in the absence of fertilization.

Additional MEA functions also can be deduced based on the post-fertilization phenotype. Seeds from fertilized *mea* female gametophytes undergo endosperm overproliferation, embryo arrest, and eventual abortion. *mea* endosperm nuclei continue to proliferate after the wild-type endosperm has ceased to replicate, resulting in large, balloon-like developing seeds in which endosperm cellularization is delayed. Compared with that in the wild type, the *mea* CZE is specifically enlarged and expanded to more anterior regions.

Thus, another function of MEA is to restrict endosperm proliferation after fertilization. Also, each morphogenetic stage is lengthened in the embryo, and it arrests at the heart stage. Eventually, the endosperm collapses around the embryo and the seed aborts. It is unknown whether the embryo and endosperm phenotypes are both direct consequences of the *mea* mutation or whether one is a primary defect and the other a downstream event.

Other experiments have suggested that *MEA* also is silenced paternally in the embryo. Using a RT-PCR approach designed to amplify only RNA expressed from the mutant *mea-1* allele and not from the wild-type *MEA* allele, *mea-1* RNA was detected in whole siliques at 52 to 56 h after pollination (HAP) when a *mea-1* female was crossed to a wild-type male.

No *mea-1* expression was detected from siliques of the reciprocal cross. Because siliques contain whole seeds, the authors concluded that the paternal *MEA* allele is silenced in both the embryo and the endosperm. Thus, it is possible that a number of genes are not expressed paternally at 52 HAP and that the inability to detect paternal *MEA* expression in whole siliques at this time reflects a genome-wide phenomenon not specific to the *MEA* locus itself.

Imprinting of *MEA* in the endosperm agrees with the parental conflict theory of parent-of-origin–specific gene expression. The *mea* phenotype resembles paternal genomic excess: endosperm mitotic activity is prolonged, cellularization is delayed, and the chalazal endosperm is enlarged.

The wild-type function is to restrict the growth of the embryo-nourishing endosperm. Growth restriction is in the maternal interest, not in the paternal interest, so *MEA* is expressed maternally but not paternally.

Additionally, MEA, a putative chromatin-modifying enzyme that maintains gene repression, might establish imprinting of genes that promote endosperm proliferation. In accordance with the parental conflict theory, these genes would be silenced maternally through the action of MEA and expressed paternally.

2

Seed Treatment

INTRODUCTION

Seed treatment is a term that describes both products and processes. The usages of specific products and specific techniques can improve the growth environment for the seed, seedlings and young plants. Seed treatment complexity ranges from a basic dressing to coating and pelleting.

In agriculture and horticulture, a seed treatment or seed dressing is a chemical, typically antimicrobial or fungidal, with which seeds are treated (or "dressed") prior to planting. Less frequently insecticides are added. Seed treatments can be an environmentally more friendly way of using pesticides as the amounts used can be very small. It is usual to add colour to make treated seed less attractive to birds if spilt and easier to see and clean up in the case of an accidental spillage.

One seed treatment, imidacloprid, from the neonicotinoid family of insecticides, is controversial and was banned in France for use on maize, due to that government's belief that the chemical was implicated in recent dramatic drops in bee counts, and possibly in the so-called Colony Collapse Disorder. Dust from treated seed is known to have caused at least some problems particularly from crops such as maize drilled during the main honey flows. Improvements to pneumatic drills to reduce dust release, and improvements to seed treatment compounds to prevent the compound breaking up into dust have been introduced in Europe led by Germany and Holland from 2009 to 2012.

Information on seed treatments including the information above can be seen on the registration authority databases.

Successful seed treatment is a complex story as it depends on a number of interacting factors. Full compliance with all of them is essential to achieve a high quality seed treatment. It all starts with the seed being treated needing to be clean and of high quality itself.

The seed treatment formulation has to be stable and have basic adhesive properties; the film coating has to complete the adhesiveness in order to ensure even distribution of the seed treatment product on the seeds and to provide good seed flowability. The slurry recipe has to be adapted to the seeds, and the volume needs to be adjusted to achieve good coverage and a homogeneous distribution among the seeds. Finally, the equipment must be able to be adjusted precisely and has to work reliably. The treatment will be of high quality only if the operator applies exactly the right dose of both seed treatment product and film coating.

Seed treatment technology has come a long way since the use of salt brine in the mid-1600s. Today, seed treatments deliver clear environmental, economic and social benefits, making the technology a perfect tool for sustainable agriculture.

With a growing world population, now more than ever it's critical that farmers have access to the tools that will help them grow more food while protecting the environment. And seed treatments are one of these tools.

"By utilizing modern agriculture technologies, farmers will be able to boost yields, conserve water usage and protect biodiversity," says Keith Jones, director of stewardship and sustainable agriculture for CropLife International based in Brussels, Belgium. "Seed treatments represent one tool on which many sustainable agriculture technologies rely upon. By protecting seeds from planting to emergence, seed treatments can improve stand establishment and increase potential yield."

Helmut Schramm, head of the seed treatment business at Bayer CropScience in Monheim, Germany, agrees, saying seed treatments deliver clear environmental, economic and social benefits, making this technology a perfect tool for sustainable agriculture.

"Innovative seed treatment technology represents an environmentally-sound approach to crop protection," he says. "Treating the seed provides a targeted and effective means of application that helps increase yields, safeguard our environment and ensure a sustainable means of crop production."

Over the years, seed treatments have evolved from simply protecting the seed to helping improve plant stand and early plant health.

"Seed treatments are increasingly designed to also enhance plant emergence, growth or nutrition efficiency, which, along with crop protection, lead to a more vigorous and uniform crop," says Schramm. "This forms the optimal base for a high quality crop which can fully exploit its yield potential." Greg Lamka, chair of the International Seed Federation Seed Treatment and Environment Committee, says that in order to maximize yield, it's important to have a full and uniform stand. The STEC committee was established in the 1990s to raise the seed industry's awareness about the use of different seed treatments and to promote a better understanding of how production could be improved and made more efficient. The committee consists of seed companies and crops protection companies who wish to promote the safe and effective use of seed treatment products.

"Growers are paying more for the seed, so expectations are rising about the performance of our products. Seed treatments are a way to ensure that the products will perform to their maximum, based on the environment that they're put into," says Lamka, adding the vast majority of the seed used in developed countries is treated.

GENERAL PRINCIPLES OF SEED

Low moisture, low temperature, low ethylene concentration and probably low oxygen concentration are the most relevant factors to have in mind for an efficient long term seed preservation. Some interrelations among them exist.

Low moisture and low temperature have been the traditional factors considered. Still, in the practice of the past few decades, low temperature has comparatively received much more attention, while an efficient control of low moisture has been largely neglected. However, in order to guarantee an efficient long term preservation of orthodox seeds, a good control of seed moisture content is essential. At this respect, two major principles should be taken into account:

The container should be perfectly tight, preventing water vapour intake. If the container is not perfectly vapour tight, the seeds will tend to get balanced with the external air humidity. In this way, any potential benefit of low temperature will be offset by an increase in seed moisture.

It must be noticed that relative humidity inside any uncontrolled cold room is usually very high. Further to this, it should be emphasised that dried seeds are highly hygroscopic and–even if often very slowly—there is time enough ahead for moisture to enter the container. Unfortunately, the utilisation of inadequate containers has been commonplace in seed genebanks (Gómez-Campo, 2002).

Ultradrying down to a moisture content of 1–3% might extend the seed life span perhaps 4–16 times (Harrington, 1972). Ultradrying has been considered harmful by some authors (Walter & Engels, 1998; Hu, C. & al. 1998), while others have defended that it is not damaging to orthodox seeds (Ellis, 1998; Hong & al. 2005).

The debate might last for years until the subtle factors behind these opposite opinions are fully understood. However, under the conditions utilised in the Universidad Politécnica de Madrid (UPM) genebank, ultradry seeds preserved during 40 years have maintained their viability intact (average = 98,4%, not significantly different from 100%; Pérez-García & al., 2007). As far as we know, these results are unmatched by any other seed genebank and set a solid base to develop improved preservation methods. This has moved the author to prepare the present condensed guide.

When dealing with ultradry seeds, enormous savings of energy can be made by avoiding the use of too low temperatures during storage. Temperature may be the key factor when seeds are merely dried but it loses relative importance for ultradried seeds (Pérez-García & al., 2007). Independently, very low temperatures, close to that of liquid nitrogen (–196 ºC), provide an alternative method for the preservation of seeds and tissues (cryopreservation) that might play an important role in the future, especially for recalcitrant material.

A third factor—the presence or absence of oxygen–has been the source of a contradictory literature over the years and, perhaps for this reason, it has largely been neglected from the practical point of view. However, recent research by Ellis & Hong (2007) suggests that ultradry seeds may be sensitive to oxygen; therefore, it is a factor that, at least tentatively, should be taken into account. Perhaps, the success of the UPM bank partly resides in the fact that air was originally substituted by CO_2 in the atmosphere within the containers. Ethylene and other toxic gases that slowly evolve during seed aging should be removed in order to extend the seed life span.

CHOICE OF CONTAINERS

For many years it has been widely admitted that containers which are appropriate for cold drinks or to store chips for a few months, are also appropriate for long term seed storage. This is very far from reality. Results of a survey on 40 different containers—some of them widely used in many genebanks-showed that 36 of them (90%) allowed moisture inside in less than three years.

All plastic containers allowed moisture inside. The water molecule is small enough to get through the pores of the polymers used to make the plastics and in the medium and long term the fact is unavoidable–although not noticeable in the short term. Polyvinyl bags are especially permeable to humidity. An additional reason for the failure of containers with a lid is based on the fact that both pieces are usually made of different materials, therefore expanding or contracting in a different way with changes in temperature. This creates fissures allowing humidity to get inside the container. This last mechanism appears not only in plastic containers, but also in most glass containers with twist-off or screw lid. Even plasticized twist-off type jam jars often show oxidation stains inside after some time has elapsed.

A special mention should be made to aluminium foil bags because of their still widespread use in genebanks of cultivated species. It's been a long time since bi-laminated bags (coated with plastic only on one side) were rejected on the basis of their inefficiency. However, tri-laminated bags (coated on both sides) are still widely accepted. According to our experience a maximum of 80% of a given set keeps vapour tight after 3–6 years. It is evident that to base the whole operation of a genebank on a material with these characteristics is risky.

To put an example forwards, a collection of 100.000 samples where we learn that 20% of them (20.000) won't keep their tightness creates a really awful scenario which is now quite common in genebanks. Additionally, it is not possible to see from the outside in which ones the problem is. Massive germination tests or even measurements of moisture in all the samples, are not practical solutions. Let's add to it that, being sealed with plastic, they carry on the long term the disadvantages of this material.

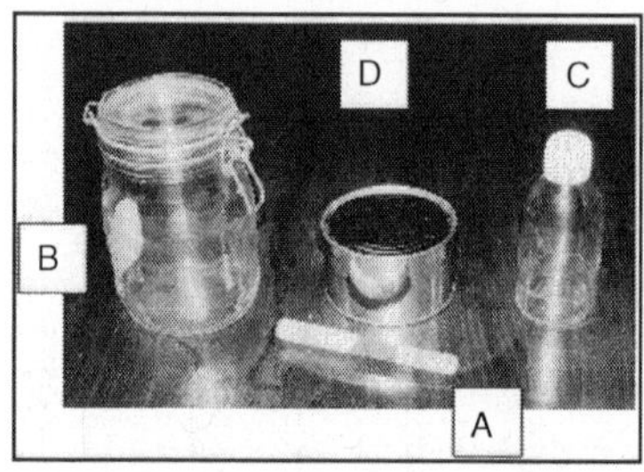

Fig. (A) Flame sealed vials (B) Kilner (Scotch) jars (C) Flask for laboratory chemical (D) Sealed metal cans

The aforementioned experiment showed that only four containers remained vapour tight:

a. Flame-sealed glass vials. Used in the UPM's genebank since 1966, they have only very exceptionally (0,3%) showed cases where moisture had come inside, due to the sealed tip having got broken. This tip can be protected with a mixture of resin and wax, as will later be explained. Undoubtedly this is the best option for small valuable accessions of wild or crop species such as rare or endangered species or genotypes, crosses, mutants, and so on.

b. "Kilner" ("Scotch") jars with a rubber joint and propped up lid, whose use is quite spread in kitchens. After 12 years, only one out of 25 units allowed moisture inside. Its size allows their use with samples of every size, including those bulky ones often kept in genebanks of cultivated species. A good adjustment of the joint should be ensured and also that nothing sticks between the glass and the joint (adhesive tape, little seeds or gel grains). It is possible that some joints need to be replaced after several years. As they are hermetic, seed samples can be stored inside in bulk, or in any type of smaller container such as paper or polyvinyl bags.

c. Some glass flasks with plastic lid, usually used in laboratories to store chemicals behaved very well during the experiment but other very similar ones behaved very badly. The difference was due to the quality of the plastic joint under the lid. Their opening being narrow, these jars are somehow uncomfortable to be used with medium sized or large seeds.

d. Heat sealed, metal cans coated with plastic on the inside. This inside plastic coating may be beneficial to prevent any possible damaging effects from the metal vapours in contact with the seeds over the years. The seal is also plastic. On the other hand, these containers are opaque and do not allow external control with coloured moisture indicators - unlike the three previous ones.

Those vials with a rubber lid fastened by an aluminium cover, used some years ago to contain penicillin, were not tested. However, users of this container refer to its behaviour positively.

It is very possible that other appropriate containers could be found in the future. However, to do reliably so, perhaps several years might be necessary. On the other hand, models very similar in appearance—case (C) and also some (B)—can behave very differently and it is advisable to test them before choosing a particular one. The method there described might serve to test other containers. Considering that most banks around the world have too often used non vapour-proof containers, this experiment probably shows the main reason of failure in seed preservation – especially for banks of cultivated species. It also shows some possible solutions. Solution (B), in particular, provides a larger volume and opens a way to the use of the silica gel method in seed genebanks of cultivated species–A method already used successfully since 1966 with wild species (Gómez-Campo, 1972).

On the other hand, it is obvious that the containers tightness is linked, in a practical way, to the need for frequent or only occasional handling of the seed lots.

For very long term collections of valuable materials addressed to the future ("black-box" collections), there is no doubt that the flame-sealed glass vials are the best option. Those can also be of help for other material used more often, especially from wild species or any small sized species, by placing smaller quantities of seeds and making more vials, so that one of them can be broken when necessary. For base (long term) collections, Kilner jars are also advisable and so are the other options already mentioned. Active collections (for distribution), require more frequent access and need containers capable of being opened and closed. Kilner jars may also be used here if being fast when taking the sub-samples and, whenever possible, doing it in a dry environment.

For active collections of species with small seeds, tubes with a screwed plastic lid might be of interest. They are easy to use but not entirely reliable. Transparency of the containers becomes a key quality when wishing to take advantage of the benefits of the silica gel method for an easy and direct monitoring of the samples from the outside.

ULTRADRYING

To this date, the most commonly used procedure has consisted of drying the seeds with different procedures until reaching a moisture content of approximately 5–7%. After this, the seeds are placed in containers whose vapour tightness is most often untested and the success is mainly trusted to low temperatures. Ultradrying has been very rarely used, perhaps by no more than 50 banks – mostly of wild species in botanical gardens—out of more than 1500 existing in the world. The UPM's results have proved beyond doubt the efficiency of ultradrying in orthodox seeds at least under anaerobic conditions and also prove the comparatively lesser importance of temperature for ultradried seeds.

It should be noted that moisture content of 4–5% may be enough for the efficient conservation of some orthodox seeds such as those of many legumes (Ellis & al. 1988). However, as this author points out, these seeds do not suffer when ultradesiccated to 1–3%. In a genebank, with thousands of accessions being handled, the need to individualise the methods should be avoided. In this light, a general benefit to the use of ultradrying can be attributed–always speaking of orthodox seeds.

ULTRADRYING WITH SILICA GEL

Ultradrying could be achieved following different ways, but perhaps the most practical one might consist of storing the seeds in balance with dehydrated silica gel. Leaving some gel afterwards with the stored seeds inside the tight container is most useful because the coloured indicator would show any accidental entry of moisture.

Advantages of the use of Silica Gel:

- It provides a practical method to desiccate the seed samples.
- It reaches moisture levels of 2-3%, lower than those obtained by most.
- Tkeeps these levels indefinitely in tight containers.
- It warns over possible anomalies (*i.e.* moisture intake) in the container—when used in combination with a coloured indicator.
- It can be regenerated after its use through a new dehydration process with heat.
- It delays also ageing by absorbing toxic gases produced during the process.

Summary of the advantages of using silica gel with an indicator. Besides those related to ultradrying itself, it is worth noting the last one (number 6), which refers to a totally independent protection mechanism. In seed samples stored within a closed container, damaging gases are generated during the aging process and as they accumulate, they prove deathly for the entire sample. The mortality, which might have been supposed a priori probabilistic for every individual seed, becomes accelerated at the end. Being capable of absorbing these gases, silica gel significantly delays aging and postpones the moment of death. Ethylene is an important component of this mixture of gases (Lee, & al. 2001).

Silica gel is simply silicic anhydride, SiO_2, though amorphous (non crystalline) obtained through an industrial process. It is granular in texture, white and very porous. It is this last characteristic which gives its absorbent properties. When it is well dehydrated and placed on a scale, it can be observed to absorb up to 20% of its own weight of water. If placed in a closed container such as a Kilner jar, it balances itself with the confined atmosphere until this reaches approximately 10–12% of relative humidity.

The name "gel" is here improperly used. We understand by gel some substance of a colloidal nature, which is not this case. However, the use of this term for this product has long been used and is now strongly established.

The colour usually shown by silica gel is due to an indicator, added to see directly when it is dehydrated and when it has absorbed moisture. For many years, cobalt chloride (Cl_2CO) has been used.

This substance gives the dehydrated gel a strong blue colour and a pale pink colour to the gel having absorbed moisture. Recently, the European Union banned its use because of considering it carcinogen through inhalation. A search for new alternatives led to some iron salts, where the change in colour can be poorly distinguished. At present, the most advisable alternative is methyl violet, which gives the dehydrated gel an orange colour and a green colour to the hydrated gel.

The gel is manufactured in granules of different sizes. This size is not irrelevant because, to be used in Kilner jars, a gel with a size > 4 mm can be more convenient, while a 2-3 mm size can be better for glass vials. By hydrating with water coarse silica gel granules (be careful, heat is produced!) the grains get fragmented, and it is possible to screen it in fractions with smaller granule sizes.

Hydrated silica gel can be regenerated through heating. When it is accompanied with cobalt or iron salts, this is done at approximately 220 °C, thus a standard cooking oven can be enough. The process may last one or two hours - depending on the mass to be regenerated. With an organic indicator (*e.g.* methyl violet) a temperature of 110–120 °C must not been surpassed and the process will require comparatively more time.

FREEZE DRYING

Freeze Drying ("liophillsation") has been successfully used to obtain ultradry levels of moisture. This procedure consists in freezing the seed water and sublimating the ice afterwards.

This technique was developed in the National Centre of Natural History (Paris) in the 1980s for long term storage of pollen grains (Schoenke & Bey, 1981. The Conservatoire Botanique National Méditerranéen (CBNM) started using it for seeds in Porquerolles in 1987.

It is assured that freeze dried seeds do not suffer any decrease in their viability because of the process. However, it does not seem to be the case for every species and the details when applying this technique must be optimized for every species (Anonymous, 2006). As it will be explained, freeze drying may have the advantage of achieving a direct ultradrying instead of one done in stages – as is generally the case with silica gel which needs to be regenerated every time it is used.

In comparison, Freeze drying needs a rather expensive machine (Hu, X. & al. 1998). In turn, silica gel reaches lower water contents. Also, considering that silica gel is going to play an essential role afterwards in keeping a low moisture and monitoring the samples, it may be more practical a direct drying with gel. In other words, while freeze drying only ultradries, silica gel not only ultradries but also provides an efficient monitoring during subsequent storage.

THE LOW TEMPERATURE FACTOR

The use of low temperatures has been widespread in long term seed preservation. Standards set by the Food and Agriculture Organization of the United Nations (FAO) and the International Plant Genetic Resources Institute (IPGRI, now Biodiversity) advise a storage temperature of -18 0C or lower (FAO/IPGRI, 1994) and almost everybody now planning a genebank instinctively thinks of a cold room capable of at least reaching -20°C. Localisation of the new installations of the Nordic Seed Bank in a region of "permafrost" responds to the same idea.

Unfortunately, emphasis on low temperatures and neglecting low moisture (either because of not drying enough or because of the use of inappropriate containers) has brought that a significant number of the actual genebanks are worried with the burdensome need to rejuvenate their prematurely aged material.

Seeds which had been freeze dried in the Conservatoire Botanique National Mediterranéen (CBNM) of Porquerolles and then kept at room temperature for 10 years showed a slightly better behaviour than those kept under low temperature. In a much longer term, ultradry seeds with silica gel kept for 40 years at room temperature inside a closet (Pérez-García & al., 2007) did not behave very differently from those kept in the cold room. It appears hence clear that the main role attributed to low temperature during decades should now be replaced by a much more thorough attention to the moisture factor.

However, it is obvious that low temperature helps, and it would be irresponsible not to take it into consideration. In the last mentioned work, it is suggested that temperatures between –5 °C and +5 °C could very well be enough for orthodox ultradry seeds, thus saving considerable amounts of energy.

APPLICATION OF SEED TREATMENT

Fungicide seed treatment products come in a variety of formulations and in a variety of packaging sizes and types. Some are registered for use only by commercial applicators using closed application systems, others are readily available for on-farm use as dusts, slurries, water soluble bags, or liquid ready-to-use-formulations.

Whatever the formulation used or application method chosen, some precautions should be taken to assure applicator safety and appropriate seed coverage.

CAUTIONS

Follow label directions when handling seed treatment chemicals. These products are potentially poisonous if mishandled or misused. Extreme caution must be used when handling seed treatment chemicals: some are toxic, others may be irritating. An approved chemical respirator and goggles are recommended even if not specifically required by the fungicide label.

The rate of application prescribed by the label must be used: over treatment may injure the seed and under treatment may not provide good disease control. To apply the correct rate, it is essential to calibrate application equipment carefully and to check calibration frequently. Metering cups of commercial applicators should be cleaned daily to prevent a buildup of chemical that might result in reduced application rates.

An auger which has been used to treat seed cannot be cleaned up sufficiently for use in augering grain for food or feed. Once an auger is used for seed treatment, it should be used only for treatment or augering seed for planting. It should not be used to auger grain used for food or feed.

Treated seed should not be used for food or feed, and treated grain should not contaminate grain delivered to elevators or be placed in bins or in trucks delivering to elevators.

Containers should be triple rinsed with the rinse water added to the treatment mixture. The rinsed containers should be punctured and crushed for disposal in an approved landfill.

CEREAL SEED TREATMENTS

Fungicidal seed treatments help control soil-borne pathogens that cause seed decay, seedling blight and root rot. Control of these diseases may result in better stands, more vigorous seedlings, and increased yields. Protectant fungicides such as captan, maneb, PCNB, thiram, or fludioxonil (Maxim) help control most types of soil-borne pathogens, except for common root rot and take-all.

Protectant fungicides containing captan, maneb, or thiram are sold under various trade names. Fungicidal seed treatments to protect against the soil-borne fungi that cause common root rot and take all will be discussed under barley seed treatment. Fungicidal seed treatment controls most, but not all, of the seed-borne diseases of small grains. Seed treatment should be used when the seed is contaminated with smut, scab, or black point fungi. Specific recommendations for these diseases are discussed under barley and wheat. No seed treatment is a substitute for good seed.

Barley

Barley has three smuts: covered smut, black semi-loose smut (nigra smut), and loose smut. Covered smut and black semi-loose smut are surface-borne fungal pathogens that infect the emerging seedling. These two smuts can be controlled by various protectant fungicides. Loose smut infects the embryo of the seed before harvest.

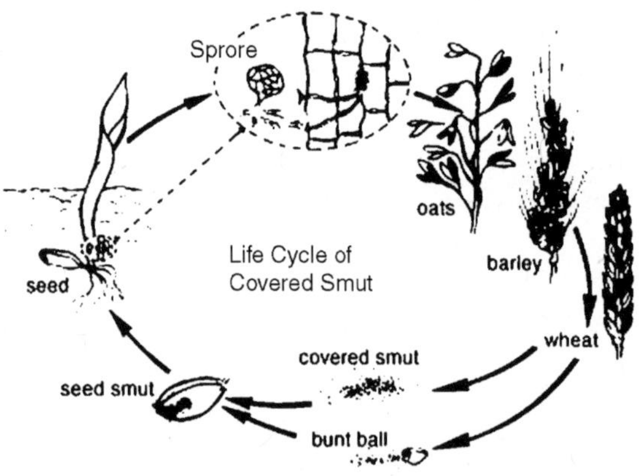

Fig. Life cycle of covered smut and bunt.

Protectant fungicides do not control loose smut—only the systemic fungicides carboxin (Vitavax or Enhance), triadimenol (Baytan) or tebuconazole (Raxil) will control loose smut in barley. These same products also will control the covered and semi-loose smuts of barley.

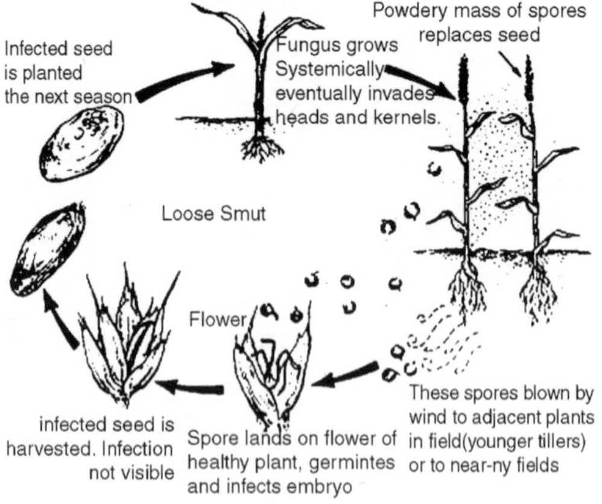

Fig. Life cycle of loose smut.

Covered Smut

Seed of barley varieties susceptible to covered smut should be treated with a protectant fungicide or with carboxin, triadimenol or tebuconazole. At present, all barley varieties recommended for production in North Dakota must be considered susceptible.

Black Semi-loose Smut

Many varieties are susceptible to this smut. It is indistinguishable from loose smut in the field, but it is borne on the seed surface and infects seedlings just as covered smut does. It is not detected by the embryo test, because it does not infect the embryo. It is controlled by seed treatment with a protectant fungicide or with carboxin, triadimenol, or tebuconazole.

Loose Smut

Many of the barley varieties released in the late 1970s and early 1980s were resistant to the races of the loose smut fungus present at that time. In the mid-1980s, a new race of the loose smut fungus was detected in North Dakota.

Seed treatment for loose smut in barley planted for seed production is especially important. Seed samples of susceptible barley varieties should be sent to the North Dakota State Seed Department for an embryo test to determine the percentage of loose smut infection.

Losses from loose smut are about equal to the percent infection; 5 percent loose smut represents a yield loss of about 4.3 percent. If the embryo test shows 2 percent or greater infection, the seed should be treated with carboxin, triadimenol or tebuconazole, or loose smut-free seed should be used for planting.

Common Root Rot

The fungus that causes common root rot is soil-borne and is widespread in North Dakota soils. The common root rot fungus increases under barley or wheat culture, and it often causes mature plant root rot. The fungus also may cause severe root infection in the seedling stage and if splashed to the leaves and head causes spot blotch and black-point infections, respectively.

A number of seed treatment fungicides are now labeled for wheat and barley for suppression of mature plant root rot and also for suppression of seedling blight due to the common root rot fungus. Imazalil is available under several trade names. Triadimenol (Baytan) also is registered for suppression of seedling blight due to common root rot. Tebuconazole (Raxil) is registered for wheat and barley, and difenoconazole (Dividend) is registered only for wheat at the time of publication of this circular.

The conditions in which seed treatment for common root rot would be most beneficial are: where continuous wheat or barley is grown, or if short rotations between these susceptible crops are practiced, and in soils where moisture stress is very likely.

Barley Stripe

This fungus disease is rarely seen in North Dakota but occasionally appears in two-row barley if contaminated seed is planted. The disease causes distinct yellow, then brown stripes that run the entire length of the leaves. The stripes join together and the leaves become split and frayed or shredded in appearance.

The fungus is seed-borne in the hull and seed coat. If seed from a suspected disease source is used, seed treatment with a carboxin + thiram or tebuconazole product provides some control of the barley stripe fungus. The systemic fungicide imazalil provides good control of barley stripe.

CORN

Corn seed is especially susceptible to attack by soil-borne pathogens when sown in cold wet soil, when the seed is in poor condition, when it is mechanically injured, or if it has been stored for two years or more.

Seed treatment will protect against seed rot and reduce the danger of seedling blight. Sweet corn is more susceptible to attack than field corn, but both should be treated. Most field corn is already treated when purchased. A number of protectant and systemic fungicides are registered for control of seedling blights and seed rot in corn.

WHEAT

Bunt or Stinking Smut

The fungus causing this disease adheres to the seed surface and then infects the emerging wheat seedling. The bunt fungus has a fishy odor and imparts the same odor to flour made from bunted kernels. The price of bunted wheat is discounted for this reason. Bunt is not known to occur in North Dakota at the present time, nor has it occurred in recent years. However, it occurs in other wheat-producing states in the Great Plains.

Growers who purchase seed from out of state should use a seed treatment fungicide, since out-of-state seed could be infested with (carrying) bunt. Growers who have their crop custom combined and then save their own seed should use a fungicide, since the combine could be contaminated with bunt spores or bunt balls from states to the south. Many protectant, systemic, or combination products are available to control bunt.

Loose Smut

Loose smut of wheat infects the embryo, as with loose smut of barley. However, the wheat loose smut pathogen does not infect barley and the barley loose smut pathogen does not infect wheat. No embryo test is available for detection of loose smut in wheat.

Counts of loose smut-infected heads can be made in wheat and durum fields to estimate the percent infection in the crop. Counts should be made at flowering time, as loose smut heads are hard to detect later. The counts provide information on the percent infection in the seed that was planted but are not a reliable estimate of the amount of infection in the seed of the harvested crop.

If weather conditions are favorable, *i.e.* cool and wet, infection in the harvested seed could be considerably higher than in the seed that was planted. Carboxin, difenoconazole, tebuconazole, and triadimenol seed treatments control wheat loose smut.

Scab and Black-point

Scab and black-point are fungal diseases that attack both wheat and barley seed. The scab fungus also may infect oats. The scab fungus infects the kernels during flowering if warm, wet conditions prevail. Black-point infection occurs from heading to maturity and also is favored by warm, wet weather.

Scabby seed is shriveled, light in test weight, and often has a chalky white or pink discolouration. When planted, scabby seed has poor germination and poor vigor. The scab fungus does not grow systemically from the seed through the plant to cause subsequent head scab infection. The source of head infection is spores from infected small grain or corn residue.

Black-pointed seed has a black to brown discolouration of the embryo or germ end, a discolouration which may extend around the kernel into the crease. When planted, black-pointed seed also may have low germination; if the seed germinates, the seedling roots are often infected by the fungus.

Protectant seed treatments or protectants in combination with systemics, or systemic products alone have provided significant improvement in stand and vigor. Yield increases also have occurred with treatment of scabby and black-pointed grain. Seed treatment helps assure good stands and seedling vigor; it will not prevent possible scab or black-point infections of the head during the growing season.

Take-all

Take-all is a root disease of barley and wheat caused by a fungus that thrives in very wet soils. Take-all generally is not a serious or common disease in barley or wheat in North Dakota, unless the crop is grown under irrigation. Triadimenol (Baytan), difenoconazole (Dividend) and tebuconazole (Raxil) are registered for suppression of this disease.

PYTHIUM

Damage to wheat from the *Pythium* fungus has not been well documented in North Dakota, but the fungus is common in agricultural soils and generally does more damage where wheat is planted into crop residues and soil temperatures are cool and soil moisture is high. Several seed treatment fungicides are registered for control of *Pythium* spp. in wheat, including difenoconazole + mefenoxam (Dividend XL), mefenoxam (Apron XL), and metalaxyl (Allegiance).

OATS

Smuts on oats have not been reported as a recent problem in North Dakota, although they have been serious in other states in the region. Oat smut can cause severe losses. Distinguishing between covered and loose smuts of oats in the field is not easy.

Oat loose smut is more difficult to control with protectant fungicides than oat covered smut, because the loose smut spores are lodged under the hulls where they are difficult to reach. Carboxin and tebuconazole are registered for oat loose smut control.

LEGUMES

Alfalfa and Small Seeded Legumes

Alfalfa, clover and other small seeded legumes may be treated for seedling blight diseases and damping off. The extent of occurrence of these disease problems in North Dakota has not been documented.

Captan or thiram products are registered for seedling blight control, while mefonaxam, metalaxyl, and oxadixyl products are registered for control of *Pythium* damping off and early season infections by *Phythophthora.*

Soybeans

Generally, seed treatment is not required for soybeans. However, it may pay to treat seed if: the germination is below 85 percent; the seeds are contaminated with fungi; the seeds are badly weathered or injured; or the seed coats are broken. If captan or PCNB treated seeds are to be inoculated with *Rhizobium* bacteria, the inoculant should be applied in-furrow.

Thiram, fludioxonil, metalaxyl and mefenoxam have little or no effect on *Rhizobium* bacteria. Generally, inoculant is not used on soybeans planted on land that was previously cropped to soybeans.

In fields with *Rhizoctonia* problems, a seed treatment containing carboxin, chloroneb, or PCNB should be used. Seed infested with Sclerotinia should be treated with fludioxonil, which will control infection of seedlings by the Sclerotinia fungus.

Dry Edible Beans

Most dry bean seed is treated prior to sale, often with a combination of fungicide, insecticide and bactericide. A fungicide protects against seed- and soil-borne fungi. The bactericide streptomycin controls surface-borne blight bacteria but will not control internally-borne blight bacteria.

Chickpeas

Several seed treatment fungicides are registered for use on chickpeas. Fludioxonil (Maxim) gives broad spectrum protection against soil-borne fungi; mefenoxam and metalaxyl (Allegiance) provide very specific and highly effective protection against *Pythium* seed rot and seedling blight and early season *Phytophthora* root rot.

Chickpea growers in some parts of western North Dakota have experienced severe difficulty with seed rot of untreated seed and should treat seed with mefanoxam or metalaxyl.

The most serious foliar disease of garbanzo beans is *Ascochyta* blight. *Ascochyta* can be seed-borne, usually at low levels. Wet weather may cause rapid spread of *Ascochyta* from a few infection centers. Thiabendazole (TBZ) is registered for use on garbanzo beans for eliminating seed-borne *Ascochyta*. Purchase of western grown seed that has been laboratory checked for *Ascochyta* also is desirable to minimize the danger of losses from this serious disease, but it should also be treated with thiabendazole.

Lentils

Lentils should be treated with captan, fludioxonil (Maxim), mefenoxam (Apron XL) or metalaxyl (Allegiance). Captan and fludioxonil provide broad spectrum protection against *Rhizoctonia* and *Fusarium* seedling blights. Mefenoxam and metalaxyl (Allegiance) provide excellent control against *Pythium* seedling blight, but do not provide protection against *Rhizoctonia* or *Fusarium*.

The two most serious diseases of lentils are anthrac-nose and *Ascochyta* blight. Both can be seed-borne at low levels but seed transmission to developing seedlings has not been deomonstrated for anthracnose. Ascochyta can spread rapidly in wet weather from a few infection centers.

The *Ascochyta* that infects lentils is a different species than the one that infects chickpeas. No fungicide currently registered is effective for elimination of either of these disease fungi on the seed. Purchase of western grown seed that has been laboratory tested for *Ascochyta* is desirable to help minimize the danger of losses.

Flax

Flax seed can be attacked by seed- and soil-borne pathogens, especially when the weather is unfavorable for germination and growth or when the seed coats are cracked and split. Seed treatment with protectant fungicides captan, mancozeb, maneb or thiram will reduce the amount of seed rot and seedling blight.

Yellow-seeded varieties are more susceptible to seed coat damage than are brown-seeded varieties. All varieties currently being grown are brown-seeded varieties except Omega, which is yellow-seeded.

Peas

Common seedling blights of peas can be controlled with captan, fludioxonil, PCNB or thiram. The water mold fungi *Pythium* and *Phytopthora* are common problems on peas and can be controlled with mefenoxam, metalaxyl or oxydixyl treatments. No seed treatment is registered for Aschochyta blight on peas, but data from Manitoba indicates that thiram provides good suppression of seed-borne Ascochyta.

Safflower

Safflower rust is borne on the surface of the seed and produces infections on the hypocotyl of the emerging seedlings. Seed-borne rust can result in poor stands and reduced vigor of the seedlings. Fungicidal seed treatment of safflower with one of several available fungicides is recommended to control seed-borne safflower rust. Carboxin, mancozeb, and thiram are labeled for seed treatment use on safflower.

The winter spores of safflower rust can survive from one season to the next in the soil but will not survive to the second season. Control of safflower rust requires seed treatment with a fungicide and crop rotation. Never plant safflower on land that had safflower the year before.

Sugarbeets

Pythium, Aphanomyces, and *Rhizoctonia* are fungi that may cause stand establishment and seedling disease problems in sugarbeets. The *Pythium* fungus occurs in most sugarbeet soils. It causes seed rot, pre-emergence damping off, and post-emergence damping off.

Post-emergence damping off caused by *Pythium* may occur when the seedlings are so tiny that they dry up and blow away within a day or two. Consequently, Pythium-induced seedling death is seldom noticed by the grower. The only thing noticeable may be an unusually poor emergence and stand.

Aphanomyces and *Rhizoctonia* cause death of seedlings at a later growth stage, with plants progressively dying from the two to the eight leaf stage. Later in the season, both fungi may cause a root rot which weakens the plant and also reduces the weight and quality of beet roots. Both diseases are favored by warm soil conditions. *Aphanomyces* is favored by heavy soils with poor drainage, resulting in saturated or puddled soils. *Rhizoctonia* is favored by moist soils.

Most sugarbeet seed is sold treated, but different treatments vary in their effectiveness against these three fungi. Certain fungicides have specific activity: mefenoxam, metalaxyl (Allegiance) and oxadixyl (Anchor) are highly effective against *Pythium;* thiram is moderately effective against *Pythium;* PCNB, chloroneb and fludioxanil (Maxim) are effective against *Rhizoctonia.*

Growers planting in fields with known *Rhizoctonia* problems may wish to request a special or supplemental seed treatment from their seed supplier or else use a planter box overtreatment. Hymexazol (Tachigaren) is effective against *Aphanomyces* and *Pythium.*

Canola

Blackleg is a fungus disease of canola that can cause severe losses. It is seed-borne and also spread by wind-borne as well as rain-splashed spores. The wind-borne spores come from canola crop refuse. Most long distance spread of the blackleg fungus is on seed. Most spread within a field or between fields is from wind-borne spores. The severe (highly virulent) strain of blackleg is common.

Benomyl seed treatment provides excellent control of the seed-borne phase of blackleg. Registration of other seed treatments effective against blackleg may occur in 2000 or 2001. Seed treatment is highly recommended for all canola seed planted in areas that do not yet have the severe strain of blackleg. In areas where the severe strain is prevalent, it may be necessary to plant tolerant varieties of canola. Other seedling blights may be controlled with captan, fludioxanil or thiram

Sunflower

Downy mildew is a soil-borne disease that can cause severe losses if excessive rains occur shortly after planting. The downy mildew fungus is widespread across North Dakota. It survives many years in the soil and infects emerging or recently emerged seedlings when the soil is saturated. Several new races have occurred in recent years and at present only one hybrid is resistant to all races.

The downy mildew fungus has developed resistance to mefenoxam, metalaxyl and oxadixyl. Seed treatment with these products provides poor to fair control of downy mildew, depending on what percent of the downy mildew population is resistant. No suitable replacement fungicide was available for the 2000 growing season.

BIOLOGICAL CONTROL

Kodiak

Kodiak concentrate contains *Bacillus subtilis* bacteria which colonize the developing root system, suppressing disease organisms such as *Fusarium, Rhizoctonia, Alternaria* and *Aspergillus* that attack root systems. When used with a chemical seed treatment, the combination of chemicals and Kodiak provides protection to the root for a much longer time than with chemicals alone. As the root system develops, the bacteria grow with the roots extending the protection throughout the growing season. As a result of this biological protection, a vigorous root system is established by the plant, which often results in more uniform stands and greater yields. Registered for seed and pod vegtables, soybeans, wheat and barley, and corn plus all other agricultural seeds.

CHEMICAL TREATMENT OF PHYSIOLOGICAL DORMANCY

A Wide range of chemicals have been tested experimentally in an attempt to overcome internal dormancy. They include gibberellic acid, citric acid, hydrogen peroxide and a number of other compounds. Some have given a degree of improvement, *e.g.*

Bachelard (1967) found that the germination of dormant seeds of *Eucalyptus delegatensis*, *E. fastigata* and *E. regnans* could be improved by treatment with gibberellic acid (GA). 24 hours' immersion in either GA 3 or GA 4/7 of *Nothofagus obliqua* has given rapid and complete germination in 14 days, although this normally dormant species otherwise requires 28–42 days' stratification (Gordon 1979). Shafiq (1980) found that the strength of the gibberellic acid had only a small effect, 200 ppm giving 100% germination in 8 days, 50 ppm the same germination in 12 days.

The best stratification treatment (42 days at 3–5 °C) yielded 70% in 14 days and 88% in 28 days and the control (24 hours' soaking in distilled water and no prechilling) only 20% in 28 days. The saving in time effected by the GA treatment (1 + 12 days compared with 42 + 28 days) is considerable. Results reported later by Rowe and Gordon (1981) showed that GA 4/7 was more reliable than GA 3 as it was less sensitive to temperature during the germination period. Excellent germination was obtained throughout the temperature range 15–30 °C, whereas GA 3 required temperatures over 21 °C for comparable results.

These successes, however, are the exception. In general, for cheapness and reliability chemical treatments cannot compete with stratification or moist prechilling and they are unlikely to play a major role in normal nursery practice in the foreseeable future.

TREATMENTS FOR ENDOGENOUS DORMANCY

X-rays, Gamma rays, light rays in the red region of the spectrum and high frequency sound waves have all been used experimentally to try to overcome dormancy and stimulate germination. Improvement has been reported in some species including *Tectona* (Bhumibhamon 1973), but it has proved difficult to achieve consistent results and the treatments may induce chromosome damage and other abnormalities (Kemp 1975c). Lynn (1967) concluded that there was far more evidence of unfavourable effects from irradiation of seed than beneficial effects. None of these methods are suitable for practical application at the present time.

MOIST PRECHILLING METHODS

Similar Results to stratification in layers may be obtained with many species by storing the seeds moist in polythene bags. As with indoor stratification, seeds should be soaked in several times their volume of water before prechilling, a 48 hour soak at 3–5 °C is suitable for many temperate broadleaved species (Gordon and Rowe 1982).

After soaking, the water is drained off and the moist seeds are then prechilled at 3–5 °C for the period appropriate to each species. Prechilling may be "naked" *i.e.* without any medium, or the seed may be mixed with 2–4 times its volume of a medium such as moist sand, moist peat or a mixture of the two Polythene bags of about 100 micron thickness make suitable containers since they are moisture proof but somewhat permeable to oxygen.

They should be lightly tied and opened weekly when the seed should be mixed and, if necessary, remoistened. A smell of alcohol on opening a bag indicates that anaerobic respiration is taking place because of inadequate oxygen; in this case frequency of opening and mixing should be increased (Bonner *et al.* 1974).

Naked prechilling has the advantage that it is easier to check the condition of the seeds throughout the prechilling operation and there is no need to separate seeds from medium at thve end of treatment. On the other hand, there is evidence that germination in some species benefits from use of a medium.

Gordon and Rowe (1982) reported less than 30% germination in 50 days from seeds of *Sambucus racemosa* treated "naked", compared with 60% in 20 days from seeds in peat/soil mixture, other elements in the treatments being identical. These authors give detailed prescriptions for pretreatment of a large number of temperature broadleaved species; as a general rule "naked" prechilling is satisfactory for species which need only a few weeks' prechilling, while use of a medium is advisable for those which need a longer prechilling period and for all species which need warm moist pretreatment. Periods of prechilling vary considerably from species to species and, to some extent, from seed lot to seed lot within a species.

For *Abies* a period of 3 weeks at 3–5 °C has proved satisfactory (Aldhous 1972). The same temperature and period is effective for most of the cold temperate eucalypts, but some provenances of *E. delegatensis* need 4–8 weeks for fast, uniform germination. For *Nothofagus obliqua* and *N. procera* naked prechilling at 3°–5 °C for six weeks, and surface drying before sowing, gave high germination (usually over 80% in 28 days) under nursery conditions (Rowe and Gordon 1981). But, treatment with gibberellic acid was a reliable and simpler alternative. In species having deep physiological dormancy the prechilling period may be as long as 20 weeks *e.g.* Liriodendron tulipifera(Bonner *et al.* 1974, Gordon and Rowe 1982).

For Fagus sylvatica in poland, after storage for several years at –5 °C and 10% MC, the following pretreatments are recommended (Suszka 1979, Suszka and Kluczynska 1980):

1. Allow to Defrost.
2. Moisten by sprinkling of water and thorough mixing of wetted nuts twice a day for 6 days, at 3 °C, until MC rises to 31%.
3. Leave wet nuts in unsealed containers without any storage medium at 3 °C for a period equal to two weeks longer than the minimum required to produce 10% germination in a sample within two weeks of its transfer to a moist germinating medium. Recurrent sampling and germination testing is needed to estimate this period, which may vary considerably from seed lot to seed lot. MC of 31% to be maintained meanwhile by periodic weighing of containers and remoistening of nuts to restore loss of weight.
4. Sow in moist germinating medium at 3 °C and leave for two weeks. This should initiate germination of radicles.
5. Transfer to temperature of 20 °C to promote hypocotyl and epicotyl elongation and seedling emergence, which are inhibited in this species at 3 °C.

For Large-scale operational sowings, the ideal experimental conditions described above can be simulated to some extent in the nursery by timing the spring sowing so that the nuts experience an adequate period of cold temperature first, followed by the higher temperatures of late spring and early summer.There is Some evidence that Steps (2) And (3) Of the pretreatment described above can be carried out before, as well as after, storage. The advantage is that the nuts are ready for sowing as soon as they come out of store, without the need for a subsequent cold, moist treatment covering several weeks. Nuts pretreated before storage have been successfully stored for 15 months in France (Muller 1982).

TREATMENTS DESIGNED TO OVERCOME DOUBLE DORMANCY

Some species combine more than one form of dormancy at the same time. Pretreatment to break one type of dormancy alone will be largely ineffective, unless it is followed by a second pre-treatment to overcome the other type.

Physical dormancy of the seedcoat may be combined with physiological dormancy of the embryo. In this case seedcoats should be treated first *e.g.* by scarification and cold moist prechilling applied afterwards to overcome embryo dormancy. *Cercis canadensis* provides a good example; single treatments gave less than 10% germination in all cases, while successive applications of seedcoat and internal dormancy treatments gave rapid germination of 45% (mechanical scarification) or 65% (acid scarification)

(Bonner *et al.* 1974). In a Few cases treatment of the seed covering after stratification has proved effective. Recent trials with *Fraxinus pennsylvanica* showed that cold stratification at 4 °C for 88 days produced 35% germination in 3 weeks, compared with 2% in unstratified seed.

But removal of the pericarp after stratification further increased germination to 56%, while the best, but laborious, treatment of stratification followed by removal of embryos from pericarp and endosperm gave 88% (Marshall 1981). Treatment to increase permeability of the pericarp in acorns after stratification benefits germination. Tuskan and Blanche (1980) found that 6 hours of mechanical shaking in distilled water, after 3 months' storage at 0 °C and 30–40% MC gave 90% germination in *Quercus shumardii,* compared with 70% from 6 hours' simple soaking after storage. The shaking treatment also increased MC over that from the soaking treatment, indicating increased permeability.

Fraxinus excelsior combines morphological dormancy (underdeveloped embryo) with physiological dormancy. Warm moist treatment to remove the morphological dormancy should be followed by cold moist treatment to remove the physiological dormancy. For UK conditions Gordon and Rowe (1982) recommend 8–12 weeks warm, followed by 8–12 weeks cold, while in Poland, Suszka (1978 a) recommends 16 weeks of each treatment.

A Number of species, especially in the family Rosaceae, combine mechanical dormancy, caused by a tough, thick pericarp, with physiological dormancy. The same combination of warm moist followed by cold moist pretreatment is effective. In the case of a resistant species, *Crataegus monogyna,* the recommended periods are 4–8 weeks warm, followed by 12–16 weeks cold (Gordon and Rowe 1982).

In some species, germination of the radicle takes place readily at warm temperature, but the epicotyl will not start to grow until,

- The radicle has already started to germinate and
- The seed has been exposed to a period of low temperature.

Examples are *Viburnum opulus* and *Carpinus caroliniana* (Bonner *et al.* 1974). Here again the same sequence of warm moist treatment at 20–25 °C followed by cold moist treatment at 3–5 °C should induce satisfactory germination.

Thus the same combination of warm moist plus cold moist treatments applied in that order is capable of removing several different combinations of double dormancy

SPECIAL TREATMENTS FOR MECHANICAL DORMANCY

The Thick, tough but water-permeable coverings of seeds exhibiting mechanical dormancy prevent embryo growth even when water can be freely imbibed. This mechanical obstruction to germination can be overcome by a period of "warm moist" treatment the length of which varies according to species.

The treatment recommended by Gordon and Rowe (1982) for temperate species is:

1. Soak Seeds in several times their volume of cold water at approximately 3–5 °C for 48 hours.
2. Drain off the water and mix the seeds with two to four times their volume of a moistened, water-retaining medium such as sand, sand/peat mixture, vermiculite.
3. Store at a warm temperature. A constant 20–25 °C or alternating 20° and 30 °C is suitable for many species.
4. Open containers weekly, mix seeds and, if surfaces show signs of drying out, remoisten with water spray.

The period of "warm moist" treatment can be shortened in some species by a preliminary treatment in sulfuric acid. This requires more care and expertise than the use of acid for physical dormancy. Seeds or fruits must be thoroughly dried before treatment and the process should be limited to the partial digestion of the outer layers only, leaving the final weakening of the inner layers to be done by the subsequent warm moist treatment (Gordon and Rowe 1982). In most cases it is preferable to accept the safer but slower method of warm moist treatment alone. Periods of treatment vary from 2 weeks for some species of *Prunus* to 16 weeks for some species of *Crataegus*.

At the end of the appropriate period seeds which possess only mechanical dormancy are ready for sowing. Many species in this class also have physiological dormancy of the embryo.

It May be noted that the warm moist treatment which removes mechanical dormancy is identical with the warm moist treatment which removes morphological dormancy (underdeveloped embryos).

3

Field Management in Seed Production

INTRODUCTION

Careful field management in seed production can significantly mitigate seed-borne diseases in seed crops. Many of the same practices that prevent disease in field crops are used to prevent disease in seed crops. Disease promoting conditions are avoided by not using overhead irrigation under moist conditions, using drip irrigation, and cutting water to allow seed crops to dry before harvesting.

Spacing plants adequately and orienting rows with the prevailing winds aids in increasing air flow—an important factor in minimizing disease. Rotating crops is important to avoid disease build-up in the soil. Of course, starting with clean, disease-free seed for planting stock is also crucial. Additionally, seed growers must use care in harvesting and cleaning seed to avoid post-harvest infection.

Seeds of Change works with our seed growers to select varieties for disease resistance. In this programme disease is intentionally allowed to build up in an isolated field for breeding and selection purposes. Seed crops are selected under the disease conditions for resistance to the disease. In this case care is taken to ensure the stock seed is not harboring seed-borne diseases when used as planting stock for seed production.

Many of Seeds of Change beet varieties have been selected in this manner for resistance to Rhizoctonia, a bacteria that causes scarring and lesions on beet roots. Many of the lettuce varieties offered have also been selected for resistance to Downy Mildew and Sclerotinia. Seeds of Change growers regularly practice rouging (selection) in seed production fields by removing diseased plants from the field.

Over 1,500 microorganisms have been shown to be related to seeds of all types. However only a small percentage of these organisms pose a threat of disease to the following crop. The majority of common diseases are not seed-borne, but initiate from the local environment. As most diseases are regionally based it is valuable to check with local extension services and growers to identify which seed-borne diseases pose a threat in your ecosystem and then place extra care in sourcing varieties of these crops. If there are crops of particular concern, ask your seed company for recommendations and information about their disease management programme. And always use good organic disease management practices in the field.

CAUSE

Black rot is caused by a bacteria, *Xanthomonas campestris* pv. *campestris,* that can infect most crucifer crops at any growth stage. This disease is difficult for growers to manage and is considered the most serious disease of crucifer crops worldwide.

The disease can cause significant yield losses when warm, humid conditions follow periods of rainy weather during early crop development. Late infections can provide a wound for other rot organisms to enter and cause significant damage during storage.

Fig. Black rot Infected Cabbage.

SYMPTOMS

Symptoms of black rot vary considerably depending on the host, cultivar, plant age and environmental conditions. The bacteria can enter plants through natural openings and wounds caused by mechanical injury on roots and leaves. Seedborne bacteria infect the emerging seedlings through pores on the margin of the cotyledons and then spread systemically through the seedling.

Infected seedlings grown in the greenhouse under cool conditions (below 15–18 °C) frequently do not show any symptoms of the disease. When infected seedlings are transplanted to the field and temperatures rise to 25–35 °C during periods of high relative humidity (80–100%), they become stunted with dead spots on the cotyledons and will eventually wilt, and die.

In regions with temperate climates (where temperatures remain cool), disease symptoms on infected seedlings may not always be obvious or appear severe. Infected seedlings grown under cool conditions may ooze bacteria from pores and lesions, which then serve as a source of the pathogen for neighbouring plants.

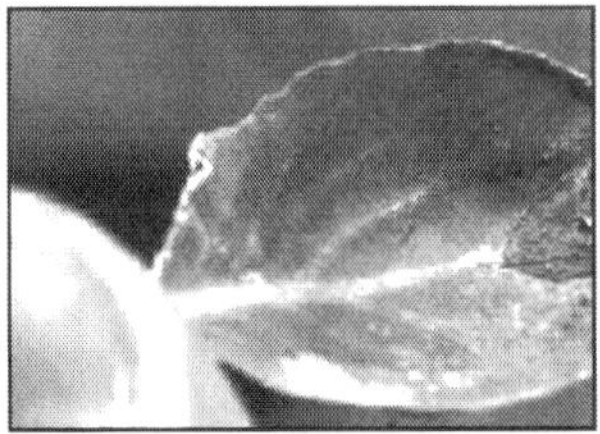

Fig. Young Cabbage leaf with V-shaped Lesion Characteristic of black rot Symptoms.

On older plants, the disease symptoms often appear as yellow or dead tissue at the edges of leaves, similar to tip burn, except the lesion frequently progress into a V-shape with the base of the V usually directed along a vein. Close inspection of infected leaves and stems may reveal black veins running through the infected tissue from which the disease gets its name. Lesions on leaves can expand down towards the base of the leaf causing the leaf to wilt and die. The bacteria produce a sticky polysaccharide called xanthan that eventually plugs the vascular tissue inside the veins causing them to collapse and turn black. The tissue above the plugged, collapsed xylem eventually turns yellow, wilts and dies. During hot humid environmental conditions, the bacteria can move from the leaf into the stem through the xylem.

Once inside the stem, the bacteria can move up or down to other parts of the plant including the roots. Systemically infected plants may produce chlorotic areas anywhere on the leaf. Severely infected leafy cole crops such as kale and cauliflower tend to shed their leaves from the bottom up leaving only a tuft of distorted leaves separated from the root system by a scarred barren stem.

Fig. Black rot Symptoms Appear as dead Tissue at the tips of (a) Kale, (b) Cauliflower, and (c) Cabbage Leaves. Note the V-Shaped Lesion Progressing from the tip along the vein of the black rot Infected Cabbage leaf.

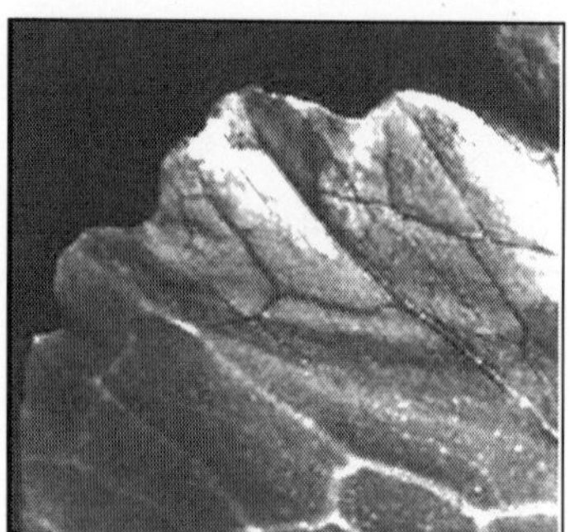

Fig. Black Rotting Veins Running through Black Rot Lesion on tip of Cauliflower Leaf.

Symptoms on cauliflower often appear as black flecks or scorched leaf margins. The curds of infected cauliflower heads often become blackened.

Foliar symptoms may not be visible on infected root crops such as rutabaga and radish but blackened vascular tissue can appear inside the edible root tissue rendering the plants unmarketable. Although some infected plants may appear healthy, cutting across infected stems will reveal characteristic blackened vascular tissue. This is a simple method of determining the presence of the disease.

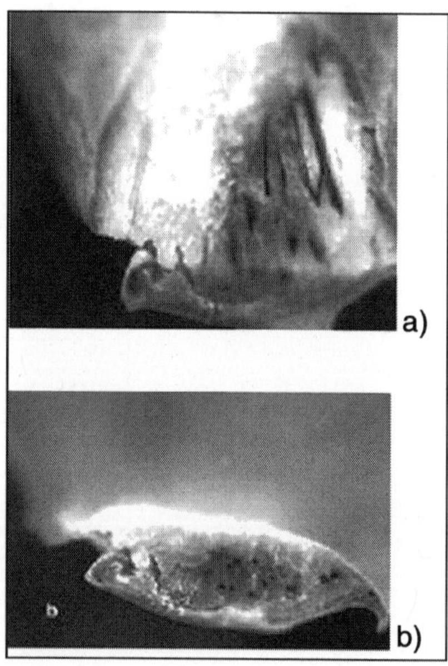

Fig. Cross Section of the base of a Black rot Infected Cabbage (a) Stem and (b) Leaf Revealing the Black, Collapsed Xylem.

Some symptoms of black rot closely resemble those caused by *Fusarium* yellows, which causes the vascular tissue to turn brown. Most commercial crucifer cultivars are resistant to *Fusarium*.

Fig. Leaf Symptoms of Fusarium Yellows Sometimes Appear Similar to black rot Except the Vascular Tissue turns brown Instead of black.

DISEASE SPREAD

Seed contaminated with black rot bacteria is considered the most important source of the pathogen and significantly contributes to the spread of this disease worldwide. As few as 3 infected seeds per 10,000 (0.03% infected seeds) can result in a black rot epidemic. Seed should be tested and certified to be disease free with less than 1 in 30,000 infected seed.

The organism survives in infected crop tissue left on the soil until the crop tissue rots. However, the bacteria do not survive very long in soil as unprotected free living organisms. The black rot bacteria can also infect and survive on many crucifer weeds. This also contributes to the persistence and spread of the disease. It can grow and multiply on host tissue without infecting or causing disease.

Rain splashed bacteria from contaminated plant residue left on the soil or from neighbouring diseased plants is the primary method of disease spread throughout a field. The bacteria enter and exit through water-secreting glands called hydathodes located at the edges and tips of leaves. Hydathodes often produce a drop of water during periods of high humidity early in the morning. The pathogen spreads very quickly when rain droplets contaminated with bacteria splash onto healthy leaves and enter the hydathodes.

Fig. Hydathodes are Special Glands or Pores at the end of Vascular Tissue on leaves through which Water Exudes and are a Natural Opening for Black Rot Bacteria to Infect.

The bacteria move into the leaf veins through hydathodes and begin to multiply, rot and plug the veins. Contaminated water droplets that exude out of hydathodes of infected leaves can then be rain- splashed to other plants. Black rot is more severe and widespread in fields that receive frequent early morning rains, particularly in May and June. Equipment, people, animals and overhead irrigation can further spread the disease. Insects can also spread the bacteria; however, their contribution to the spread of black rot is limited.

VEGETABLE SEED PRODUCTION

Vegetables are harvested for some component of their vegetative (celery, carrot) or reproductive (tomato, broccoli) structure. In most cases, the harvest of the market-use crop requires environmental conditions that are very different from those for reproductive and development of a mature seed crop. For example, ambient temperature, photoperiod, length of growing season, and precipitation amount and distribution may vary for the optimum production of market-use and seed crops.

Ambient temperatures have very different effects on vegetables and their end use. Those vegetables that are harvested for market-use such as carrot, onion, radish, brassicas, and lettuce are typically grown in areas with lower temperatures from those required for a seed crop in order to maintain good vegetative growth and minimize bolting. Seed production of these crops often requires higher temperatures to encourage bolting although extremely high temperatures also hinder pollen growth and fertilization which lead to reduced seed number per floret or fruit.

This is true in onion, tomato, and lettuce. In contrast, beet and carrot market-use crops have been selected to avoid bolting in mild climates but they still require adequate periods of cool temperatures (vernalization) for flowering and seed production. The Pacific Northwest of the United States provides such sustained cool temperatures and this is the reason why beet and carrot seed production occurs in this region while market-use production of these crops is geographically more diverse.

The requirement of some vegetables for specific photoperiods before induction of flowers also limits those areas where vegetable seed production is successful. This effect is dependent on the latitude and the length of the photoperiod which greatly affect seed yield. For example, onion is a vegetable crop that exhibits a wide diversity in photoperiod response and this influences where seed production occurs.

Seed production of long-day onions occurs at latitudes greater than 45° N and short-day onions at latitudes no greater than 35° N. Limited seed production also can be accomplished in glasshouses where supplemental light is provided or light excluded to promote flowering.Even when appropriate photoperiods are present, temperatures must be suitable for the completion of the entire life cycle of the crop during the growing season. Successful market-use crops such as onion, radish, broccoli, and cabbage are harvested before flowering occurs. As a result, these crops are grown in regions with cooler temperatures not suitable for seed crops which require longer periods and warmer temperatures for maturation of seed.

Precipitation amount and distribution also varies for market-use and seed crops. Many vegetable crops harvested for their vegetative structure require large amounts of early season rainfall for most rapid development. Seed crops, in contrast, demand uniform precipitation during development followed by dry conditions during seed maturation. Under these environments, dependable conditions for seed harvest are possible and pathogen infestations minimized.

This is the principal reason much of the vegetable seed production has shifted to the western United States. The Mediterranean climate of the Pacific Northwest assures cool, dry summers and warm, wet winters for optimum development of many biennial seed crops. Southern California and much of Arizona are hot and dry during the summer which is ideal for vegetable seed production. The dry condition of these locations permits seed crops to be irrigated with the proper amount of moisture at the proper time so that optimum seed quality is achieved.

SEED MARKET PRODUCTION

The following represents general requirements and considerations utilized in vegetable seed production.

TILLAGE

The preparation of a soil site for vegetable seed production differs little from that for fresh market production. Soils should have a high water holding capacity to encourage uniform vegetative growth. Depending on the crop and water availability, planting is done on raised, shaped, or flat planting surfaces which require appropriate soil preparation. In all cases, the seed bed must be uniform because most vegetable seed crops are precision planted and uniform emergence and seedling development necessary for optimum performance.

PLANTING

The planting of vegetable seeds for seed production is generally done by direct drilling or transplanting of glass house grown seedlings. Biennial vegetable seed production, however, can be achieved using two planting techniques.

The "seed-to-seed" method is done by planting biennial seed, allowing the resulting plant to overwinter without being transplanted, and harvesting the seed crop the following season. This method does not permit the selection or roguing of root or other genotypic characters. A majority of carrot and onion seed crops are produced by this method which is less expensive and complex than the "root-to-seed" method. The "root-to-seed" method is done by planting biennial seed, removing the resulting plant from the soil (in some cases stored), replanting, and harvesting the seed crop the following season. The replanted plant is referred to as a "steckling." This method permits the selection or roguing of root or other genotypic characters. It also decreases the time required for varietal development. For example, onion bulbs adapted to dry regions can be planted in the spring in the desert regions of the southwest and then shipped for seed production to central California or the Pacific Northwest for planting in the early fall of the same year. This is the most expensive method of vegetable seed production due to transportation, handling, and replanting costs, but the benefits of checking for minimum genetic quality standards before replanting are important in vegetable seed production.

Vegetable seed crops must be planted so that sufficient vegetative development occurs to support optimum fruit and seed development. In addition, the increasing emphasis on hybrid seed production necessitates planting of male and female lines at times that assure successful pollination, a process known as nicking. This may require that genetically different male and female lines be planted at differing times so that synchronous flowering occurs.

Most hybrid broccoli is transplanted and staggered plantings are necessary for the male and female lines to ensure proper nicking. Hybrid cauliflower and squash are direct seeded but because of the differences in time required for the male and female lines to flower, they are planted in a relay fashion where the earlier flowering line is planted first followed by the later flowering line. Vegetable seed production also is unique from most agronomic crops because vegetables such as beets, carrots, and cabbage are biennials and must develop sufficient vegetative growth prior to cool temperature exposure in order that vernalization successfully induces flower formation the following season.

Planting such crops too early causes winter kill or late season pest infestations. Planting too late results in a lack of vernalization which limits flowering and reduces seed yield. In some instances, vegetable seed producers opt not to start the seed crop by seed. Rather, to ensure optimum vegetative growth for vernalization, they plant some developed portion of the vegetative structure. This is commonly done with carrot and beet stecklings and onion bulbs which allows for maximum production of plant biomass before vernalization.

Vegetable seeds also vary greatly in their tolerance to soil temperatures at planting. Pea, radish, and spinach require cool soil temperatures for optimum seedling emergence. Beets, cabbage, carrot, and onion are tolerant of cool soil temperatures although they do better in warmer soils. Squash and melons require warm soil temperatures for optimum seedling emergence. In regions with high summer temperatures, vegetables such as tomato, pepper, eggplant, and cucurbits are planted in the early spring to optimize flowering and seed set in early summer. Later plantings of these seed crops are possible by transplanting young plants previously started in glass houses. This avoids the delayed establishment of plants from direct seedings due to cooler night temperatures and wetter soils encountered in the early spring.

Row spacings and planting densities of vegetable seed crops differ from those for fresh market production. Sufficient space for flower development, air movement to reduce pathogens, unrestricted access to inflorescences by pollinators, mechanical cultivation, and harvest operations are necessary for seed crops.

In some cases (e.g., eggplant, pepper, tomato, muskmelon, watermelon), row spacings used for seed production are the same as those for fresh market production. In most cases, row spacings are different for vegetable seed production. In lettuce, the size of the head will be larger than the fresh market head in order that bolting is promoted.

Large-head types are planted 25 to 30 cm apart while small-head and loose-leaf types are planted 15 to 20 cm apart. Similar considerations are encountered for cabbage. Broccoli and cauliflower produce sizable seed stalks requiring rows 40 to 60 cm apart. Planting densities are often different as well in vegetable seed crops from fresh market vegetable crops. Beet seed production is accomplished at higher planting densities than beets harvested for their roots. Carrot, celery, and parsnip seed production is planted at densities much less than for root or stalk production. Higher planting densities of onions grown by the bulb-to-seed method reduces lodging of the large seed stalk scapes.

The method (wind or insect) of pollination affects row spacings and planting densities. Vegetables that are wind pollinated include sweet corn and spinach. Insect pollinated vegetables include most of the cole crops, carrot, and onion. Row spacings and planting densities are also important considerations in the production of vegetable seed hybrids. In order to maximize the amount of hybrid seed produced, the optimum ratio of female to male rows must be determined.

FERTILIZATION

Vegetable seed crops have similar nutrient requirements as fresh market crops. Where possible, nutrients that stimulate reproductive development can be supplemented with normal fertility regimes. Boron, for example, increases the number of fruiting sites and seed produced per pod in legumes such as pea and bean. Molybdenum also is essential for optimum nitrogen fixation of legume seed crops. Another approach to stimulating reproductive development in vegetable seed crops is to split the application of fertilizers at planting and before flowering. This avoids luxury consumption of the elements at the time of application, improves crop uptake efficiency, and increases the formation of reproductive structures. Generally, lower nitrogen and phosphorus levels are used in seed production because these encourage continued vegetative growth. Because most vegetable seed crops are grown in rows, fertilizers are typically applied either in bands in the planting row or as a side-dressing beside the row after establishment.

WEED AND PEST CONTROL

The management of weeds and pests in vegetable seed production poses additional problems due to their longer growing season not encountered for vegetables grown for market use. This extended growing period necessitates an increase in residual pesticide activity or additional pesticide applications. In some cases, governmental regulations do not permit the use of the same pesticide for a vegetable seed crop that is used for the market use crop.

This is because the approval process is directed at the more economically important value crop. Some pesticides which control insects also detrimentally affect insect pollinators essential for cross pollination or hybrid seed production and cannot be used in a seed production field. Such restrictions on the use of pesticides in vegetable seed production imply that the safe use of pesticides will be more difficult in the future and greater reliance on crop rotations and biological control will be necessary.

Insects and the timing of their infestations affect vegetable seed production. The Lygus bug prefers legume crops and safflower as its primary host. However, once these crops are harvested, it commonly flies to adjacent carrot seed fields and rapidly creates epidemic infestations.

There, it feeds on developing seed embryos and subsequently reduces seed yield and quality. Cabbage looper is a significant pest of brassicas during vegetative development but does not affect seed stalk and pod development. Thus, control of this pest is important during vegetative growth if maximum yield and seed quality are attained. Aphids affect the harvest and subsequent conditioning of brassica seed crops by the development of "honeydew" which adheres to the seed pods and makes separation difficult.

Vegetable seed crops are susceptible to the same diseases as fresh market crops. However, the need to complete their entire life cycle for seed production means that other diseases may also be encountered and require control. Many significant diseases of vegetable seed crops can be controlled by growing the crop in differing geographic locations. For example, the low relative humidity and high temperatures of the southwestern United States reduce the incidence of the seedborne pathogen Cercospora carotae in carrot seed fields commonly found in the high relative humidity, low temperature Pacific Northwest. The incidence of downy mildew is reduced by growing long-day onion seeds in the low humidity areas of the desert southwestern United States.

Diseases in vegetable seed fields can also be minimized by separating the seed crop from disease pests either physically or over time. Physical separation occurs by producing the seed crop in an isolated region from fresh market crops or from the disease causing organism. Celery seed fields are isolated from celery stalk production fields to reduce the incidence of late blight caused by Septoria.

Lettuce seed fields are isolated from leaf and head production fields to reduce the incidence of seed-borne lettuce mosaic virus. Isolation from the disease causing organism is achieved by maintaining weed-free irrigation canal banks and roadsides which serve as alternate hosts or growing the seed crop where the disease causing organism does not exist. To meet phytosanitary requirements for export, onion seed crops must be grown in regions where certain disease and nematodes are certified to be absent.

The lettuce mosaic virus is another serious disease that results in unmarketable lettuce heads. To control this disease, an integrated pest management strategy has been established which focuses on the planting of disease-free seed and control of insects and weeds. Disease-free seed is achieved by planting stock seed in glass houses free of such virus vectors as the green peach aphid (Myzus persicae). Commercial seed lots are certified to have not more than one infected seed in 1,000 seeds that have been screened. Other effective measures include control of common weeds that serve as alternate hosts to insect vectors and frequent roguing of lettuce seed fields to remove any plants suspected of carrying the virus.

Some disease causing organisms have relatively short life spans. In these cases, separation of the seed crop from the organism over time reduces disease problems. Alternating cabbage and other cruciferous seed crops in a three year rotation with non-cruciferous crops decreases the incidence of club root caused by Plasmodiophora brassicae. Nematodes can similarly be controlled in cucurbit and solanaceous seed crops by three year rotations with crops not from these families.

In other cases, diseases can be controlled by storing the seed for a period during which the disease causing organism dies. An example is soft rot (Erwinia carotovora) on celery seeds which dies after two years of seed storage. Hot water baths or chemical treatments also control diseases. Hot water treatment is used on a variety of vegetable seeds to control a number of seedborne bacteria and fungi Many disease causing organisms are carried on the seed coat and most of these are controlled by various topical seed treatments consisting of fungicides, insecticides, and bactericides.

HARVESTING, THRESHING, AND DRYING

Vegetable seed crops are divided into two categories based on the type of fruit harvested. Dry-seeded fruits include the brassicas, legumes, and onion and wet-seeded fruits include the cucurbits, melons, and tomatoes. The methods of harvest and extraction of seeds vary dependent on the type of fruit harvested. Of great concern in dry-seeded fruit vegetable seed production is the potential for loss of yield due to seed shattering when harvesting occurs after physiological maturity. To minimize seed loss, two approaches are employed. Since many dry-seeded vegetables achieve physiological maturity before harvest maturity, they are cut while the plant is still green and placed into rows where the wind further dries the plant and its reproductive structure.

This permits more uniform maturation of all dry-seeded fruits and seed quality is enhanced compared to the more costly option of sequential harvest of mature fruits. Later, after the seeds have reached harvest maturity, they are extracted from the seed heads either by hand or mechanically using a combine. The second approach is to visually monitor the development of the dry-seeded fruit and harvest the crop before shattering occurs.

For example, onion seeds are very susceptible to shattering once the florets have dried and seed germination and quality is reduced if the scape stalks are cut too early. A common harvest criterion has been to examine umbels for the first mature seeds. In indeterminate crops such as carrot, seeds mature in sequential order with those on the primary umbel maturing first followed by those on secondary and tertiary umbels. Secondary umbels produce about 50% the quantity of pure live seed of the primary umbel. Thus, harvest of this dry-seeded crop is complicated by the umbel maturation sequence. Seeds produced from the first lettuce flowers also tend to be larger and of higher quality than seeds produced from later flowers.

Seed harvest from wet-seeded fruits is often determined by fruit colour. Seed quality in pepper as determined by percentage germination and rate of germination is higher in seeds harvested from red fruits compared to those harvested from green fruits. In tomatoes, maximum seed germination is attained when the fruits are at a mature green stage compared to those which are at the red or fully-ripe stages.

After the fruits are determined to be at the correct stage for maximum seed quality, they are gathered and the seed extracted. In some cases, this is done in one mechanical operation where the fruit is picked from the row, conveyed to a crushing mechanism, and the pulp removed from the seed by washing with water in a rotating drum. For high moisture vegetables such as melons and tomatoes, water is periodically drained from the equipment.

For low moisture vegetables such as squash and peppers, water is periodically added to the equipment. In most cases, vegetable seed companies mechanically harvest crops such as tomatoes, peppers, and cucumber in the field, transport the fruit to a central extraction plant, and remove the seeds from the fruit. In tomato, cucumber, and some melons, a mucilaginous layer around the seed must be removed prior to cleaning. This can be done by two methods.

The first, and most common, is the retention of seeds in the crushed fruit pulp for a few hours to more than a day so that fermentation occurs. This process is closely monitored because the fermentation creates heat and mechanical injury to the seeds if continued too long. The second approach is to add sodium carbonate or hydrochloric acid to the crushed fruit extract to degrade the mucilaginous layer. After the fruit is crushed and any mucilaginous layer removed, the seed is separated from the pulp and passed along a water sluice containing gates along its entire length. Mature high quality seeds sink behind the gates and are separated from the pulp by the running water.

After the pulp is removed, the sluice gates are opened and the mature seeds captured on a screen. At this stage, the wet seed must be promptly dried to avoid heating and loss of seed quality with increasing imbibition. To accomplish this, the wet seeds are collected and placed into a spin dryer to remove excess water and the moist seeds are evenly spread on screens and exposed to large volumes of heated air. After drying is completed, the seeds are handled and conditioned in the same way as employed for dry-seeded vegetable crops. Examples of conventional harvest methods for dry- and wet-seeded vegetable seed crops are provided in

CLEANING AND STORAGE

The physical quality of the seed lot following harvesting, threshing, and drying must be evaluated to determine whether precleaning is necessary. If the seed lot possesses large amounts of inert material such as stones or plant debris, it should be passed over a scalping machine. If the seed possesses undesired appendages which cause the seeds to clump together such as carrot, then it must be passed through a debearder to ensure free-flow of individual seed units.

Impact debearders are also used in multigerm beet seeds to split the fruit into single seed units. After precleaning, vegetable seed crops are conditioned in the same way as other seed crops using the same principles and equipment Some additional improvements can be made. For example, greater seed density improves the seed quality of lettuce, tomato, and onion crops. This can be accomplished by placing the seed mass into a solution of specific osmotic concentration. The more dense seed fractions are retained and the more buoyant seeds removed. After cleaning is completed, vegetable seeds are stored using the principles of successful seed storage Approximate seed moisture contents of vegetable seeds held at differing relative humidities in storage are provided

LIQUID SEED PRODUCTION

TOMATO (LYCOPERSICON ESCULENTUM)

Introduction

Tomatoes originated in Peru and Bolivia as small, wild fruits. Migration of local natives brought the tomato to Central America and Mexico where large, domesticated fruits were developed. Archaeological evidence reveals that tomatoes were a popular fruit of the Aztec and Toltec Indians who planted them with their corn. Spanish settlers to the New World carried tomatoes back to Europe where they quickly became established throughout the continent. Initially, tomatoes were grown as ornamentals because of their colourful, bright red fruits and the assumption that the fruits were poisonous. By 1750, however, tomatoes were cultivated in England and were a popular component of soups.

Tomatoes came to the United States in the late 1700's by European colonists. Seed catalogues offered tomato seeds for sale as early as 1812 but tomatoes were not considered a major food crop until the 1860's. During that period, ambitious breeding programs were established to expand the number of tomato varieties. Today, continuing efforts at improvement of extra early varieties, development of disease resistant varieties, and improvements in plant type and fruit quality have led to tomatoes being accepted as one of the world's most versatile food crops.

Tomatoes are the second most consumed vegetable in the world behind only the potato. They are eaten fresh or processed and can be stewed, fried, baked, or used as juice. In addition to this versatility, tomatoes are also nutritional. They are low in calories (20 calories per average size fruit) and an excellent source of iron and vitamins A and C. They also contain small amounts of the B complex vitamins thiamin, niacin, and riboflavin.

Plant Development

The tomato is a warm-season herbaceous perennial that requires three to four months from seeding to production of fruit. It grows best at temperatures between 18 and 29°C (65 and 85F), is often killed at temperatures below freezing, and does not grow at temperatures above 35°C (95°F).

Seed

Tomato seeds are 3 to 5 mm (0.12 to 0.20 inches) in size, flat, reniform, and have a grey, slightly hairy seed coat. The embryo is large and coiled in a limited, fleshy endosperm

Vegetative

Tomato plants are typically viny, prostrate, and are either determinate or indeterminate based on whether the apical stem terminates in an inflorescence. Most shoots form in the axils of leaves. The tomato plant produces a deep tap root with extensive secondary roots.

Reproductive

Tomato flowers are yellow and formed on racemes. The stamens are fused with the lower part of the corolla forming a staminal cone. Generally, pollen release occurs prior to elongation of the style through the staminal cone resulting in self-pollination. However, temperature has a major effect on this process. If temperatures are below 15°C (60°F) or above 29°C (85°F), pollen release is restricted resulting in incomplete fertilization of ovules. This causes collapsed fruit walls and formation of deep indentations in the fruit, a phenomenon called "catface." Hot, drying winds also cause tomato flowers to abscise from the plant. The best shaped and fullest tomato fruits result from optimum fertilization of ovules. The mature tomato fruit is botanically a berry. Fruit colour varies from the common red to yellow, pink, and orange.

Self-pollinated and hybrid tomato seeds are produced and marketed. Production of hybrid tomato seeds requires the planting of male (pollen) and female (hybrid fruit) parents. A ratio of 1:5 male to female plants is adequate for hybrid seed production.

The male parent is typically planted earlier than the female in order that sufficient flowers are available for pollen collection prior to flowering of the female. Pollen is collected from the male plants by a mechanical hand-held vibrator that is placed on opened flowers. The vibration causes the anthers to release their pollen which is collected on a receptive plate, disk, or tube attached to the vibrator.

Female plants are often supported in the field so that the flowers and developing fruit are off the ground. Hybridization is accomplished by selecting flowers from the female plant in the morning for emasculation. Flowers must be in the late bud stage of development. The bud is opened with a pair of forceps and the staminal cone removed. Care must be taken to ensure that the receptive sigma is not damaged during this process. Usually, transfer of the male parent pollen is accomplished by dabbing the pollen onto a fine brush that is then placed on the stigma of the emasculated flower within a day or two after pollen isolation. After pollination, the flower is tagged and half of the calyx removed to document that the developing fruit is a hybrid.

Seed Production

Tomatoes are grown for three principal purposes: processing, field shipping, and greenhouse production. The processing areas are located in the eastern (Ohio, Indiana, and Ontario) and western (California) parts of North America. The major production of tomatoes for local shipping are produced in Florida, California, and South Carolina. Greenhouse tomatoes represent 65 to 75% of the vegetable production that occurs in greenhouses. The leading areas are Michigan, Pennsylvania, Virginia, Illinois, and Ontario.

Seed production in greenhouses has occurred on a limited scale. As a result, most of the following discussion will consider tomato seed production as practiced under field conditions.

Tillage

Crop rotations are useful in tomato seed production because of the crop's susceptibility to a variety of diseases. Rotations with soybeans, sugar beets, wheat, and beans have been successful. One of the most common rotations in the California San Joaquin Valley is with cotton.

Fields used in tomato seed production should be Fall plowed to a depth of 20 cm (8 inches). This practice not only improves the physical structure of the soil but also encourages the decay of roots and other plant parts thereby enhancing the level of soil organic matter. If organic matter remains low, manure can be applied at a rate of 13,440 to 17,920 kg per hectare (6 to 8 tons per acre) and incorporated into the soil at the time of Fall plowing.

Manure applications should not be applied in the Spring because microorganism activity is increased and competes with the tomatoes for available nitrogen. After plowing, beds are prepared 137 to 167 cm (54 to 66 inches) apart with furrows 20 to 15 cm (8 to 10 inches) deep depending on the soil type. The beds allow more uniform fruit ripening and better drainage following heavy rains.

Planting

Tomatoes can be planted either by direct seeding or transplants. When 152 cm (5 foot) rows are used, seed is planted to give a spacing of 20 to 30 cm (8 to 12 inches) or 30 to 40 cm (12 to 16 inches) if double rows are used. Planting should not begin until the soil temperature has reached 10°C (50°F) since germination will not occur below this temperature. The optimum range of soil temperatures for seed germination is between 15°C and 29°C (60°F to 85°F).

Direct seeding is accomplished by precision planters that equally space the desired number of seeds in the row. Seeds should be planted at a depth of 1.3 to 2.0 cm (1/2 to 3/4 inch) to obtain a population of 24,700 to 49,500 plants per hectare (10,000 to 20,000 plants per acre) in single rows. Planters should be modified to allow the application of anticrustants (keep the soil from crusting and thereby facilitating uniform seedling emergence) and starter fertilizers applied either in a solid or liquid form.

Tomato seeds can also be pelleted with clay/fertilizer coatings that allow the use of traditional plate planters. Because rapid emergence and seedling establishment are desirable in tomatoes, another approach has been to pregerminate tomato seeds in a solution of anticrustants and fertilizers. The germinated seeds are planted by a planter which makes a furrow, the germination medium including the seeds flows into the furrow, and the furrow is closed.

Many seed companies are using a variation of this approach for marketing tomato seeds. They osmocondition the seed (a prehydration process) in an osmoticum such as a salt or in a solid matrix such as vermiculite. The seeds are hydrated to a moisture content below that required for germination and then redried. Following osmoconditioning, the dried seeds are handled and packaged in traditional ways. Transplanting is another important planting approach used in tomato seed production. Transplants can be obtained either from local greenhouses or can be from transplants produced using grower seed.

They are at the correct stage for planting when the first pair of seedling leaves appear. When possible, transplanting should be done either during the afternoon or on a still, cloudy day to avoid transplant shock. The transplants should be thoroughly watered before planting and a starter fertilizer high in phosphorus added. Transplants every 40 to 45 cm (16 to 18 inches) in rows 168 cm (66 inches) apart provide a population of 14,800 to 19,700 plants per hectare (6,000 to 8,000 plants per acre).

The choice of direct seeding or transplants is made by the seed grower. Advantages of direct seedling over transplants are:

1. Reduced cost per hectare (acre)
2. Less introduction of plant diseases, and
3. Greater flexibility of planting time, variety selection, and plant population. Transplanting, when properly done, offers the advantage of more rapid plant establishment and higher seed yields.

Fertilization

While tomatoes grow on almost any soil, soils that are sandy, light, well-drained, and high in organic matter and mineral nutrients produce the best quality and highest fruit yields. Optimum soil pH should range from 6.0 to 6.5.

Tomatoes require an abundance of the three major elements: nitrogen, phosphorus, and potassium. Adequate nitrogen is important to enhance foliage growth which has a major bearing on crop maturity and protects fruits from sunscald. Phosphorus influences fruit quality by stimulating vigorous root growth that enables more nutrients to enter the plant thereby promoting sturdy stem growth and healthy leaf formation. Tomatoes use large amounts of potassium. This element is important in stimulating early plant growth and regulating normal carbohydrate and protein metabolism.

Fertilizers are generally applied broadcast before Fall plowing. All the phosphorus and potassium and one-half the nitrogen are added at this time. Between the time the plants have two to three leaves and when the flowers are first observed, a side-dress nitrogen application is useful. Nitrogen should not be applied after fruit formation because it stimulates vegetative growth.

Because fields vary in the absolute quantities of fertilizers to apply for optimum tomato growth, it is difficult to make specific recommendations without a soil analysis. However, the following general guides of fertilizer ratios (N-P-K) may be useful for certain situations. 1) tomatoes grown on light soils in rotation with other crops (1-2-1 or 1-2-2), 2) tomatoes grown on sandy soils (1-2-2), and 3) tomatoes grown on loam, silt loam, and clay loam soils

(1-3-1 or 1-4-1).Tomatoes are sensitive to deficiencies in the minor elements manganese, boron, copper, zinc, and molybdenum. Lack of rapid calcium movement from vegetative to fruit tissues under conditions of high moisture often results in a condition known as blossom-end rot of the fruit.

Weed and Pest Control

Mechanical and chemical control of weeds is practiced in tomatoes. Cultivation to eradicate weeds begins soon after seedling emergence. Early cultivations are close to the seedling and 2.5 to 5.0 cm (1 to 2 inches) deep. Later cultivations are further from the plants and shallower to minimize injury to rapidly spreading surface roots. Chemical control also involves application of an array of pesticides specific to the weed problem. Weeds that are difficult to control in tomato fields include nightshade, pigweed, lambsquarters, green foxtail, wild oats, nutsedge, smartweed, and morningglory.

Tomatoes are subject to a variety of disease and insect pests. Crop rotations on a two to three year cycle with nonsolanaceous species reduce pest problems and are a sound seed production practice. Very few of these diseases infect the seed directly but they can reduce seed yields and cause decreases in seed quality as a result of reduction in vegetative growth.Common tomato diseases include early blight, septoria leaf spot, bacterial canker, bacterial wilt, fusarium wilt, damping-off, and tobacco mosaic virus.

Early blight (Alternaria solani) affects the plant at any stage of development but the greatest plant injury is observed when the fruit begins to mature. Injury is first observed with the formation of small, irregular dead spots that enlarge and commonly show a ridged, concentric ring pattern. The pathogen is believed to be soil-borne and can be carried by infected seeds to new locations. Septoria leaf spot (Septoria lycopersici) is one of the most destructive tomato leaf diseases. It appears first as small water-soaked spots on the underside of leaves, usually after fruit formation. Fruits are rarely affected but severely infected leaves completely dry out.

Bacterial canker (Corynebacterium michiganese) causes a downward curling of the lower leaves which eventually wilt and die. A diagnostic feature of this disease is the presence of a shiny reddish-brown streak in the vascular tissue of a stem cut lengthwise. The fruit is also invaded and snowy white raised spots appear on its surface. Developing seeds are infected and can carry the disease to new fields.

Bacterial wilt (Pseudomonas solanacearum) causes a rapid wilting and death of the entire plant unaccompanied by any yellowing of leaves or spotting of fruit. Freshly cut stems reveal a grayish, slimy exudate in the water soaked pith. Two important bacterial diseases which invade the seed are bacterial spot (Xanthomonas vesicatoria) and bacterial speck (Pseudomonas syringee). Fusarium wilt (Fusarium oxysporum f. sp. lycopersicum) is a common tomato disease that is manifested by a yellowing, wilting, and death of tomato leaves that begins at the base of the plant and moves upward. Damping-off (Pythium spp., Phytopthera spp.) attacks small seedlings at the soil surface. Initially the stems shrivel, then the plant topples over and dies. Other important fungal diseases of tomato are corky root (Phyrochaeta lycopersici), verticillium wilt (Verticillium dahliae), and powdery mildew (Leveillula taurica). Tobacco mosaic virus causes mottled, distorted foliage that reduces fruit yields. Insects spread the diseases after feeding on infected plants.

The principal insects that damage tomatoes are cutworms, flea beetles, aphids, hornworms, grasshoppers, and wireworms. Of these, only the cutworms and hornworms are consistent problems in tomato producing areas. Cutworms are smooth, plump caterpillars that feed at night on young tomato plants. They ultimately cut the plant off at the soil surface. Hornworms are green caterpillars characterized by distinctive V-markings on the side and a large horn on the end. They are voracious feeders and one or two hornworms can completely defoliate a mature plant.

Harvesting

Tomato fruits can be hand or machine harvested. In the case of hybrid seed production, the marked or tagged fruits are always hand harvested. All mechanical tomato harvesters are based on a "once-over" principle in which the entire tomato plant is harvested. They have four components: pick-up mechanism, fruit and vine separating area, fruit sorting area, and container loading mechanism.

The pick-up mechanisms cut the vine at or slightly below the ground surface and the plant and fruit are moved by a conveyor to the fruit and vine separating area. There the plants are transferred to a reciprocating mechanism that begins a shaking action which allows the fruit to be separated from the plant and to be conveyed to sorting belts. The remaining plant is discharged into the field. Unacceptable fruits are removed by hand sorters or electronic colour sorters. Acceptable fruits continue to the discharge area where they are loaded into containers. After harvesting, the tomato fruits are transported to an area dedicated to seed extraction, washing, drying, cleaning, and storage.

Extraction

The tomato fruit is placed into a crusher that pulverizes the fruit and separates the gelatinous seed from the remaining fruit tissues by pressing them through screens. The extract containing the gelatinous seed material must still be separated from the remaining pulp. This is accomplished naturally by fermentation or with chemicals. When fermentation is used, the seeds remain in the extract until the gelatinous coating has been degraded by microorganisms; a process that can take up to three days under normal field conditions. The fermentation mixture is frequently stirred and inspected until this occurs. Excessive fermentation reduces seed quality. Chemical methods include the use of hydrochloric acid or sodium carbonate. Enough concentrated hydrochloric acid is added to the fermenting extract to make a 5% concentration.

The acid rapidly degrades the gelatinous seed coating resulting in the production of a clean seed sample. Similar cautions concerning excessive digestion time on seed quality and concern about worker safety when using concentrated acids need to be emphasized with this technique. Sodium carbonate accomplishes the same task as hydrochloric acid without the potential danger to workers.

However, it darkens the seed coats and makes the seeds less attractive as a commercial product. After seed extraction, the seeds are immediately washed to remove digested materials and any remaining chemicals. This is accomplished by placing the seeds and extraction in long, slightly angled water troughs that contain a series of riffles at specific intervals.

Conditioning

Tomato seeds are generally clean and free of unwanted material and other crop seeds following washing. They are further cleaned and sized on an air-screen cleaner.

Storage. When stored under cool, dry conditions and at a seed moisture content of 6%, tomato seeds are relatively long lived. They can be successfully stored under these conditions for 4 to 6 years.

CUCUMBER

(Cucumis sativus)

Introduction

Cucumber likely originated in India because a close weedy relative, Cucumis hardwickii, has been located there. Evidence exists that the cucumber was cultivated in western Asia over 3,000 years ago. From there, it spread westward into Egypt and was introduced throughout Europe by the Greek and Roman empires. Columbus and Spanish settlers brought cucumbers to North America where they quickly became established as a popular crop.

Cucumbers are used in two principal ways:

1. Fresh when they are sliced, diced, and cubed in salads and
2. Pickled in sweet, sour, and dill brines. Nutritionally, the cucumber is low in calories and low in mineral and vitamin food value. Its excellent flavor and texture, however, continue to make cucumbers among the most prized of vegetables.

Plant Development

Cucumbers are considered a warm-season crop which grows best at average daily temperatures from 18 to 24°C (65 to 75F). Young plant development can be severely retarded if exposed to frost. Cucumbers are noted for requiring the shortest interval from planting to fruit production of any major vegetable crop.

Seed

The seed is flat, oval-ovate in shape, 10 to 15 mm (0.4 to 0.6 inches) long, pointed at the hilum end and rounded at the opposite end. The seed coat is smooth, white to black, with a marginal groove on each side near the base. Internally, the embryo is surrounded by a thin perisperm. The embryo is large, having completely digested the endosperm, and is composed of two leafy cotyledons, each with a prominent midrib, attached to a small, straight embryonic axis. The 1,000 seed weight is 25 g (18,000 seeds per pound).

Vegetative

The cucumber is an herbaceous, prostrate, annual vine that produces from three to eight angled stems covered with short, stiff hairs. The leaves are simple, large, triangular and palmately five-lobed with the middle lobe being sharply pointed. The root system produces a taproot that penetrates more than a meter (39 inches) into the soil with numerous shallow, lateral secondary roots. The lateral root extension is often equal to that of the above ground shoot growth.

Reproductive

Cucumbers are typically monoecious plants that produce imperfect yellow male and female flowers. The flowers are yellow and 2 to 3 cm (0.8 to 1.2 inches) in diameter. Male flowers generally appear first and in clusters in the leaf axils.

Female flowers occur after the male flowers appear with usually one flower found in each leaf axil. The appearance of either male or female flowers can be altered by the environment. Male flowers are generally formed under long days while female flowers occur under short days. The flowers are cross-pollinated by insects. Bee hives are often placed around the periphery of cucumber seed fields to enhance pollination. One hive may pollinate up to 1.2 hectares (3 acres) of cucumbers.

Hybrid cucumbers have been developed with the recent discovery of plants that possess a gynoecious flowering habit. Such plants produce primarily female flowers. This unique genetic trait is retained by treating the gynoecious plant with gibberellic acid (1,000 ppm) or silver nitrate (600 mg/L) to induce the formation of male flowers for self-fertilization for production of stock seed of the female parent. Hybrid seed production is achieved by planting gynoecious varieties treated with ethylene to suppress male flower formation. If ethylene is not used, male flowers must be removed by hand from the seed plants every morning. The female:male row ratio is typically 4:1 arranged in a pattern of 8 female rows between two male rows. After flowering and pollination, the male parent row is plowed under so that only hybrid fruit are harvested. Hybrid cucumbers are noted for being earlier, producing a concentrated set of fruit, and being higher yielding than open-pollinated varieties.

The cucumber fruit is oblong and contains three seed-bearing placentae. Developing seeds enhance fruit growth. Unevenly shaped fruit can be attributed to incomplete pollination or environmental stresses that hinder uniform seed development.

Seed Production

Cucumbers can be grown almost anywhere. They require warm, dry weather for maximum production and are particularly sensitive to frost. Major cucumber producing states are Florida, Texas, Michigan, and California. Most cucumber seed is produced in the Sacramento Valley of California, and in Colourado and Michigan.

Tillage

Cucumber seed fields should be rotated from other cucubits such as melons. Cucumbers can be grown on almost any soil. Cucumber fields should be plowed deeply and early so that the field is settled prior to planting. Additional disking and harrowing operations may be necessary to smooth the seedbed. If irrigation is used or if the soil is poorly drained and heavy, raised beds are necessary. This is accomplished using a lister (a double moldboard plow) that makes the furrows 20 to 25 cm (8 to 10 inches) deep and the beds are formed by back-furrowing with a turn plow. A raised seed bed offers two important benefits: 1) the vines and fruits remain above the water level and 2) the raised beds are warmer; providing additional protection against frost injury.

Planting

The time of planting is delayed until the danger of frost is over. Because of the irregular, flat seed shape, most cucumber seed is coated so that precision planting can be accomplished. Direct seeding using drills into rows that are 91 to 122 cm (3 to 4 feet) apart at a rate of nine seeds per meter (three seeds per foot) is practiced. After emergence, the seedlings are thinned by cutting (pulling creates injury to the developing roots of the plant to be retained) to one plant per foot. When the initial vegetative leader stem has produced three to five leaves, it is removed by pinching to encourage growth of the remaining two lateral stems and their fruits. This results in the development of four or five fruits per parent plant.

Fertilization

Optimum growth of cucumbers requires soils with high organic matter and a pH of 6.5 or above. The organic matter is often supplied by green manure crops that are turned under before they reach maturity and become woody. It is difficult to provide a specific fertilization recommendation because of the diversity of soils on which cucumbers successfully grow. However, a fertilization regime of 1-2-2 is suitable before planting under non-irrigated conditions. Under irrigated conditions, the ratio should be 2-1-1 with half of the nitrogen applied a month after emergence to to replace that lost from leaching.

Weed and Pest Control

Chemical weed control in cucumber is expensive and uncertain so the most effective practices have concentrated on land management and mechanical cultivation. One approach has been to use a stale seedbed in which the field is left untouched until two weeks prior to planting. At that time, the site is sprayed with a contact herbicide to kill all plants followed by direct seeding of cucumber. Another approach has been to plow the killed plants under the soil following herbicide application to increase organic matter. Cultivation 5 to 10 cm (2 to 4 inches) deep begins as soon as seedlings have emerged. Later cultivations, however, must be shallow to avoid damage to subsurface roots and not close to the plant in order to minimize damage to the vine. After vines have covered the ground, further cultivation is not possible.

Cucumber is exposed to a variety of diseases and these are more prevalent in the humid eastern areas than the dry western areas of North America. Downy mildew, bacterial wilt, anthracnose, and root knot are common diseases that are destructive to cucumber as well as a number of other vegetable crops.

Three specific diseases that affect cucumber are discussed because of their relationship with seed production: angular leaf spot, cucumber mosaic virus, and scab. Angular leaf spot (Pseudomonas lachrymans) causes small, angular, water-soaked spots on the leaves and fruits. The causative bacterial organism over winters in the soil and on the seed. Seeds should be tested to determine the presence of this organism. Cucumber mosaic virus causes a dwarfing, mottling, yellowing, and wrinkling of leaves, a warting of fruits leading to loss in yield.

The virus over winters on the roots and seeds of common weeds including pokeweed, milkweed, and groundcherry. It is transmitted to the cucumber crop by aphids and the striped cucumber beetle. Control includes eradication of the host weeds and use of resistant cucumber varieties. Scab (Cladosporium cucumerinum) produces sunken, dark-brown spots on the fruits and is most damaging in cool, moist weather. The causative fungus over winters in the soil and on the seed. Control includes the use of seed treatments such as ziram and maneb.

Insects also reduce cucumber yields. The most destructive insects are cucumber beetles, aphids, and spider mites.

Irrigation. Cucumbers require high levels of water during vigorous vegetative and reproductive growth. Even in areas where rainfall is plentiful, periodic droughts substantially reduce yields and irrigation is necessary in these instances. While furrow irrigation is preferred, when overhead irrigation is used, it should be applied early in the day to permit the vegetation to dry out prior to nightfall and thereby minimize fruit rooting and foliar diseases.

Harvesting and Extraction

Fruit maturity affects seed quality in cucumber. For maximum seed quality, cucumber fruits must reach full maturity. This is signaled by a change in colour from green to yellow and by a withering of the fruit stalk. Fruits can also be examined to determine seed maturity. Seeds that separate easily from the flesh of a sliced cucumber fruit are ready for harvest. At this stage, the fruits are typically harvested by hand and carried to a central seed extraction, drying, and storage facility for stock seed or small seed lots. For commercial seed, the plants are windrowed and then machine harvested. The slurry is augered into steel bins and then taken to the plant for seed extraction.

Seed extraction is accomplished by natural fermentation or acid treatment. In the former process, fruits are first sliced and macerated and the surrounding pulp, juice, and seeds allowed to ferment for about 4 to 6 days under normal field conditions. In the latter process, commercial hydrochloric acid (3 fluid ounces) or sulfuric acid (1 fluid ounce) added to and stirred with 11.3 kg (25 pounds) of the sliced, macerated fruit tissue. After 15 to 30 minutes, water is added to the mixture and the mature seeds sink to the bottom while the digested pulp floats to the top. Whether fermentation or acid extraction is used, the seeds must be immediately washed to preserve seed quality. This is done by collecting the seeds on screens and adding the seeds to water troughs with riffles as described for tomato seeds. The remaining pulp material and acid float off while the cleaned seeds are retained on the riffles.

Drying

After extraction, the seeds are immediately dried. This is done either in the sun with frequent stirring and turning of the seed or with artificial batch driers that have a rotary paddle to expose the seed to heated air. The initial drying temperature should not exceed 40C (104°F) and the seed should be dried until it reaches a moisture content of 6% prior to cleaning and storage.

Cleaning

Cucumber seeds may still contain slight amounts of fruit debris following drying. This can be removed by an air-screen machine. Light and immature seeds can be separated from mature seeds on a gravity table.

Storage

Cucumber seeds stored at 6.5% moisture content under proper temperature and relative humidity conditions can remain viable for 5 to 7 years.

TECHNICALITY REQUIREMENT IN SEED PRODUCTION

The basic requirement of seed production is availability of improved varieties/hybrids and their demand among the farmers. Preferably the variety should be realsed and notified for certified seed production, however, any kind/variety can be multiplied for Truthfully Labeled Seed (TFL).

ISOLATION REQUIREMENTS

The cucurbits are cross pollinated in nature and honeybees are major pollinator, thus for pure seed production an isolation distance all around seed field is necessaryto separate it from fields of other varieties, fields of the same variety not confirming to varietal purity requirement. The isolation distance of 400 m for C.S. and 800 m for F.S. and at least1000 m isolation is required for breeder seed production. It is important to mention that muskmelon, longmelon and snapemelon (phoot) can cross with each other. Similarly, species of genus Cucurbita have the risk of crossability among them. The cucumber can cross easily with its wild relative Cucumis hardwickee found in wild form in sub-mountanious regions of Himalayas.

CHOICE OF SEASON AND AREAS OF SEED PRODUCTION

Seed crop should be raised in such a seasons which remain dry at the time of seed maturity and seed extraction. Rainy season is preferred over summer season for raising seed crop. Locations are also important in seed production with reference to seed yield and quality of seed. To harness the advantage of climate, private sector seed companies are organizing their seed production in these areas.

MAJOR CUCURBITS SEED PRODUCTION REGIONS IN INDIA

Crop	Areas	Region
Muskmelon, Longmelon	Punjab & Haryana	Northern Region
Bittergourd	Eastern U.P., Faizabad and Jaunpur	Northern Region
Cucumber and Muskmelon	Azamgarh, Ballia and Gonda in U.P. Jalana (Maharashtra)	Northern Region Central Region
Pumpkin and Ridge gourd	West Bengal (South West)	Eastern Region
Sponge gourd, Watermelon, Cucumber	Telingana, Karool and Vijayawada (North A.P.) Ranibenur (Karnataka)	Southern Region Southern Region
Watermelon, Bottle gourd and Bitter gourd	Costal districts of A.P.	Southern Region

Roguing

Seed crop is to be monitored at various stages of crop growth for removal of off-type and obviously should be carried out before flowering to avoid natural cross-pollination.

However, fruit set and complete fruit development stages are also important.

The crop-wise main features described here under for effective roguing:

- *Muskmelon:* Fruit shape, colour, rind colour, skin (netted/plain), flesh colour (orange/red), TSS and cavity size.
- *Watermelon*: Fruit shape, colour, rind colour, flescolou(red/yellow/white).
- *Longmelon*: Fruit shape, colour, bitterness.
- *Cucumber*: Fruit shape, colour, presence of spines, spines colour, colour of ripen fruit (green, yellow, white or orange)
- *Pumpkin (Squash, summer)*: Fruit shape, colour, flesh colour.
- *Gourds (Bottle gourd, Bitter gourd and Luffa etc.)*: Fruit shape, colour, stripe, neck etc.

MATURITY OF FRUIT

Cucurbits takes fairly long time to attain harvestable maturity. The maximum period is required in crops like pumpkin (Cucurbita moschata), ashgourd and watermelon, however, muskmelon, round melon and bitter gourd take relatively lesstime.

The maturity also influenced by the environmental factors and crop management (trailing etc.). The maturity period is shorter in summer season than rainy season.

HARVESTABLE MATURITY IN CUCURBITS IN INDIA

Besides days to maturity, some other parameters like change in colour also used as crietaria of maturity index.

- *Cucumber and summer squash*: Fruit turn pale yellow to golden yellow and attached with plant.
- *Pumpkin*: Fruit reddena and seeds inside the shell breake readily from pulp.
- *Muskmelon*: Full slip stage.
- *Watermelon*: Fruits are ready for harvest when they reach edible maturity, fruit colour change from green/white to pale yellow of under side of the fruit.
- *Bitter gourd and snake gourd*: Fruit turn to bright yellow.
- *Bottle gourd*: At maturity fruit colour fade to starw freen or pale yellow.
- *Luffa*: Complete drying/fruit turn to gray colour.

Crop	Variety	Period in Days (Seed to Seed)
Bitter gourd	Priya, Pusa Vishesh	65 days
Snake gourd	TA-19	65 days
Bottle gourd	Arka Bahar,	75 days
	Pusa Naveen	80 days
Pumpkin	Ambali	90 days
	Pusa	(Rainy season)
	Vishwas/Vikas	100 days
Ash gourd	Local	90 days
Cucumber	Mudikode Local	70 days
Watermelon	Sugar Baby	85 days
Muskmelon	Pusa Madhuras	80 days

SEED EXTRACTION

There are two method of seed extraction employed in cucurbits:

1. *Dry method*: The dried fruits are cut from one side and the seeds comes out from the fruit *e.g.*, sponge gourd, ridge gourd, snake gourd.
2. *Wet method*: This method is employed for seeds extraction of cucumber, muskmelon, watermelon, ash gourd, bitter gourd, round melon and long melon. The fruit of cucumber and bitter gourd, summer squash and long melon are cut longitudianlly and seed is scooped out while fruit of muskmelon and pumpkin are cut into two piece and seed is scooped out from cavity. However, in case of watermelon and ash gourd whole central portion are manually scooped out and macerated to separate the seed from pulp. In wet method, the seed extraction done by three ways:
   - *Mechanical Extraction:* In this method the fruits are cut into pieces and macerated by machine. The seeds are separated out from pulp by floating with water. This method is quick, less expensive and seeds retain good lusture, but require good amount of water. This method is applicable in bottlgourd, watermelon, roundmelon and ash gourd.

     Natural Fermentation: The scooped material kept in wooden/ plastic or steel vesel for 48 hours at room temperature and stired 2-3 times and then seed is washed throughly with water 2-3 times. The main problem with this method are discolouration and poor lusture of seed.
   - *Chemical Extraction*: 25-30 ml. of HCL or 8-10 ml. of commercial H_2SO_4 added per 5 kg of pulp and some quantity of water is mixed, stiring of pulp is done to enhance to separation and left

for 30 minutes. The impurities will float and seed will sink. The seed should be washed throughly with clean water. This is quick method but accuracy of acid and time is important.

CONQUEST OF PRODUCTION

Even today we see men and women denying themselves necessaries to acquire mere trifles, to obtain some particular gratification, or some intellectual or material enjoyment. A Christian or an ascetic may disapprove of these desires for luxury; but it is precisely these trifles that break the monotony of existence and make it agreeable. Would life, with all its inevitable sorrows, be worth living, if besides daily work man could never obtain a single pleasure according to his individual tastes?

If we wish for a Social Revolution, it is no doubt in the first place to give bread to all; to transform this execrable society, in which we can every day see robust workmen dangling their arms for want of an employer who will exploit them; women and children wandering shelterless at night; whole families reduced to dry bread; men, women, and children dying for want of care and even for want of food. It is to put an end to these iniquities that we rebel.

But we expect more from the Revolution. We see that the worker compelled to struggle painfully for bare existence, is reduced to ignorance of these higher delights, the highest within man's reach, of science, and especially of scientific discovery; of art, and especially of artistic creation. It is in order to obtain these joys for all, which are now reserved to a few; in order to give leisure and the possibility of developing intellectual capacities, that the social revolution must guarantee daily bread to all. After bread has been secured, leisure is the supreme aim.

No doubt, nowadays, when hundreds and thousands of human beings are in need of bread, coal, clothing, and shelter, luxury is a crime; to satisfy it the worker's child must go without bread! But in a society in which all can eat sufficiently the needs which we consider luxuries to- day will be the more keenly felt. And as all men do not and cannot resemble one another (the variety of tastes and needs is the chief guarantee of human progress) there will always be, and it is desirable that there should always be, men and women whose desire will go beyond those of ordinary individuals in some particular direction.

Everybody does not need a telescope, because, even if learning were general, there are people who prefer examining things through a microscope to studying the starry heavens. Some like statues, some pictures. A particular individual has no other ambition than to possess an excellent piano, while another is pleased with an accordion. The tastes vary, but the artistic needs exist in all. In our present, poor capitalistic society, the man who has artistic needs cannot satisfy them unless he is heir to a large fortune, or by dint of hard work appropriates to himself an intellectual capital which will enable him to take up a liberal profession. Still he cherishes the *hope* of some day satisfying his tastes more of less, and for this reason he reproaches the idealist Communist societies with having the material life of each individual as their sole aim.— "In your communal stores you may perhaps have bread for all," he says to us, "but you will not have beautiful pictures, optical instruments, luxurious furniture, artistic jewelry—in short, the many things that minister to the infinite variety of human tastes. And in this way you suppress the possibility of obtaining anything besides the bread and meat which the commune can offer to all, and the grey linen in which all your lady citizens will be dressed".

These are the objections which all communist systems have to consider, and which the founders of new societies, established in American deserts, never understood. They believed that if the community could procure sufficient cloth to dress all its members, a music hall in which the "brothers" could strum a piece of music, or act a play from time to time, it was enough.

They forgot that the feeling for art existed in the agriculturist as well as in the burgher, and, notwithstanding that the expression of artistic feeling varies according to the difference in culture, in the main it remains the same. In vain did the community guarantee the common necessaries of life, in vain did it suppress all education that would tend to develop individuality, in vain did it eliminate all reading save the Bible. Individual tastes broke forth, and caused general discontent; quarrels arose when somebody proposed to buy a piano or scientific instruments; and the elements of progress flagged. The society could only exist on condition that it crushed all individual feeling, all artistic tendency, and all development.

Will the anarchist Commune be impelled by the same direction? Evidently not, if it understands that while it produces all that is necessary to material life, it must also strive to satisfy all manifestations of the human mind.

SEED YIELD

Seed yield depends upon the crop, variety, location, season and management of the seed crops. Seed attributes of selected cucurbits varieties of IARI.

Crop	Variety	Seed Yield/ (gm)	Fruit (gm)	No. of 1000 Seed/Fruit wt.(gm)	Seed
		Summer Season	Rainy Season		
Bottle gourd	Pusa Naveen	63.00	73.33	634.93 and 654.69 in summer	120-130 gm

				and rainy season	
Sponge gourd	P. Supriya	25.78	–	248.44	–
	P. Chikni	–	9.93	–	–
Pumpkin	P. Vikas	52.83	–	270.00	200 gm
Bitter gourd	P. Vishesh	4.052	–	21.77	160-170 gm

ROLE OF SOCIETY

Is it a dream to concieve a society in which—all having become producers, all having received an education that enables them to cultivate science or art, and all having leisure to do so—men would combine to publish the works of their choice, by contributing each his share of manual work? We have already hundreds of learned, literary, and other societies; and these societies are nothing but voluntary groups of men, interested in certain branches of learning, and associated for the purpose of publishing their works. The authors who write for the periodicals of these societies are not paid, and the periodicals are not for sale; they are sent gratis to all quarters of the globe, to other societies, cultivating the same branches of learning.

This member of the society may insert in its review a one-page note summarizing his observations; another may publish therein an extensive work, the results of long years of study; while others will confine themselves to consulting the review as a startingpoint for further research. It does not matter: all these authors and readers are associated for the production of works in which all of them take an interest. It is true that a learned society, like the individual author, goes to a printing office where workmen are engaged to do the printing.

Nowadays, those who belong to the learned societies despise manual labour; which indeed is carried on under very bad conditions; but a community which would give a generous philosophic and *scientific* education to all its members, would know how to organize manual labour in such a way that it would be the pride of humanity. Its learned societies would become associations of explorers, lovers of science, and workers—all knowing a manual trade and all interested in science.

If, for example, the society is studying geology, all will contribute to the exploration of the earth's strata; each member will take his share in research, and ten thousand observers where we have now only a hundred, will do more in a year than we can do in twenty years. And when their works are to be published, ten thousand men and women, skilled in different trades, will be ready to draw maps, engrave designs, compose, and print the books. With gladness will they give their leisure—in summer to exploration in winter to indoor work And when their works appear, they will find not only a hundred, but ten thousand readers interested in their common work.

This is the direction in which progress is already moving. Even to-day, when England felt the need of a complete dictionary of the English language, the birth of a Littré, who would devote his life to this work, was not waited for. Volunteers were appealed to, and a thousand men offered their services, spontaneously and gratuitously, to ransack the libraries, to take notes, and to accomplish in a few years a work which one man could not complete in his lifetime.

In all branches of human intelligence the same spirit is breaking forth, and we should have a very limited knowledge of humanity could we not guess that the future is announcing itself in such tentative co-operation, which is gradually taking the place of individual work.

For this dictionary to be a really collective work, it would have required that many volunteer authors, printers and printers' readers should have worked in common; but something in this direction is done already in the Socialist Press, which offers us examples of manual and intellectual work combined. It happens in our newspapers that a Socialist author composes in lead his own article. True, such attempts are rare, but they indicate in which direction evolution is going.

They show the road of liberty. In future, when a man will have something useful to say-a word that goes beyond the thoughts of his century, he will not have to look for an editor who might advance the necessary capital. He will look for collaborators among those who know the printing trade, and who approve the idea of his new work.

Literature and journalism will cease to be a means of money-making and living at the cost of others. But is there any one who knows literature and journalism from within, and who does not ardently desire that literature should at last be able to free itself from those who formerly protected it, and who now exploit it, and from the multitude which with rare exceptions pays it in proportion to its mediocrity, or to the ease with which it adapts itself to the bad taste of the greater number? Letters and science will only take their proper place in the work of human development when, freed from all mercenary bondage, they will be exclusively cultivated by those that love them, and for those that love them.

CONTRACT FARMING IN SEED PRODUCTION

Seed is important in sunflower production. However, quality seeds are not available to meet the growing areas of sunflower in Karnataka. To meet the sunflower seeds, KSSC,KOF, UAS etc. agencies are involved in seed production. The seed production targets are being fulfilled by several ways including contract farming. However, there are several issues raised by different stakeholders about the profitability and feasibility of seed production under contract farming system. Hence, the present study was carried out to address these issues.

Generally, there are three types of contract in agriculture viz."

1. Procurement contracts, under which only sale and purchase conditions are specified
2. Partial contracts, wherein only the contracting firms supply some of the inputs and produce is bought at preagreed prices and
3. Total contracts, under which the contracting firm supplies and manages all the inputs on the farm and farmer is just a supplier of land and labour.

The relevance and importance of each type varies from product to product and these types are not mutually exclusive. Whereas, the first type is generally referred to as marketing contracts, the other two are of production contracts. But, there is a systematic link between product market and factor markets under the contract arrangements as contracts requires a definite quality of produce. Different types of production contracts allocate production and market risks between the producer and the processor in different ways.

The scenario of contract farming can be traced to sugar mills in Karnataka were practicing the contract farming since many decades, where farmers were growing sugarcane at the specified pre-agreed price. In the same manner, corporate sector has introduced it to horticultural and vegetable crops. Similarly, the contract farming of oilseeds for seed production was introduced in Northern Karnataka during 1985-86. It was first introduced in the taluks of Haveri, Ranebennur, Hanagal and Hirekerur of Haveri district by the different firms like KOF, KSSC and private seed companies *viz.*, Mahyco and Sungro. The farmers of this district had engaged since long time in the production of oilseed for seed purpose.

IMPORTANCE OF THE PRESENT STUDY

The KOF in seed production resort to contract farming mainly to have assured supply of genuine seed material in required quantity at the right time, which has been produced under their supervision. On the other hand, the farmers are interested to enter into contract mainly to minimize the price risk, and also to reap higher profits out of this seed production activity over commercial production of crops.

The present investigation is an integrated effort to study all socio-economic aspects and profitability of production of hybrid sunflower seeds in addition to identifying the constraints in its production with an overall view of exploring the possibility of bringing about required improvement. The results obtained from this study would be used to overcome the present limitations in production of hybrid sunflower seeds.

The information on cost and returns structure will guide the producer in readjustment and proper management of resources and to bring down the cost of production at the farm level without affecting the output. The results would help the planners and policy makers in formulating suitable policies for grant of loans and fixation of prices and also throw further light on the avenue for future research in the area of hybrid sunflower seed production.

PRODUCTIVITY OF RESOURCES IN SEED PRODUCTION

Kshirsagar studied the resource use efficiency of different inputs in hybrid cotton seed production using Cobb-Douglas production function. Out of four selected variables in Varalaxmi, the human labour input was significant at one per cent level, while fertilizers and plant protection measures were significant at five per cent level of probability. In the case of H-4 cotton seed, except fertilizer, all selected variables *viz.*, human labour, plant protection measures and irrigation were significant at five per cent level of probability. Patel studied the resource use efficiency of various inputs used in seed production of hybrid cotton in Gujarat. The ratios of MVPs of different factors to respective cost of factors were calculated. It was found that 9.14, 1.23, 5.48, -1.08 and 6.83 ratios of MVP to MFC of bullock labour, human labour, plant protection chemicals, irrigation and other expenditures, respectively.

Pathak studied economics of production and marketing of certified seeds of high yielding varieties of jowar and wheat. He concluded that the factor product relationship, between the output and human labour, bullock labour and manures and fertilizers was positive in jowar seed production. In wheat seed production, it was seen that there was a positive relationship between the output and human labour, manures and fertilizers. Muralidharan studied the resource use efficiency in rice production in Kerala employing the Cobb-Douglas type of production function. Adjusted R2 (0.84) indicated that per cent of the variation in yield of paddy could be explained by the estimated production function. The coefficient of land and human labour were positive and significant at one per cent probability level.

Radha *et al.*evaluated the resource-use efficiency in rice-rice and rice-pulse farming systems in Krishna district of Andhra Pradesh. The results indicated that manures and fertilizers and irrigation were quite productively used in both the farming systems. The sum of production elasticities indicated the operation of constant returns to scale in both farming systems.

Karisomanagoudar employed Cobb-Douglas type of production function in Gadag taluk of Dharwad district to study resource use efficiency in rainfed onion roduction. It was observed that land and labour inputs significantly increased the gross revenue. The seed variable exercised a significant negative influence on earnings from onion. The variables included in the production function explained 96 per cent of the variation in output.

Vasudha assessed the relationship between agricultural output and selected explanatory variables in Karnataka by using Cobb-Douglas production function. It was found that the gross cropped area, fertilizer consumption, bullock labour and human labour were important variables in explaining variation in agriculture output. The study also indicated that the agricultural production was increased at the rate of 3.11 per cent per annum during the period between 1956 and 1983 and fertilizer and irrigation were the predominant contributors for growth.

EVOLUTION OF SEED POLICY IN INDIA

The question arises as to what was the need for such a bill to come into existence. For this it is essential that the history of Indian seed regulation be briefly looked into. Two decades after India's independence, during the 1960s, the formal seed sector in India was dominated by the Public sector. It was in 1961 that the National Seeds Corporation (NSC) was established under the Ministry of Agriculture and was at the centre of seed production of breeders, foundation and certified seeds and their quality control. In furtherance of their control in the seeds sector, the National Seeds Project (NSP) was undertaken by the Indian Government in 1967 along with the assistance of the World Bank.

While the NSP did set up huge seed processing plants in order to provide 'certified' seeds of food crops, mainly self-pollinating to farmers, most of these plants operated well below capacity and were stated to be examples of 'faulty technology being pushed into India'. This was, of course, not the only big public sector project that failed to achieve the desired goals and gradually a distrust of the public sector set in. However, the private sector has not proved much more efficient or beneficial either for the cultivator or for the consumer. This led to the National Seed Policy of 1988, which involved a US $ 150m loan from the World Bank to help privatize the Indian Seed Industry. At this time, the import of seeds was still restricted but this sector was gradually opened up, to allow more private participation. Further, after India signed the GATT agreement and joined the WTO, these agreements required that India make some changes to its law, especially regarding

INTELLECTUAL PROPERTY RIGHTS

These requirements were met through the Protection for Plant. In 2002, a new National Seed Policy was released, and as mentioned above, to meet the goals of this policy the new Seed Bill was drafted and tabled in Parliament in 2004.

The objective of the new seed policy seems to be to reduce the direct involvement of government in seed production and marketing, and to actively encourage the private sector to engage in research and development of new varieties. The dominance of the public sector is blamed for the backwardness of Indian agriculture, and one of the stated aims of the National Seed Policy, 2002 was to encourage more private participation in agriculture and seed production, specifically, to complement the existing structures and toreplace them, when necessary. However, this is not to say that the Seed Policy is an insidious conspiracy to betray small and medium-scale farmers. Liberalization has been targeted towards certain components of national seed policies, retaining regulation of some components to safeguard national interests. As the Food and Adulteration

Organisation has noted, the change in the structure from one geared towards the public sector to a more private sector – centric approach is an extremely complex one and would take time. Public and private sectors need to complement each other, perhaps on the basis of a division between cash crops and essential food crops. It is a fact that neither the private nor the public sector can fulfil India's agricultural requirements by itself.

Only effective cooperation and coordination will allow farmers to have access to quality seed and thus contribute to sustainable agriculture and food security. The National Seed Policy, 2002 clearly identifies the twin aims of encouraging the seed industry, especially the domestic industry and of ensuring maximum prosperity and security for farmers. A number of the National Seed Policy's recommendations have been addressed in the Protection of Plant Varieties and Farmers' Rights Act, 2002, including the establishment of a National Gene Fund and a Plant Varieties' authority to regulate the quality of seeds in the country. The further aims of the National Seed Policy that include building up infrastructure, ensuring good quality of seeds and facilitating international trade in seeds, are sought to be addressed through the proposed Seeds Bill.

THE SEED BILL

At the very outset, the preamble of the bill makes clear its intention. The Bill states that it is to "provide for regulating the quality of seeds for sale, import and export and to facilitate production and supply of seeds of quality and for matters connected therewith or incidental thereto".

RATIONALE FOR THE SEED BILL

The new seed bill is a tool to address the grievances and the concerns which the Seed Bill of 1966 does not cover. Even though a large majority of our population depends on agriculture for their livelihood, agriculture in India remains, to a large extent, backward and relatively unproductive. The much touted Green Revolution did have a positive impact but the lot of marginal and small farmers has not improved much over the past few decades. As such, a need was felt for using new techniques and methods to increase the productivity of Indian agriculture. At the same time the Bio-Technology sector came up with promises of extremely productive Genetically Modified (GM) Crops.

These new scientifically manipulated crops caught the imagination of the Indian Government, and to some extent, that of farmers as well. On one hand, there were reports of farmers' suicides owing to failure of crops, including Bt Cotton, and on the other, stories of farmers who found the highway to prosperity through the use of the same Bt Cotton. It is believed that this new technology has the potential to improve living standards, and as such, several powerful groups support the commercialization of such GM crops (especially Bt cotton), including the Indian Council of Agricultural Research, Ministry of Environment and Forests,

About 50% of the total investment in agricultural research in developed countries is contributed by private sector. The idea behind the Seeds Bill, 2004 is that effective implementation of this new law can be expected to promote private plant breeding in the country in the long run. The major rationale behind the policy is the hope that these developments would provide Indian farmers multiple choices and increased access to improved seeds. As such, the Seed Bill 2004 also seeks to address the concerns of the Seed Industry. The Seed Association of India and the Association of Seed Industries raised certain demands at the National Seeds seminar organized by them in 2005. They demanded a level playing field for the private sector, for subsidies and support to the private sector for R&D (specifically to facilitate exports). Another major demand was that seeds be taken out from the purview of the Consumer Protection Act, 1986 and that a scientific system of scrutinizing claims, along with a system of crop insurance, should be developed to study the causes of crop failure.

While the Seeds Bill, 2004 does take care of the infrastructural demands but it retained the right of the farmer to go to the Consumer Courts under the Consumer Protection Act, 1986. One cannot be sure if, in the present situation in India, where a majority of farmers are not only small landholders, but they are also illiterate and/or uneducated, a system of crop insurance can work to their advantage. It was in this atmosphere that the National Seed Policy was formulated in 2002 with the National Seed Bill being drafted.

SALE OF SEEDS

The Seed Bill aims to regulate the quality of seeds sold in India. It requires registration of all varieties of seeds sold in India. The Bill states in S.1 (3) that it applies to every dealer and every producer of seed except when the seed is produced by him for his own use and not for sale. Thus, the only exception to the rule is the exemption granted to farmers to use and sell seeds from their own farms, as long as such seeds are sold unbranded. However, such seeds will also have to meet the minimum standards set for registered seeds, a requirement which will obviously be hard to fulfill for a small farmer and probably as hard to detect for the enforcement authorities. This has been criticized as an assault on the freedom of farmers, a denial of their time-honoured rights, as an undermining of seed sovereignty of farming communities. Farmers care for their own seed quality more than a centralized authority can.

Regulation of farmers' own seed varieties needs to be left to farmers. That is why we have established Community Biodiversity Registers and Jaiv Panchayats. Further, in India, where up to 66% of the seed requirements are met by farmers, declaring all of this to be non–commercial does not make much sense. Farmers' Rights have to be recognized for one major reason; they are the original source of the germ-plasm. Under most IPR Regimes all over the world, farmers are not recognized as innovators, and it is the providers of technology who acquire Intellectual

Property Rights. For the sake of conservation and to give the farmers a fair share of the economic benefits arising out of usage of their seeds, there should be provisions for recognizing and awarding community rights. The Seeds Bill, 2004 has no such provision although the Plant Variety Protection and Farmers' Rights (PVPFR Act) does include provisions for compensation for communities for their role in research and development.

A feature of the policy and the bill is that they start off with complete acceptance of GM crops. There is no longer space for argumentation and debate on that particular issue. India, as mentioned above, has taken to GM crops in a big way. GM crops have not found the same level of acceptance everywhere. Several EU countries continue to block entry of GM products based on concerns for their environment and public health. The seed policies of the European Union have been described as 'control-oriented' seed legislations. Seed quality control and certification agencies in Europe are run by the Government in countries like Germany or by an independent foundation like in the Netherland. These independent foundations are usually a representative of farmers' societies, and owing to the sheer magnitude of India's population, may not be the ideal solution for Indian agriculture.

Also, in Europe, especially in the horticultural sector, certification agencies can perform a full control or they may concentrate on certifying the internal quality control operations of the seed companies, rather than certifying each and every seed lot throughout the seed production-marketing-user chain. All these systems monitor GM crops and their effects from the stage of production of seeds till food processing and consumption.

Once GM crops are allowed entry into the market in a major way it becomes necessary to protect the Intellectual Property Rights (IPRs) of the organizations and industries which have patented such seeds. The Plant Variety Protection and Farmers' Rights (PVPFR) Act was passed in 2001 to take care of this need. The government has provided sufficient protection to the Intellectual Property Rights of the seed developers. It is possible that the implementation of the IPR regime will increase private sector activity. This would invariably lead to an increase in prices, which would adversely affect resource-poor farmers in marginal areas. Thus, the government will have to monitor the seed sector very closely and should effectively intervene if the market fails to serve the farmers. This requires more decentralization and flexibility in operations of public seed agencies.

Although, registration and certification of seeds under the Seeds Bill do raise certain questions regarding Intellectual Property Rights, the issues regarding Intellectual Property Rights are dealt with by the PVPFR Act. Through certification and registration, the Seed Bill only safeguards such rights acquired under the PVPFR Act. Under both Statutes, a degree of leeway is given to farmers. Under the Seed Bill, farmers are free to use and sell their own unbranded seeds while under the PVPFR Act, under Section 42, farmers have been given protection in cases of 'innocent infringement', wherein if the farmer is not aware of breeder rights, at the time of infringement, she will not be liable for infringement.

It is perhaps a testament to the lack of progress on the PVPFR Act, that a need is still felt for a comprehensive database of seeds in India. The PVPFR Act itself has allowed for a National Register of Plant Varieties to be set up. Now, the proposed Seed Bill incorporates the requirement for a National Register of Seeds. The Bill proposes to establish a Central Seed Committee (CSC), which is roughly on the lines of the National Seeds Board envisaged by the Seed Policy, 2002. The CSC will appoint a Registration Sub-Committee, which shall maintain the National Register of Seeds. Further, State Seed Certification Agencies will be set up in each State. These Agencies will have the authority to certify seeds. Further, certain accredited agencies may also be allowed self-certification.

This is a risky proposition at best. Neither the public nor the private sector enjoy an unimpeachable reputation in India. It would be an unsurprising development should wake one up to newspaper headlines of companies and agencies having fudged data to grant certification to certain varieties. The State needs to be the ultimate overseer and guardian of farmers' rights and public health. In the United States, for example, seeds and the crops that they grow into have to secure approval from one, two or three government agencies, as the case may be, before being marketed. The Food and Drug Administration has to review whether the food that will ultimately be produce will be safe to eat, the U.S.

Department of Agriculture's Animal and Plant Health Inspection Service reviews whether the crop is safe to grow, and the Environmentl Protection Agency studies whether the new crop is safe for the environment. Another aim of the National Seed Policy was to streamline the monitoring mechanisms regarding seeds and crops. As such, under the Seed Bill any type of seed for sale will have to be registered with the Registration Sub-Committee. The registration will be valid for 15 years for annual/biennial crops and 18 years for long duration perennial crops. On expiry, the registration can be renewed for a similar period. This is a problematic proposition as usually IPRs are granted for a specific period, whereas in this case, the Registration sub–committee can renew the registration ad infinitum. All registered seeds will need to meet minimum standards with respect to the proportion of seeds that must germinate, levels of genetic and physical purity, and the permitted proportion of diseased seeds.

No transgenic variety of seed can be registered under the proposed Bill unless the applicant has obtained clearance under the provisions of the Environment (Protection) Act, 1986. Any variety of seed that contains any technology considered harmful or potentially harmful cannot be registered in India under the proposed scheme.

It is hoped that this terminology excludes the use of terminator seeds in India, as such seeds cause great harm, especially to small farmers. Also, a major lacuna under the Bill is that there is requirement for disclosure of 'parental lines' at the time of registration. Such a disclosure is essential for the sake of information and safety.

Under the proposed Bill, it is the responsibility of the Central Seeds Committee to set minimum standards for seeds and to decide which seeds are harmful to the environment or dangerous to public health. Further, Central and State Seed Testing Laboratories will be set up. These Laboratories will analyse all varieties of seeds.

Lastly, the respective State Governments will need to appoint Seed Inspectors. The Seed Inspectors would be given powers of search and seizure relating to an offence under the proposed Bill. This is quite a worrying component. Under these provisions, Seed Inspectors have been given wide powers and unlike under the Code of Criminal Procedure, no warrant is necessary nor are procedural safeguards applicable. It is not necessary to outline how this particular provision develops scope for abuse of power, but the safeguards against harassment by functionaries of the State exist for a very good reason, and such safeguards should be included here as well.

REMEDIES AVAILABLE TO FARMERS

Apart from the success stories publicized by the Bio-tech industry to illustrate the success of their seeds, there are numerous stories of failed crops and economic ruin as well. As such, one of the most important provisions in the Bill as far as Farmers' Rights are concerned is regarding the remedy available to farmers in case of defective seeds.

Under the Seed Bill, in case the seed supplied to the farmer is found to be of inferior quality and does not give the desired results, the farmer can approach the Consumer Courts under the Consumer Protection Act, for compensation and redressal.

Also, under the PVPFR Act, under Section 39 (2) of that Act, a farmer or a farmer's organization can claim compensation if a variety fails to give the expected performance under given conditions. If the performance of the variety is not found to be as claimed by the breeder, the Authority can deal with claims of failure of performance and could decide about such claims independently, instead of the courts.

However, it is not clear whether the quantum of compensation awarded would cover only the cost of the seed but that of the crop as well.

WIDESPREAD OPPOSITION TO THE SEED BILL

The Seed Bill, 2004 has met with opposition from farmers' organizations, political parties as well as large swathes of society. The All India Kisan Sabha (AIKS) has come out openly against this Bill. The AIKS has decided to organise widespread campaigns and mass actions at village, mandal and district levels throughout the country in May and June on the issues concerning peasants in particular and people in general.

Political parties have also come out against this Bill. The Communist Parties have denounced this bill as being anti-farmer. On the other end of the political spectrum, B.S. Yediyurappa, Deputy Chief Minister of Karnataka and a member of the Bharatiya Janta Party (BJP), has also condemned the new Seed Bill. Bharat Krishak Samaj (BKS), the ruling Congress party's farmers' outfit, has also opposed the legislation.

Apart from concerns expressed by prominent activists such as Vandana Shiva and Devinder Sharma, the Bill has also been blasted by mainstream newspapers. Dr. Suman Sahai, a noted geneticist and President of the Gene Campaign, has pointed out that the PVPFR Act, drafted to comply with India's obligations under the TRIPS Agreement, does a much better job of regulation and protection of farmer's rights. She notes, "Key differences between the Seed Bill and the PVPFR relate to declaring the origins (parentage) of the variety, the conditions for multi-location testing and who will conduct these tests, the level of transparency maintained on grant of registration, price control and the treatment of farmer varieties.

While the PVPFR requires the declaration of the origin of the variety with pedigree details, the Seed Bill does not. The Seed Bill does not grant any recognition to the contribution of farmers as conservers of agricultural gene pools, or as breeder of successful new varieties This violates the provisions of the Convention on Biological Diversity and the PVPFR according to which if farmer varieties have been used in breeding a new variety, farmers are entitled to share in the profits through the benefit sharing mechanism.

The industry gets its way with seed pricing in the Seed Bill; there is no mechanism to regulate seed supply or seed price unlike the PVPFR which has a clear provision for compulsory licensing. Compulsory licensing safeguards the interests of the farming community since it places the responsibility of ensuring an adequate seed supply at reasonable price on the government. As a result of the Seed Bill, we could have high cost of seeds fixed arbitrarily by the seed companies, leaving the government with no means to control the price.

It could also mean that seed providers are under no obligation to ensure a reasonable seed supply to farmers." This is the most damning indictment of the proposed Seed Bill. The fact is that a legislation that adequately protects farmers' rights as well as the Intellectual Property Rights of the Industry is already in place. The National Gene Fund and the recognition of Community

Rights under S. 41(1) of that Act are path breaking developments and go a log way in convincing skeptics of the fairness of the National Seeds Policy. Now, after hectic lobbying by the Seed Industry the new Seed Bill actually seeks to dilute the PVPFR Act. The PVPFR Act was drafted to ensure compliance with Article 27.3 (b) of the TRIPS agreement. After that agreement, India's obligations to the WTO as regard intellectual property rights of seed companies have ended. There is no need to change the law once again. The government needs to take another look at the proposed bill to ensure that there is no overlap with the PVPFR Act, and especially that the beneficial provisions of that Act are not diluted.

THE MODEL LAW

A Model Law for Developing Countries The African Model Law was drafted by the Organization for African Unity (OAU), now the African Union as a model law for developing countries where the major and immediate issue is not profitability but security, both for the producers and the consumers. The African model law expressly recognises that local communities possess certain rights over their biological resources and the technologies that have evolved over generations and that rights of a collective nature take precedence over rights based on private interests.

The African model law seeks to implement the relevant provisions of the Convention for Biological Diversity and includes protection of farmers' rights, breeders' rights as well as protection of collective and traditional knowledge. Access to such biological resources, knowledge, or technologies of local communities is granted by the simple act of submitting an application for prior informed consent and a written permit. The application to the Competent National Authority has to disclose the full details of the project for which the resource is required. We can have similar systems in India as well, especially with the envisaged structure under the Seed Bill.

The Central Seeds Committee can be the regulator not only of seed quality but of the quality of research as well. The arguments against the Seed Bill have been that it gives too much freedom to industries while curtailing that of farmers. However, the Bill can easily be changed to accommodate such concerns and instead of focusing on encouraging the private sector, it can focus on encouraging co-operation and participation of local communities with the capital and expertise brought in by private companies.

Under the Model Law, the researcher (or the collector, the term used by the Model Law) has to obtain prior informed consent of the community before embarking on research. An access permit is granted through a signed written agreement among the three parties the National Competent Authority, the community or communities concerned, and the applicant or collector. This agreement requires the collector to contribute financially to the efforts of the state and communities concerned in the regeneration and conservation of the biological resource. The collector can only apply for intellectual property protection of the biological resource, or parts or derivatives thereof, or for community knowledge or technology with the additional prior informed consent of the original providers.

Article 9 then goes on to provide that patents over life forms and biological processes will not be recognized and cannot be applied for, but does provide for plant breeders' rights. The model law drafted by the AU, along with the provisions of the Convention on Biological Diversity and associated law instruments serves to provide a more just and effective regime than the one under the TRIPS regime. It is understood that India has certain obligations under the TRIPS agreement. However, many of the beneficial provisions in the African Model Law can be incorporated within the existing framework without India reneging on its international commitments.

4

Seed Germination

INTRODUCTION TO SEED

A seed is a small embryonic plant enclosed in a covering called the seed coat, usually with some stored food. It is the product of the ripened ovule of gymnosperm and angiosperm plants which occurs after fertilization and some growth within the mother plant. The formation of the seed completes the process of reproduction in seed plants (started with the development of flowers and pollination), with the embryo developed from the zygote and the seed coat from the integuments of the ovule.

Seeds have been an important development in the reproduction and spread of flowering plants, relative to more primitive plants like mosses, ferns and liverworts, which do not have seeds and use other means to propagate themselves. This can be seen by the success of seed plants (both gymnosperms and angiosperms) in dominating biological niches on land, from forests to grasslands both in hot and cold climates.

The term *seed* also has a general meaning that predates the above — anything that can be sown i.e. "seed" potatoes, "seeds" of corn or sunflower "seeds". In the case of sunflower and corn "seeds", what is sown is the seed enclosed in a shell or hull, and the potato is a tuber.

SEED STRUCTURE

A typical seed includes three basic parts:

1. An embryo,
2. A supply of nutrients for the embryo,
3. A seed coat.

The embryo is an immature plant from which a new plant will grow under proper conditions. The embryo has one cotyledon or seed leaf in monocotyledons, two cotyledons in almost all dicotyledons and two or more in gymnosperms. The radicle is the embryonic root. The plumule is the embryonic shoot. The embryonic stem above the point of attachment of the cotyledon(s) is the epicotyl. The embryonic stem below the point of attachment is the hypocotyl.

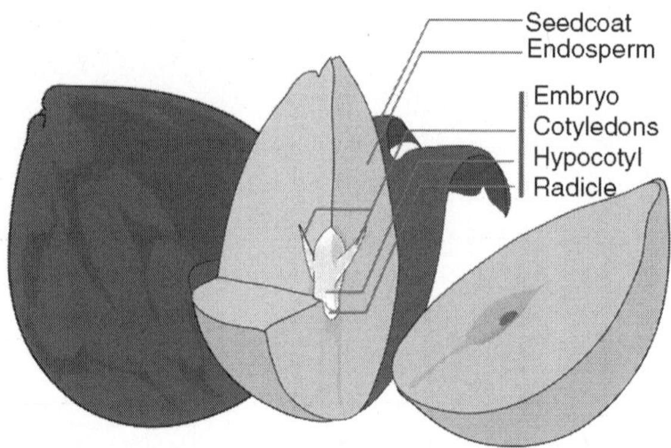

Fig. The parts of an avocado seed (a dicot), showing the seed coat, endosperm, and embryo.

Within the seed, there usually is a store of nutrients for the seedling that will grow from the embryo. The form of the stored nutrition varies depending on the kind of plant. In angiosperms, the stored food begins as a tissue called the endosperm, which is derived from the parent plant via double fertilization. The usually triploid endosperm is rich in oil or starch and protein.

In gymnosperms, such as conifers, the food storage tissue is part of the female gametophyte, a haploid tissue. In some species, the embryo is embedded in the endosperm or female gametophyte, which the seedling will use upon germination. In others, the endosperm is absorbed by the embryo as the latter grows within the developing seed, and the cotyledons of the embryo become filled with this stored food.

At maturity, seeds of these species have no endosperm and are termed exalbuminous seeds. Some exalbuminous seeds are bean, pea, oak,.walnut, squash, sunflower, and radish. Seeds with an endosperm at maturity are termed albuminous seeds.

Most monocots (e.g. grasses and palms) and many dicots (e.g. brazil nut and castor bean) have albuminous seeds. All gymnosperm seeds are albuminous. The seed coat (or *testa*) develops from the tissue, the integument, originally surrounding the ovule. The seed coat in the mature seed can be a paper-thin layer (e.g. peanut) or something more substantial (e.g. thick and hard in honey locust and coconut). The seed coat helps protect the embryo from mechanical injury and from drying out.

In addition to the three basic seed parts, some seeds have an appendage on the seed coat such an aril (as in yew and nutmeg) or an elaiosome (as in Corydalis) or hairs (as in cotton). There may also be a scar on the seed coat, called the *hilum*; it is where the seed was attached to the ovary wall by the funiculus.

SEED PRODUCTION

Seeds are produced in several related groups of plants, and their manner of production distinguishes the angiosperms ("enclosed seeds") from the gymnosperms ("naked seeds"). Angiosperm seeds are produced in a hard or fleshy (or with layers of both) structure called a fruit that encloses the seeds, hence the name.

In gymnosperms, no special structure develops to enclose the seeds, which begin their development "naked" on the bracts of cones. However, the seeds do become covered by the cone scales as they develop in some species of conifer.

Kinds of Seeds

There are a number of modifications to seeds by different groups of plants. One example is that of the so-called *stone* fruits (such as the peach), where a hardened fruit layer (the endocarp) surrounds the actual seed and is fused to it. Many structures commonly referred to as "seeds" are actually dry fruits. Sunflower seeds are sold commercially while still enclosed within the hard wall of the fruit, which must be split open to reach the seed.

Seed development

The inside of a *Ginkgo* seed, showing a well-developed embryo, nutritive tissue (megagametophyte), and a bit of the surrounding seed coat.

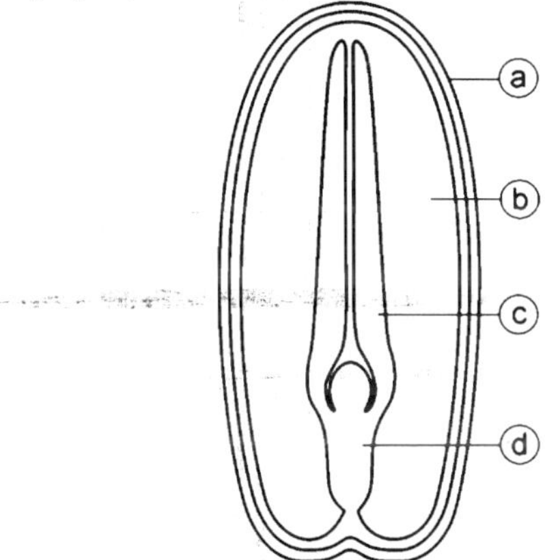

Fig. Diagram of the internal structure of a dicot seed and embryo. (a) seed coat, (b) endosperm, (c) cotyledon, (d) hypocotyl.

The seed, which is an embryo with two points of growth (one of which forms the stems the other the roots) is enclosed in a seed coat with some food reserves. Angiosperm seeds consist of three genetically distinct constituents:

1. The embryo formed from the zygote,
2. The endosperm, which is normally triploid,
3. The seed coat from tissue derived from the maternal tissue of the ovule.

In angiosperms, the process of seed development begins with double fertilization and involves the fusion of the egg and sperm nuclei into a zygote.

The second part of this process is the fusion of the polar nuclei with a second sperm cell nucleus, thus forming a primary endosperm. Right after fertilization the zygote is mostly inactive but the primary endosperm divides rapidly to form the endosperm tissue. This tissue becomes the food that the young plant will consume until the roots have developed after germination or it develops into a hard seed coat.

The seed coat forms from the two integuments or outer layers of cells of the ovule, which derive from tissue from the mother plant, the inner integument forms the tegmen and the outer forms the testa. When the seed coat forms from only one layer it is also called the testa, though not all such testa are homologous from one species to the next.

In gymnosperms, the two sperm cells transferred from the pollen do not develop seed by double fertilization but instead only one sperm fertilizes the egg while the other is not used. The seed is composed of the embryo (the result of fertilization) and tissue from the mother plant, which also form a cone around the seed in coniferous plants like Pine and Spruce.

The ovules after fertilization develop into the seeds; the main parts of the ovule are the funicle; which attaches the ovule to the placenta, the nucellus; the main region of the ovule were the embryo sac develops, the micropyle; A small pore or opening in the ovule where the pollen tube usually enters during the process of fertilization, and the chalaza; the base of the ovule opposite the micropyle, where integument and nucellus are joined together.

The shape of the ovules as they develop often affects the finale shape of the seeds. Plants generally produce ovules of four shapes: the most common shape is called anatropous, with a curved shape. Orthotropous ovules are straight with all the parts of the ovule lined up in a long row producing an uncurved seed. Campylotropous ovules have a curved embryo sac often giving the seed a tight "c" shape. The last ovule shape is called amphitropous, where the ovule is partly inverted and turned back 90 degrees on its stalk or funicle.

In the majority of flowering plants the zygotes first division is transversely orientated in regards to the long axis and this establishes the polarity of the embryo. The upper or chalazal pole becomes the main area of growth of the embryo, while the lower or micropylar pole produces the stalk-like suspensor that attaches to the micropyle. The suspensor absorbs and manufacturers nutrients from the endosperm that are utilized during the embryos growth.

The embryo is composed of different parts; the epicotyle will grow into the shoot, the radicle grows into the primary root, the hypocotyl connects the epicotyle and the radicle, the cotyledons form the seed leaves, the testa or seed coat forms the outer covering of the seed. Monocotyledonous plants like corn, have other structures; instead of the hypocotyle-epicotyle, it has a coleoptile that forms the first leaf and connects to the coleorhiza that connects to the primary root and adventitious roots form from the sides. The seeds of corn are constructed with these structures; pericarp, scutellum (single large cotyledon) that absorbs nutrients from the endosperm, endosperm, plumule, radicle, coleoptile and coleorhiza - these last two structures are sheath-like and enclose the plumule and radicle, acting as a protective covering. The testa or seed coats of both monocots and dicots are often marked with patterns and textured markings, or have wings or tufts of hair.

Seed size and Seed Set

Seeds are very diverse in size. The dust-like orchid seeds are the smallest with about one million seeds per gram. Embryotic seeds have immature embryos and no significant energy reserves. They are myco-heterotrophs, depending on mycorrhizal fungi for nutrition during germination and the early growth of the seedling, in fact some terrestrial Orchid seedlings spend the first few years of their life deriving energy from the fungus and do not produce green leaves. At over 20 kg, the largest seed is the coco de mer.

Plants that produce smaller seeds can generate many more seeds while plants with larger seeds invest more resources into those seeds and normally produce fewer seeds. Small seeds are quicker to ripen and can be dispersed sooner, so fall blooming plants often have small seeds. Many annual plants produce great quantities of smaller seeds; this helps to ensure that at least a few will end in a favorable place for growth. Herbaceous perennials and woody plants often have larger seeds, they can produce seeds over many years, and larger seeds have more energy reserves for germination and seedling growth and produce larger, more established seedlings.

SEED FUNCTIONS

Seeds serve several functions for the plants that produce them. Key among these functions are nourishment of the embryo, dispersal to a new location, and dormancy during unfavorable conditions. Seeds fundamentally are a means of reproduction and most seeds are the product of sexual reproduction which produces a remixing of genetic material and phenotype variability that natural selection acts on.

Embryo Nourishment

Seeds protect and nourish the embryo or baby plant. Seeds usually give a seedling a faster start than a sporling from a spore gets because of the larger food reserves in the seed.

Seed Dispersal

Unlike animals, plants are limited in their ability to seek out favorable conditions for life and growth. As a consequence, plants have evolved many ways to disperse their offspring by dispersing their seeds. A seed must somehow "arrive" at a location and be there at a time favorable for germination and growth.

When the fruits open and release their seeds in a regular way, it is called dehiscent, which is often distinctive for related groups of plants, these fruits include; Capsules, follicles, legumes, silicles and siliques. When fruits do not open and release their seeds in a regular fashion they are called indehiscent, which include these fruits; Achenes, caryopsis, nuts, samaras, and utricles.

Seed dispersal is seen most obviously in fruits; however many seeds aid in their own dispersal. Some kinds of seeds are dispersed while still inside a fruit or cone, which later opens or disintegrates to release the seeds. Other seeds are expelled or released from the fruit prior to dispersal. Milkweeds produce a fruit type, known as a *follicle,* that splits open along one side to release the seeds. Iris capsules split into three "valves" to release their seeds.

By Wind (Anemochory)

Dandelion seeds (achenes) can be carried long distances by the wind:

- Many seeds (e.g. maple, pine) have a wing that aids in wind dispersal.
- The dustlike seeds of orchids are carried efficiently by the wind.
- Some seeds, (e.g. dandelion, milkweed, poplar) have hairs that aid in wind dispersal.

By Water (hydrochory)

- Some plants, such as *Mucuna* and *Dioclea,* produce buoyant seeds termed sea-beans or drift seeds because they float in rivers to the oceans and wash up on beaches.

By Animals (Zoochory)

- Seeds (burrs) with barbs or hooks (e.g. acaena, burdock, dock which attach to animal fur or feathers, and then drop off later.
- Seeds with a fleshy covering (e.g. apple, cherry, juniper) are eaten by animals (birds, mammals) which then disperse these seeds in their droppings.
- Seeds (nuts) which are an attractive long-term storable food resource for animals (e.g. acorns, hazelnut, walnut); the seeds are stored some distance from the parent plant, and some escape being eaten if the animal forgets them.

Myrmecochory

Myrmecochory is the dispersal of seeds by ants. Foraging ants disperse seeds which have appendages called elaiosomes (e.g. bloodroot, trilliums, Acacias, and many species of Proteaceae). Elaiosomes are soft, fleshy structures that contain nutrients for animals that eat them. The ants carry such seeds back to their nest, where the elaiosomes are eaten. The remainder of the seed, which is hard and inedible to the ants, then germinates either within the nest or at a removal site where the seed has been discarded by the ants. This dispersal relationship is an example of mutualism, since the plants depend upon the ants to disperse seeds, while the ants depend upon the plants seeds for food. As a result, a drop in numbers of one partner can reduce success of the other.

In South Africa, the Argentine ant (*Linepithema humile*) has invaded and displaced native species of ants. Unlike the native ant species, Argentine ants do not collect the seeds of *Mimetes cucullatus* or eat the elaiosomes. In areas where these ants have invaded, the numbers of *Mimetes* seedlings have dropped.

Seed Dormancy and Protection

One important function of most seeds is delaying germination, which allows time for dispersal and prevents germination of all the seeds at one time when conditions appear favorable. The staggering of germination safeguards some seeds or seedlings from suffering during short periods of bad weather, transient herbivores or competition from other plants for light and nutrients. Many species of plants have seeds that germinate over many months or years, and some seeds can remain in the soil seed bank for more than 50 years before germination.

Seed dormancy is defined as a seed failing to germinate under environmental conditions optimal for germination, normally when the seed's environment is at the right temperature with proper soil moisture conditions. Induced dormancy or seed quiescence occurs when a seed fails to germinate because the external environmental conditions are inappropriate for germination, mostly in response to being too cold or hot, or too dry.

True dormancy or innate dormancy is caused by conditions within the seed that prevent germination under normally ideal conditions. Often seed dormancy is divided into four major categories: exogenous; endogenous; combinational; and secondary.

Exogenous dormancy is caused by conditions outside the embryo including:

- Hard seed coats or physical dormancy occurs when seeds are impermeable to water or the exchange of gases. In some seeds the seed coat physically prevents the seedling from growing.
- Chemical dormancy includes growth regulators etc.

Endogenous dormancy is caused by conditions within the embryo itself, including:

- Immature embryos where some plants release their seeds before the tissues of the embryos have fully differentiated, and the seeds ripen after they take in water while on the ground, germination can be delayed from a few weeks to a few months.
- Morphological dormancy where seeds have fully differentiated embryos that need to grow more before seed germination, the embryos are not yet fully developed.

- Morphophysiological dormancy seeds with underdeveloped embryos, and in addition have physiological components to dormancy. These seeds therefore require a dormancy-breaking treatments as well as a period of time to develop fully grown embryos.
- Physiological dormancy prevents seed germination until the chemical inhibitors are broken down or are no longer produced by the seed, often physiological dormancy is broken by a period of cool moist conditions, normally below (+4C) 39F, or in the case of many species in *Ranunculaceae* and a few others,(-5C) 24F. Other chemicals that prevent germination are washed out of the seeds by rainwater or snow melt. Abscisic acid is usually the growth inhibitor in seeds and its production can be affected by light. Some plants like Peony species have multiple types of physiological dormancy, one affects radical growth while the other affects shoot growth.
  - Drying; some plants including a number of grasses and those from seasonally arid regions need a period of drying before they will germinate, the seeds are released but need to have a lower moister content before germination can begin. If the seeds remain moist after dispersal, germination can be delayed for many months or even years. Many herbaceous plants from temperate climate zones have physiological dormancy that disappears with drying of the seeds. Other species will germinate after dispersal only under very narrow temperature ranges, but as the seeds dry they are able to germinate over a wider temperature range.
  - Photodormancy or light sensitivity affects germination of some seeds. These photoblastic seeds need a period of darkness or light to germinate. In species with thin seed coats, light may be able to penetrate into the dormant embryo. The presence of light or the absence of light may trigger the germination process, inhibiting germination in some seeds buried too deeply or in others not buried in the soil.
  - Thermodormancy is seed sensitivity to heat or cold. Some seeds including cocklebur and amaranth germinate only at high temperatures (30C or 86F) many plants that have seed that germinate in early to mid summer have thermodormancy and germinate only when the soil temperature is warm. Other seeds need cool soils to germinate, while others like celery are inhibited when soil temperatures are too warm. Often thermodormancy requirements disappear as the seed ages or dries.

Combinational dormancy also called double dormancy. Many seeds have more than one type of dormancy, some *Iris* species have both hard impermeable seeds coats and physiological dormancy. Secondary dormancy is caused by conditions after the seed has been dispersed and occurs in some seeds when none dormant seed is exposed to conditions that are not favorable to germination, very often high temperatures. The mechanisms of secondary dormancy is not yet fully understood but might involve the lose of sensitivity in receptors in the plasma membrane.

Many garden plants have seeds that will germinate readily as soon as they have water and are warm enough, though their wild ancestors may have had dormancy, these cultivated plants lack seed dormancy. After many generations of selective pressure by plant breeders and gardeners dormancy has been selected out.

For annuals, seeds are a way for the species to survive dry or cold seasons. Ephemeral plants are usually annuals that can go from seed to seed in as few as six weeks. Not all seeds undergo a period of dormancy. Seeds of some mangroves are viviparous, they begin to germinate while still attached to the parent. The large, heavy root allows the seed to penetrate into the ground when it falls.

DATABASEJ OF SEED GERMINATION

The following data is provided by Thompson and Morgan Successful Seed Raising Guide. This guide is out of print. A seed is an embryo plant and contains within itself virtually all the materials and energy to start off a new plant. To get the most from one's seeds it is needful to understand a little about their needs, so that just the right conditions can be given for successful growth. One of the most usual causes of failures with seed is sowing too deeply; a seed has only enough food within itself for a limited period of growth and a tiny seed sown too deeply soon expends that energy and dies before it can reach the surface.

Our seed guide therefore states the optimum depth at which each type of seed should be sown. Another common cause is watering. Seeds need a supply of moisture and air in the soil around them. Keeping the soil too wet drives out the air and the seed quickly rots, whereas insufficient water causes the tender seedling to dry out and die. We can thoroughly recommend the Polythene bag method which helps to overcome this problem. Watering of containers of very small seeds should always be done from below, allowing the water to creep up until the surface glistens. Most seeds will of course only germinate between certain temperatures. Too low and the seed takes up water but cannot germinate and therefore rots, too high and growth within the seed is prevented.

Fortunately most seeds are tolerant of a wide range of temperatures but it is wise to try to maintain a steady, not fluctuating temperature, at around the figure we have recommended in our guide. Once several of the seeds start to germinate the temperatures can be reduced by about 5 degrees F and ventilation and light should be given. Some perennials and tree and shrub seeds can be very slow and erratic in germination.

This may sometimes be due to seed dormancy, a condition which prevents the seed from germinating even when it is perfectly healthy and all conditions for germination are at optimum. The natural method is to sow the seeds out of doors somewhere where they will be sheltered from extremes of climate, predators, etc. and leave them until they emerge, which may be two or three seasons later.

HINTS ON SEED RAISING

Strelitzia and Similar

Do not chip or mark the seedcoat at all but merely remove the orange tuft and soak for up to 2 hours, or even overnight. Sow the seeds in moist sand, pressing them into the sand until only a small part of the black seed is visible and grow in a temperature of 75 degrees F in the dark and ensure that the sand always remains moist.

From 7 days onwards inspect the container once a week and as soon as any bulges, roots or shoots are seen remove the germinated seed and pot up in a compost of half peat and half sand. We find that Strelitzias often produce a root without a shoot and we have also found that the young shoots and roots are susceptible to fungal attack. Therefore as soon as possible pot up and provide light and fresh air. Germination can start within 7 days and carry on for 6 months or more.

Palms; Banana; Coffee; Mini-Orange; Tea; Cycads and Similar

All these items can take several months to germinate and are very erratic in germination. Soak for at least 2 hours in warm water before sowing. (After soaking the parchment shell on the Coffee seeds should be removed with the fingernail). Sow in Levington or Arthur Bowers compost and place in the dark in a temperature of 75 degrees F, keeping the compost moist at all times, but not wet. Inspect regularly and occasionally dig around in the compost with a penknife. We normally sow our seeds just below the surface of the soil and we have found that sometimes they make a very vigourous root without producing a shoot at all. If you find a seed with a root then it should be excavated and potted up into a 3-4" pot immediately when it will produce a shoot. Cycads prefer to be potted up into a compost of half sand and half peat. The Tea requires the above treatment but in a lower temperature of 60-65 degree F.

Clivia and Similar

Sow these seeds immediately on receipt in Levington or a peat based compost, covering with a 1/2 " compost. Water and place in the dark in a temperature of 65-70 °F. Germination should occur within 3 weeks.

Ferns (Garden and Indoor)

The fern spore needs a fine film of moisture over which to swim in order to complete the process of reproduction, therefore a good peat compost, such as Levington, ought to be used pressed down very firmly and which is a lot more moist than one would normally have it in order to provide the moisture film. The spore (seed) should be sprinkled close together on the surface of the soil and not covered and the container should be covered with a piece of glass and placed in diffused light, but not darkness. It is essential to ensure that the compost remains moist at all times. Germination which commences with the appearance of a film of green jelly over the soil can take anything from 1 -5 months. You may wish to try germinating the fern spore on blotting paper which is placed in a saucer and kept moist at all times. A transparent cover is inverted over the saucer and the whole lot placed in a well lit but not sunny position.

You can actually see the fern spores developing and when you can see small plantlettes appearing along the jelly the blotting paper should be lifted and placed on the surface of a container of Levington compost and watered well. It should then be covered with a transparent cover which can remain there until the plants are quite large.

Bromeliads; Cineraria

Calceolaria; Insect Eaters (Drosera, Nepenthes, Sarracenias); Living Stones; Meconopsis; Rubber Plants; Saintpaulia; Streptocarpus; Tibouchina; Xmas Cactus; Begonia and similar These seeds should be sown on the surface of the compost and not covered. The compost should be quite moist and we would recommend that you cover the seed container with a piece of glass or clear plastic and leave in a temperature of approximately 65 degrees F in a position which receives diffused light. Once some of the seeds have germinated air should be admitted gradually otherwise the seedlings may damp off.

Alternatively the seeds can be sown on to moist blotting paper or kitchen towel placed in a saucer. Cover with a transparent cover and place on a windowsill which receives plenty of light, but not direct sunlight. Keep the blotting paper wet at all times and when the tiny seedlings are large enough to handle prick out into small pots. If the INSECT EATERS are sown using the first method described the compost requires to be both moist yet free draining. Use only pure peat with no fertilizers added to which sphagnum moss should be added if available.

Alstroemeria; Bonsai; Clematis; Hardy Cyclamen; Eucalyptus; Flower Lawn; Helleborus; Hosta; Primula; Iris and similar

Sowing October-February

Sow the seeds in John Innes seed compost, covering them with a thin layer of compost. After watering place the seed container outside against a North wall or in a cold frame, making sure they are protected against mice, and leave them there until the spring. The compost should be kept moist but not wet at all times, and if the seed containers are out in the open then some shelter has to be given against excessive rain.

In the spring bring the seed containers into the greenhouse, or indoors on to a well lit but not sunny windowsill and keep the compost moist. This should trigger off germination. If the seeds do not germinate in the spring keep them in cool moist conditions throughout the summer. As each seed germinates we would recommend that you transplant it almost immediately into its own pot.

Sowing March-September

Sow in John Innes seed compost, or something similar, and place each container in a polythene bag and put into the refrigerator (not the freezer compartment) for 2-3 weeks. After this time place the containers outside in a cold frame or plunge them up to the rims in a shady part of the garden border and cover with glass or clear plastic. Some of the seeds may germinate during the spring and summer and these should be transplanted when large enough to handle. The remainder of the seeds may lay dormant until next spring. Germination of some items, particularly Alstroemeria, Clematis, Hardy Cyclamen and Christmas Rose (Helleborus) may take take 18 months or more. An alternative method for growing PRIMULAS is to sow in a peat based compost which has already been moistened and do not cover the seed. Cover the container with a piece of glass or plastic and grow in the dark in a steady temperature of 60 °F.

This is quite adequate and over 65 °F germination will be inhibited. When the seeds start to germinate sprinkle a thin layer of fine compost over them and when the seed leaves come through this, move the box to a well lit place with a temperature of 55 °F. At no time should the seed box be in full sun. Hardy Cyclamen have been found to germinate best in total darkness at around 55-60 °F. We have had good results with the following method. Place the seeds between two pieces of damp filter paper, Kleenex tissue, etc., then put into a polythene bag and place this into an opaque container in order to exclude all light. Inspect the seeds after a month and remove and prick out as the seedlings appear, returning the ungerminated seeds to total darkness.

Freesia

Soak the seeds for 24 hours and sow in Levington compost, or something similar, and place in a temperature of 50-60 °F. Germination can sometimes be slow.

Nertera Granadensis (Bead Plant)

We recently found that this subject requires a well drained compost which is completely free from fertilizer (*e.g.* moss peat and sand in equal parts). Sow by barely covering the seed and place a sheet of glass over the container, and leave in a temperature of 65-75 °F.

Turn the glass daily as excessive condensation can kill the young seedlings. On germination the seedlings look very thin and spindly and the glass should be removed almost immediately and the seed container moved to a well lit but not sunny position.

Cactus and Similar

Make very shallow furrows in compost with a plant label and sow in these. No seed should be completely buried. Water from beneath and cover with glass and brown paper or black Polythene. Place in a dark position in a temperature of 70-75 °F and keep moist. On germinating move to a light but not sunny windowsill, give plenty of ventilation and water from beneath.

Pot up when they begin to overcrowd. During the first winter only keep warm and do not allow to get too dry. If it is not possible to grow warm then keep them drier. Subsequent years keep relatively dry through the winter. Can be planted outside, plunged to the rim, all summer if required.

Lilies

Successful germination of seeds of some lilies requires a period of warmth followed by one of cold. Method 1. Put seeds in a screw top jar in moist (not wet) peat and keep at 70-75 °F for 3-4 months. Inspect regularly, any normal seedlings (that is having root and seedling leaves) should be pricked out as they germinate. Any seeds which produce roots but not seedling leaves, sow in a pan and keep at 32-40 °F for 3 months. Seed leaves and normal growth will follow.

Method. Sow in a pan in summer (warm spell); put in a frame (or outside covered by a piece of glass) for the winter. Seeds will germinate in spring. Soil Humus rich (peat or leafmould) lime free and very free drainage (use 1/3 grit). Never overwater, keep bulbs almost dry from November to March.

Delicate Seeds

A method which has proved useful for not only small delicate seeds but for a wide range of types is the Polythene bag method. The seeds should be sown on the surface of the moist compost, covered to their recommended depth if necessary and the container is then placed inside a Polythene bag after which the end is sealed with an elastic band. The bag should 'fog-up' with condensation within 24 hours and if this does not occur place the container almost up to its rim in moisture until the soil surface glistens, then replace in the bag and reseal. The bag is not removed and normally no more watering is required until the seeds germinate. However, it is wise, if left for a long period to check the compost occasionally. The seed container, bag etc.

should be placed in a well lit place with a steady temperature. As soon as a fair number of the seedlings emerge remove the polythene bag, lower the temperature a few degrees and provide plenty of light, but not bright sunshine, to ensure that sturdy seedlings develop. It is also helpful to spray the seedlings occasionally for the first 14 days.

SPECIAL TREATMENT

Hard Seeds-Chipping

Some seeds, *e.g.* Sweet peas, lpomaea etc., have hard seed coats which prevent moisture being absorbed by the seed. All that is needed is for the outer surface to be scratched or abraided to allow water to pass through.

This can be achieved by chipping the seed with a sharp knife at a part furthest away from the 'eye', by rubbing lightly with sandpaper or with very small seed pricking carefully once with a needle etc.

Hard Seeds-Soaking

Soaking is beneficial in two ways; it can soften a hard seed coat and also leach out any chemical inhibitors in the seed which may prevent germination. 24 hours in water which starts off hand hot is usually sufficient. If soaking for longer the water should be changed daily. Seeds of some species (*e.g.* Cytisus, Caragana, Clianthus) swell up when they are soaked. If some seeds of a batch do swell within 24 hours they should be planted immediately and the remainder pricked gently with a pin and returned to soak. As each seed swells it should be removed and sown before it has time to dry out.

Stratification (Cold Treatment)

Some seeds need a period of moisture and cold after harvest before they will germinate-usually this is necessary to either allow the embryo to mature or to break dormancy. This period can be artificially stimulated by placing the moistened seed in a refrigerator for a certain period of time (usually 3-5 weeks at around 41 °F).

With tiny seeds it is best to sow them on moistened compost, seal the container in a Polythene bag and leave everything in the refrigerator for the recommended period. However, larger seeds can be mixed with 2-3 times their volume of damp peat, placed direct into a Polythene bag which is sealed and placed in the refrigerator. Look at seeds from time to time.

The seeds must be moist whilst being pre-chilled, but it doesn't usually benefit them to be actually in water or at temperatures below freezing.

Light also seems to be beneficial after prechilling and so pre-chilled seeds should have only the lightest covering of compost over them, if any is required, and the seed trays etc. should be in the light and not covered with brown paper etc.

Double Dormancy

Some seeds have a combination of dormancy's and each one has to be broken in turn and in the right sequence before germination can take place; for example, some Lilies, Tree paeonies, Taxus need a three month warm period (68-8 °'F) during which the root develops and then a three month chilling to break dormancy of the shoots, before the seedling actually emerges. Trillium needs a three month chill followed by three months of warmth and then a further three month chill before it will germinate.

Outdoor Treatment

The above mentioned methods (12-15) accelerate the germination process and help to prevent seeds being lost due to external hazards (mice, disease, etc.) but outdoor sowing is just as effective albeit longer. The seeds are best sown in containers of free draining compost and placed in a cold frame or plunged up to their rim outdoors in a shaded part of the garden, preferably on the north side of the house avoiding cold drying winds and strong sun.

Recent tests show that much of the beneficial effects of pre-chilling are lost if the seed is not exposed to light immediately afterwards. We therefore recommend sowing the seeds very close to the surface of the soil and covering the container with a sheet of glass. An alternative method especially with larger seeds, is to sow the seed in a well prepared ground, cover with a jam jar and press this down well into the soil so that the seeds are enclosed and safe from predators, drying out etc.

Germination Days

The usual time period in which a particular variety will germinate given optimum conditions.

Light/Dark

Seeds needing light should have no newspaper, brown paper etc. placed over the trays. Seeds needing dark for germination should be placed in total darkness.

Slow and Irregular Germination

This is the column with the "X". Not all seeds will show at once -prick out each seedling as it becomes large enough to handle and don't discard the container until well over the time suggested.

Temperature

A steady temperature between these limits is recommended-fluctuating temperatures can damage a seedling in its critical early stages.

Compost

Most reputable seed composts will be quite adequate and we have indicated where a loam based type such as John Innes or a peat based type such as Levington would be slightly more suitable. On no account should potting composts, which have additional fertilizers, be used.

Sowing Depth

If in doubt sow shallowly, but always ensure that the compost surface is damp. J.C. = Just cover the seed with compost or sharp sand. S = Sow on the surface and do not cover at all with compost.

Sowing in Situ

Where recommended under the heading of comments, these seeds can be sown out of doors. Moist soil worked down to a fine tilth is essential. For hardy annuals and perennials sowing can be carried out from late winter onwards as soon as the ground is workable and has warmed up and half hardy annuals after all danger of frost is passed.

GERMINATION BEGINNING OF GROWTH

Germination is the beginning of growth of a seed. The seed must have the right level of warmth and moisture to begin to germinate. First, the seed leaves absorb moisture which allows the food reserves to become available to the new plant.

It can then produce a root so that it can find its own water, followed by a shoot which develops from the plumule, which will allow it to absorb light. The plant needs both water and light to grow. Sometimes, the seed leaves, or cotyledons, remain below the surface of the soil, as in germination of a Pea, below. This is called Hypogeal Germination.

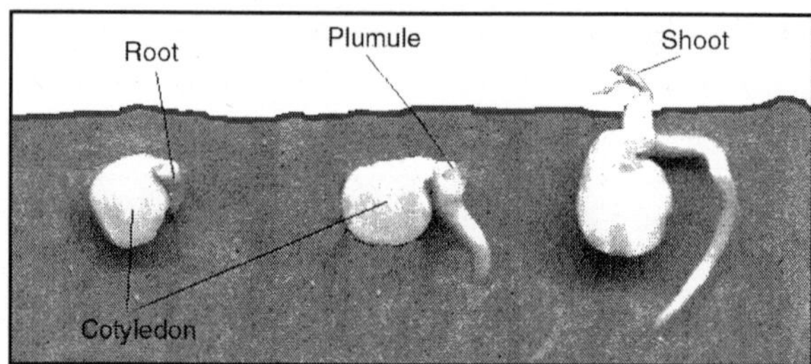

In some species, the seed leaves remain on the new shoot and are brought above the ground, as in germination of the Ash tree, below. This is called Epigeal Germination.

There is one other type of germination that is found only in a few tropical plants. In this case, one special type of leaf is formed above the ground. This special leaf is not a cotyledon, which remains under the ground. The first leaf is a round dissected or frilly circular leaf with a central stalk like an umbrella, and its purpose is to protect the proper shoot from being trampled on or otherwise damaged before the new plant has enough resources to replace it.

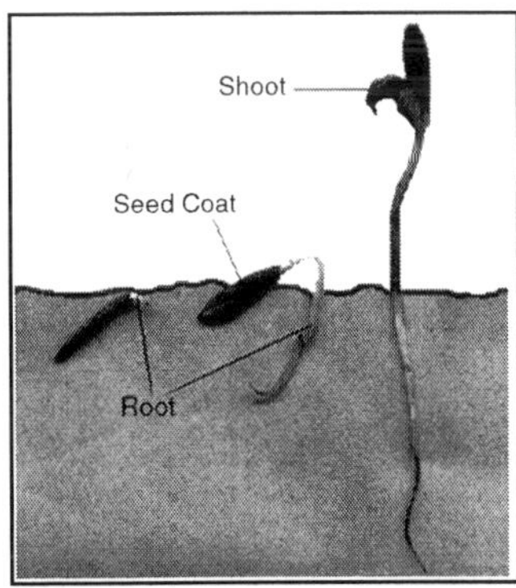

As it is green, it also gathers food by photosynthesis, which enables the root and proper shoot to grow under the ground, so that the new plant is already established when the real shoot emerges. Because the development of the shoot can not be seen, this type of germination is called Cleistogeal Germination from the Greek*cleisto-*, which means hidden.

Fig. This is the Frilly Leaf of a Combretum Seed from Zambia.

FACTORS AFFECTING SEED GERMINATION

Seed germination depends on both internal and external conditions. The most important external factors include temperature, water, oxygen and sometimes light or darkness. Various plants require different variables for successful seed germination, often this depends on the individual seed variety and is closely linked to the ecological conditions of a plant's natural habitat. For some seeds, their future germination response is affected by environmental conditions during seed formation; most often these responses are types of seed dormancy.

WATER

Water is required for germination. Mature seeds are often extremely dry and need to take in significant amounts of water, relative to the dry weight of the seed, before cellular metabolism and growth can resume.

Most seeds need enough water to moisten the seeds but not enough to soak them. The uptake of water by seeds is called imbibition, which leads to the swelling and the breaking of the seed coat. When seeds are formed, most plants store a food reserve with the seed, such as starch, proteins, or oils. This food reserve provides nourishment to the growing embryo. When the seed imbibes water, hydrolytic enzymes are activated which break down these stored food resources into metabolically useful chemicals. After the seedling emerges from the seed coat and starts growing roots and leaves, the seedling's food reserves are typically exhausted; at this point photosynthesis provides the energy needed for continued growth and the seedling now requires a continuous supply of water, nutrients, and light.

OXYGEN

Oxygen is required by the germinating seed for metabolism. Oxygen is used in aerobic respiration, the main source of the seedling's energy until it grows leaves. Oxygen is an atmospheric gas that is found in soil pore spaces; if a seed is buried too deeply within the soil or the soil is waterlogged, the seed can be oxygen starved. Some seeds have impermeable seed coats that prevent oxygen from entering the seed, causing a type of physical dormancy which is broken when the seed coat is worn away enough to allow gas exchange and water uptake from the environment.

TEMPERATURE

Temperature affects cellular metabolic and growth rates. Seeds from different species and even seeds from the same plant germinate over a wide range of temperatures. Seeds often have a temperature range within which they will germinate, and they will not do so above or below this range. Many seeds germinate at temperatures slightly above room–temperature 60–75 °F (16–24 °C), while others germinate just above freezing and others germinate only in response to alternations in temperature between warm and cool.

Some seeds germinate when the soil is cool 28–40 °F (–2—4 °C), and some when the soil is warm 76–90 °F (24–32 °C). Some seeds require exposure to cold temperatures (vernalization) to break dormancy. Seeds in a dormant state will not germinate even if conditions are favourable.

Seeds that are dependent on temperature to end dormancy have a type of physiological dormancy. For example, seeds requiring the cold of winter are inhibited from germinating until they take in water in the fall and experience cooler temperatures.

Four degrees Celsius is cool enough to end dormancy for most cool dormant seeds, but some groups, especially within the family Ranunculaceae and others, need conditions cooler than –5 C. Some seeds will only germinate after hot temperatures during a forest fire which cracks their seed coats; this is a type of physical dormancy. Most common annual vegetables have optimal germination temperatures between 75–90 °F (24–32 °C), though many species (*e.g.* radishes or spinach) can germinate at significantly lower temperatures, as low as 40 °F (4 °C), thus allowing them to be grown from seed in cooler climates. Suboptimal temperatures lead to lower success rates and longer germination periods.

LIGHT OR DARKNESS

Light or darkness can be an environmental trigger for germination and is a type of physiological dormancy. Most seeds are not affected by light or darkness, but many seeds, including species found in forest settings, will not germinate until an opening in the canopy allows sufficient light for growth of the seedling.

Scarification mimics natural processes that weaken the seed coat before germination. In nature, some seeds require particular conditions to germinate, such as the heat of a fire (*e.g.*, many Australian native plants), or soaking in a body of water for a long period of time. Others need to be passed through an animal's digestive tract to weaken the seed coat enough to allow the seedling to emerge.

BEGINNING OF GERMINATION

Germination takes place only in the first week after the beginning of the summer rains, and the seedlings are hidden in the litter under bushes or among the small stones which cover the ground.

The saguaro is a subtropical species and its seeds germinate only in the summertime. Germination occurs principally in July and August during the southwestern summer monsoon.

Thus, successful germination and early establishment depend on a short period of coincident high levels of moisture and warmth during but a few weeks of the desert year. We are not surprised, therefore, to find that this highly successful desert species has acquired adaptive strategies that fit a set of germination requirements to an environment characterised by a persistently high evaporation which is only partially offset by scanty and often uncertain rainfall.

Seedfall is completed just prior to the normal arrival of summer rains. In exceptional years, the two events—seedfall and rainfall—actually overlap. The first summer rains contribute importantly to seed survival by dispersing seeds and by washing them into locations that offer concealment from predators.

Little or no germination, however, results from these late June and early July rains. Rigid light and moisture requirements interact to inhibit germination and prevent losses that would result from germination at that time, for sporadic early monsoon rains seldom provide a sufficiently reliable supply of moisture to promote germination and support the initial establishment of seedlings. The increasing frequency of rains during the second half of July offers progressively more favourable conditions for seed germination.

While it is commonly said that "seedlings germinate," plant seedlings "sprout" and proceed to establish or die; only seeds can germinate. The highly variable spatial and temporal distribution of summer rainfall importantly affects the number of saguaro seed germinations occurring at a given location. Further, the number of germinations varies in accord with the characteristically large year-to-year variability of monsoon precipitation patterns.

Our experimental and field observations, however, indicate that the most favourable conditions for saguaro seed germination do not necessarily occur during years with the highest total summer precipitation. Rather, natural germination is associated with temporal clustering of summer rainstorms that provide continuously high moisture levels at and near the soil surface during the 2- to 3-day germination period.

Most germination takes place during the period from mid-July through the first week of August. Then, with continuing availability of high surface moisture resulting from short-spaced rainstorms and supplemented by dewfall, germination proceeds to completion within a period of 2-3 days—initial establishment of the seedlings takes place at a time when temperatures and moisture are optimum for their growth.

The high temperature limitations on saguaro seed germination operate to restrict natural germination to shaded sites where periods of high moisture are sufficiently prolonged to permit germination.

During wet periods such sites seldom reach temperatures high enough to inhibit germination. Inhibition of germination by low temperatures effectively restricts germination to the warm-wet periods of June, July, August, and September in the overall distribution of the saguaro.

Table. Per cent Soil Moisture in Shaded and Unshaded Saguaro-paloverde Habitat at Saguaro National Monument (east), July 1967

Date July	Soil moisture (%)			
	Shade		Open	
	0.5 inch (1.27 cm)	1 inch (2.54 cm)	0.5 inch (1.27 cm)	1 inch (2.54 cm)
18	15.23	15.34	6.01	8.12
19	10.06	9.63	3.13	6.69
20	3.40	8.87	2.36	5.45
21[a]	1.85	6.33	1.39	4.47
22	11.45	9.96	7.80	7.54
23	5.15	8.10	2.43	4.50

[a]Estimated by regression (log % on time).

Samples of the top 0.5 inch (1.27 cm) and top 1 inch (2.54 cm) of soil were collected at approximately 1300 hr beneath the crown of a mature foothill paloverde at 1 m (3.3 ft) from the trunk (shade), and from an unshaded (open) level site approximately S m (16.4 ft) outside the crown. Data graphed in Figure.

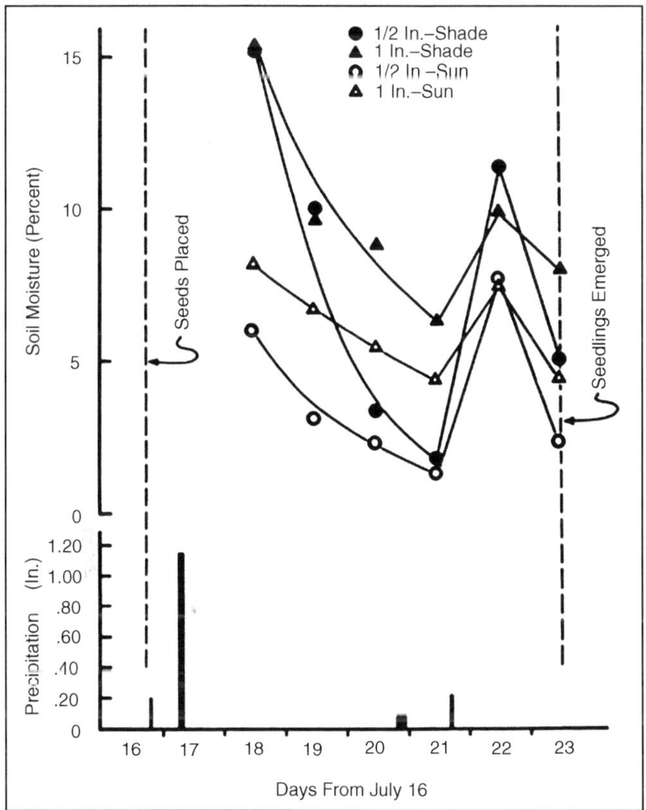

Fig. Natural Field Germination of Saguaro Seeds (on 23 July 1967) and Accompanying Precipitation (inches [Original units of Measurement; 1 inch = 25.4 mm]) and Soil Moisture at 0.5 inch (1.27 cm) and 1 inch (2.54 cm) depth.

Seeds were broadcast beneath the crown of a foothill paloverde (Cercidium microphyllum) on 16 July immediately prior to receiving 0.20 inch (5 mm) precipitation the same day. No germination occurred during the period of high soil moisture following a heavy rain the next morning, followed by clear skies (17 July, 1.5 inches; 38 mm). Germination took place after two periods of light precipitation (0.04 inch; 1 mm, 0.21 inch; 5 mm) and broken overcast skies on 21 and 22 July.

The first seedlings emerged on 23 July, 3 days after the start of these rains. Amount (bar height) and period (bar width) of precipitation is shown. Saguaro seed germination is dependent upon the continued availability of moisture at the soil surface as determined by temporal distribution of rainfall, cloud cover, and other factors that aid in continuously maintaining high relative humidity and surface moisture for a period of 2-3 days during the summer rainfall season. Generally, natural germination is associated with the occurrence of two or more rains within a 2- to 3-day period. Genetically controlled variability in the time required for completion of the germination process further insures that the entire annual seed crop will not respond to conditions that are marginal for seedling establishment. Approximately 10 per cent of seeds tested under ambient summer light and temperatures required 4 days or more of continuous exposure to free water for completion of germination. Thus, even under conditions that permit extensive germination, a substantial portion of the germinable seeds remain viable and available for subsequent germination.

The light requirement may seem to present an apparent survival disadvantage because it seemingly dictates that germination must occur at or near the soil surface where the predation pressures are greatest and requisite moisture conditions are least apt to prevail. However, in view of the characteristic development pattern of the monsoon rains, an important adaptive strategy can be recognised.

In the presence of insufficient moisture to complete germination and initial establishment of the seedling, the light requirement acts to stop the germination process without loss of seed viability. Buried seeds retain their viability, and will germinate quickly when exposed by subsequent rains, animal digging, other natural forces, or experimental exposure. The light requirement insures germination sufficiently near the soil surface for the globular seedling to emerge and immediately receive adequate light for subsequent growth. Sensitivity to far-red light may also permit better utilisation of energy from nocturnal re-radiation thus hastening completion of the germination process. Field observations indicate that in nature germination of shallowly buried seeds accounts for a substantial portion of the annual saguaro seedling crop. This observation appears to conflict with knowledge that germinating seeds must receive exposure to light after they have imbibed water, that little or no germination (<1 per cent) will occur in total darkness.

Extensive laboratory tests using a variety of environmental, mechanical, and chemical treatments have revealed no mechanism by which germination might occur naturally in the absence of light (Alcorn and Kurtz 1959; McDonough 1964). Based on successful use of acid treatments in promoting dark germination, McDonough (1964) suggested that passage through the digestive tracts of animals might promote germination in the absence of light. However, we have attained no success in the dark-germination of saguaro seeds recovered from the feces of native rodents or birds.

In fact, the natural germination of buried saguaro seeds requires no complex mechanism for explanation. In view of the brief exposure and low intensities of light required for germination as found by all investigators, the natural translucence and gravelly surface of the soils of the saguaro habitat will admit sufficient light in most instances to permit germination of seeds buried to a depth of 5-10 mm (0.2-0.4 inches), the approximate maximum depth from which the seedling can emerge. A further mechanism for satisfying the light requirement exists in the soil churning and washing action occurring during typical summer rain storms. Such action permits burial of seeds after wetting and exposure to light.

Germination of saguaro seeds requires contact with free water and exposure to light (Alcorn and Kurtz 1959; Alcorn 1961a). Under ambient summer light and temperatures in a laboratory environment, initial germination of saguaro seeds maintained in continuous contact with free water occurs approximately 48 hr after initial wetting and 50 per cent germination is reached in approximately 72 hr. However, within the natural environment of the saguaro such prolonged periods (3 days) of continuously available water are a rare, almost non-existent occurrence during July and August when most natural germination takes place.

Rather, natural germination commonly follows two or more shorter periods of water availability with interspersed drying conditions at the soil surface resulting from the occurrence of two or more distinct rainstorms within a 2- to 5-day period (Steenbergh and Lowe 1969). In a humid atmosphere, saguaro seeds hygroscopically imbibe and retain moisture required for germination. Such pre-exposure to high relative humidity effectively speeds the germination process and reduces the required period of seed contact with free water. First germination of seeds so pre-conditioned can take place after 24 hr of wet contact and 50 per cent germination can occur within 48 hr. In a near-saturated atmosphere, maximum hygroscopic imbibition (approximately 20 per cent of the air-dry seed weight) is reached in approximately 20 hr.

Table. Hygroscopic Imbibition by Saguaro Seeds in the Laboratory at High Relative Humidities (90-100 per cent).

Time (hr)	**Weight (g)**	**Uptake (%)**
I - Ambient Light		
0	2.00	0.0
1	2.10	5.0
2	2.16	8.0
4	2.23	11.5
6	2.26	13.0
17	2.35	17.5
19	2.36	18.0
21	2.32	16.0
23	2.30	15.0
25	2.29	14.5
II - Total Darkness		
0	2.00	0.0
12	2.36	17.8
15	2.37	18.5
21	2.41	20.5
23	2.41	20.5
25	2.40	20.0

Air-dry 2-g (0.07-oz) seed samples were placed in petri dishes suspended over saturated paper towels in closed chambers in full room shade (indirect light) and ambient July air temperatures (ca. 70-95°F; 21-35°C) Data are given as weight and as moisture uptake (per cent air-dry weight). Dashed line represents break in curve at maximum moisture uptake. Sample I (in clear plastic chamber) was subjected to ambient light; sample II (in a light-proof metal chamber) received light only at the times of weighing.

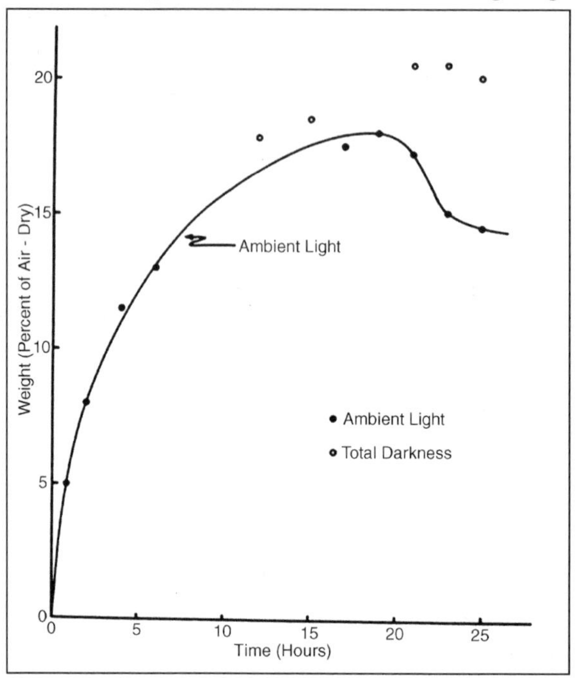

Fig. Hygroscopic Moisture Uptake by Saguaro Seeds at High Relative Humidities (90-100 per cent).

In a near-saturated atmosphere without condensation, saguaro seeds absorb atmospheric moisture up to approximately 20 per cent of their air-dry weight. Maximum hygroscopic imbibition is reached in approximately 20 hr, *i.e.*, ca. one day. Data in Table below. In an experiment to determine the effect of such treatment on the required period of contact with free water, treated seeds reached 50 per cent germination in 46.5 hr, 21.5 hr sooner than untreated (air-dry) seeds (68.0 hr).

Hygroscopic imbibition can reduce by approximately one day the required pre-germination period of seed contact with free water. Number and per cent germination of humidified (pre-treated) and of air-dry (untreated) saguaro seeds in the laboratory under ambient summer light and temperatures (full shade, ca. 65-95°F; 18.3-35.0°C) Pretreated seeds were held 5 days in open petri dishes in a near saturated atmosphere immediately prior to testing. Hours to germination of treated (2—1.00 seed lots) and untreated (3—10.0 seed lots) seeds is measured from time of initial contact with free water (15 August, 0800 hr). Data graphed in Figure.

Hours Wet	Germination			
	Humidified (20.0 Seeds)		Air Dry (300 Seeds)	
	No.	%	No.	%
24	7	3.5		
26	16	8.0		
28	27	13.5		
30	39	19.5		
42	95	47.5	11	3.7
48	116	58.0	22	7.3
54	148	74.0	88	29.3
72	167	83.5	169	56.3
120	184	92.0	254	84.0

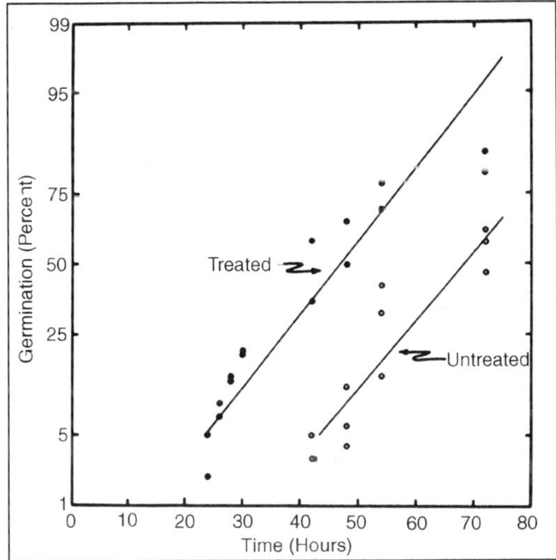

Fig. Effect of Pre-exposure to High Relative Humidity on the Rate of Saguaro Seed Germination after Contact with Free Water. The Time for Seeds Pre-exposed (Pretreated) to a Near-saturated Atmosphere to Reach 50 per cent Germination (46.5 hr) was Approximately 21.5 hr Less than the Time (68.0 hr) for Untreated (Air-dry) Seeds.

Hygroscopic imbibition can reduce by approximately one day the required period of seed contact with free water to germinate under natural conditions. Thus, in the desert environment where moisture is critically limiting, chances for germination are greatly increased by this adaptive mechanism that can reduce—by as much as a whole day—the required period of contact with a saturated soil surface. Germination 50 per cent points determined by regression analysis after probit transformation. Germination under natural conditions, therefore, need not depend upon the rare occurrence of a continuously saturated soil surface over a prolonged period of time; but rather it can be sustained by a period of high relative humidity followed by a relatively brief period of saturation. Thus, hygroscopic imbibition allows rapid germination with intermittent high moisture levels commonly associated with the characteristic summer storm patterns of the Sonoran Desert.

Requisite environmental conditions for saguaro seed germination occur far beyond the bounds of the plant's natural distribution. Only at the western boundaries of its occurrence—in the lower Colorado River Valley—where summer rainfall is seldom adequate to promote germination, does the range of the species appear to be limited primarily by germination requirements. With that exception, therefore, the factors that limit its distribution and control populations along the margins of its range must act during the post-germination stage of the plant's development.

GENERAL RULES OF GERMINATE SEED SOWING

Seeds need water and oxygen to germinate, so are best started in a light, loose soil that will not compact, get soggy, or crust over. Free flow of water & air are a must. Cover seed with 2 - 4 times their thickness of soil, unless they require light to germinate. Sow shallowly in cold wet spring, more deeply in warm dry summer. Large seed can be soaked overnight and planted singly.

Barely cover small seed, and sprinkle fine seed on the surface and water by misting. Plant flat seed edgewise and winged seed with wing uppermost or broken off. Sowing too thickly wastes seed and weakens the crowded seedlings, but some kinds sprout best if crowded. Lightly tamp soil to insure good contact with the seed, unless heavy. Keep soil moist, not soggy, and do not allow to dry out.

Common causes of failure are soil too heavy, wet or cold, or allowed to dry out, not giving slow seeds long enough to come up, pests eating the seeds or seedlings, and not giving dormant seeds the proper pretreatment. Careful attention to the instructions in the catalog and on the packet will help insure good results.

Common causes of seedling loss are damping off due to poor air circulation & overwatering, drying out or burning due to placing in full sun or outdoor conditions too quickly, transplanting shock (best done on a cool, moist day), and predation by insects, slugs and snails at night.

TEMPERATURE

Most seeds germinate best at warm (70°F) temperatures. Plants from temperate regions, the arctic, high mountains and high deserts often germinate best at cool temperatures. Plants from winter-rain areas like California, the Mediterranean, Chile, S. Africa and parts of Australia also like cool temperatures. Warm temperatures will often speed germination of these seeds, but lower vigor & survival. Warm desert plants and tropicals like warmth.

Temperatures used in the catalog are: Cold (34 - 45°F), Cool (50 - 65°F), Warm (65 - 80°F) and Very Warm (80 - 100°F). When no temperature is given, warm (70°F average) is indicated.

TIME UNTIL GERMINATION

Average range of time to germinate is usually given in weeks. A seed that takes 2 - 3 weeks will usually come up fairly evenly; one that takes 1 - 12 weeks will tend to straggle in irregularly.

Time varies with temperature, so expect considerable variation. Don't give up too soon — many who have given up and sown another seed in the pot end up with two types of plants in the same pot!

Hardy Annuals (HA)

These will stand some frost and can be sown direct to the garden as early as the ground can be worked, March to June. Prepare the soil until a smooth, fine surface is obtained. An attractive annual border can be had by planting in large, irregular drifts.

Half Hardy Annuals (HHA)

These are killed by frost and should be sown in late spring after danger of frost. For early bloom, start early indoors & plant out after danger of frost.

Tender Annuals (TA)

These need warmth and shelter and are best started in pots or flats and planted out in favored spots after the soil has warmed.

Biennials and Winter Annuals

Biennials are sown like half hardy annuals or perennials in spring or fall and planted out in September and October. Winter annuals such as some Californian and desert plants may be grown in summer, but are at their best sown in fall, even if grown in the greenhouse in cold winter areas.

Hardy Perennials (HP)

Many of these germinate readily at warm temperatures, and can be sown direct to the garden or early in the greenhouse or cold frame. If started early, they often bloom the first year. Others germinate best at cool or cold temperatures and the seedlings need cool temperatures. Many have various dormancies & need the pretreatment indicated.

Hardy Trees and Shrubs

Many will germinate readily if sown in spring. Others need cold or other pretreatment, and some are best sown in fall or winter and covered with a mulch or snow. The addition of some forest soil or litter from below both hardwoods and conifers is often very beneficial, greatly increasing seedling growth due to beneficial mycorrhizal inoculation.

Tropical and Subtropical Plants

These often need very warm temperatures. Bottom heat and constant moisture are often beneficial. Some take surprisingly long times to germinate. Many people think all tropical seeds are short-lived and perishable, but many of the longest-lived seeds are tropical, and some even need to be aged dry for a year before they will germinate!

Aquatic Plants

We have had the best results with some of these by sowing in pots, pans, or petri dishes, and submerging so that the soil is under 1/4" of water.

Fern Spores

Spores should be scattered thinly on the surface of a light, sterile, peaty mix which has been watered with clean sterile water about 4 hours before sowing. Cover pot with a pane of glass or enclose in a plastic bag and place out of direct sunlight. A light green film will form on the surface in 1 - 4 weeks or so, up to 6 months for slow species.

Botrychiums may take up to 8 years, and one *Equisetum* takes 18 years! Small round or heart-shaped prothallia will then form. Begin to mist lightly with distilled water weekly to enable fertilization of the egg cells, until true fronds develop. Transplant sporelings when large enough to handle. Good ferns can usually be had in 8 - 12 months from spores. Ferns are hybridized by sowing 2 kinds of spores together.

Soil Mixes

Many seeds do well sown direct to ordinary garden soil, but even good soil may be poor in pots or flats. These need a lighter, looser soil. Most commercial mixes are fine, but the addition of some garden soil and compost will often insure adequate beneficial micro-organisms & fungi.

A good soil mix can be made at home from 1/3 garden loam, 1/3 peat or compost, and 1/3 gritty sand. Number 1 limestone chick-grit makes a good top dressing for many alpines or slow germinating seed to discourage algae growth. Crushed charcoal also helps.

Hard Shelled Seed

These have hard impermeable shells and need nicking or scratching the coat to allow water to enter and the seed to germinate. The best results are from the least amount of nicking that will allow water to enter and the seed to swell. Many failures are due to over-nicking and damaging the seed. Different seeds need varying amounts of nicking. Most do best with lightly rubbing on sandpaper or a file until just the very outer coat is scratched. Often just scratching with a knife-point or scriber works. Others need serious nicking, sometimes with a hacksaw until the white interior shows.

Nicking seed one by one can be tedious but is most effective. Larger lots can be rubbed between two boards covered with sandpaper, or shaken or tumbled in a can lined with sandpaper. Then soak the seed overnight and any that don't swell or soften should be re-nicked and soaked again until swollen.

Nick & Soak means nick first, then soak, and re-nick any hard (unswollen) ones. Soak & Nick Unswollen means the seed is usually only partly hard and many will swell without nicking. Soak first, nick only the hard ones.

Other Methods for large lots include:

Hot Water Soak: The seeds are placed in a cup and not quite boiling (200°F) water is poured on them and allowed to cool & the seeds to swell. At least 4 - 5 times their volume of water is used. Or the seeds are dipped in boiling water for 10 seconds to 3 minutes. Produces erratic results, often killing many of the seeds and increasing fungal attack.

Dry Heat: The seeds are baked dry in an oven at 140° to 220°F for 4 - 10 hours, or are microwaved for 30 seconds to 4 minutes. Gives erratic results.

Sulfuric acid or lye soaks work with some seeds.

Cold Treatment (Prechill)

Many seeds need a cold moist period before they will sprout. The essentials are moisture, air, cold and time. Dry cold (a packet in the refrigerator) won't work. Often the seed may be sown in fall or winter and allow natural cold & snow to work.

1. Soak seed overnight until swollen or soft (up to four days for large hard nuts). Nick if needed.
2. Mix seed with about 3 times its volume of damp peat moss or vermiculite and place in a plastic bag or pill bottle. Small amounts may be conveniently layered between damp paper towels or coffee filters. Remember, air is essential; avoid sogginess. Label the bag with the name of the seed and date to be removed from cold. Mark your calendar too!
3. Store in the refrigerator (34 - 40°F) for the time specified in the catalog. Four to 12 weeks is usual
4. Remove the seed and sow. Seed is best kept cool (50°F or so) for a week after sowing, and gradually brought up to warm temperatures. Warming too quickly can be fatal for some seeds.

'Cold germinators' only sprout at cold temperatures, and are most conveniently treated by layering in damp towels and pricked off as they germinate. If using outdoor treatment, hold pots at above freezing for a few weeks before putting outside.

Snow cover is often highly beneficial.

Dormancy is highly variable. Sometimes a seed collected in the warm lowlands will germinate readily, but the same species collected at a high elevation will need cold. Dormancy even varies between individual plants at the same site, and varies with weather before harvest and conditions of storage after harvest.

Gibberellic Acid-3 (GA-3)

GA-3 is a naturally occurring plant growth regulator. Presoaking seeds in GA-3 will often cause rapid germination of many highly dormant seeds that would otherwise require cold, aging, light or other prolonged treatments. Many different types of dormancy are overcome. You may buy our 'GA-3 Primed' seed which are ready to plant, or the GA-3 powder or one of our kits. GA-3 is safe, easy to use, economical and rapidly becoming a standard tool for germinating seeds.

Other Treatments

Dry Storage or Ageing: Many seeds will not germinate when freshly harvested, but are dormant until after a period of dry storage ranging from 1 - 12 months or up to 5 years. The time varies with temperature, humidity and oxygen. We try to supply aged seed whenever necessary. Often this type of dormancy can be broken with GA-3.

Warm Moist Treatment: Many seeds need 1 - 4 months of warm moist treatment, followed by cold treatment to sprout. In some, the root sprouts during the warm period, but the shoot does not sprout until after a cold period (epicotyl dormancy). Called 'two-step' germinators by Deno.

Some of these have unformed embryos which must first develop at warm temperatures. Some seeds are '70 - 40' germinators, which need warm moist, then cold moist, and germinate at cold temperatures. Some tropicals which are slow to germinate (many palms and aralias) need prolonged warmth to first develop their embryos. Many seeds are 'multicycle' germinators, needing several cycles of cold and warm periods.

Soaking, Washing or Leaching: Many seeds contain germination inhibitors which in nature prevent the seed from germinating except during the wet season (tropicals) or only after sufficient rain has fallen (desert plants). Soaking the seed in shallow water and changing every day for several days will leach out inhibitors. Heavy daily watering of pots may work. Some have oily coats and need detergent or peroxide soaks.

Light or Darkness: Some seeds need light to germinate; others need darkness, and light prevents sprouting. If light is required, sow on the surface; if darkness is needed, cover seed well.

Oscillating Temperatures: Some seeds need the daily change from warm to cold that occurs in spring and fall, and they germinate best sown in pots and exposed to outdoor conditions.

Other Treatments: Soaking seed in potassium nitrate (KNO3), hydrogen peroxide, citric acid, sodium hypochlorite (bleach), smoke solution, charred wood leachate, or enzyme solutions have all been used to trigger germination of seeds.

Smoke Treatment

Smoke treatment often helps germination of plants from fire-prone environments, particularly Mediterranean-climate plants such as many species from California, Chile, Australia, South Africa, and the Mediterranean region.

Prepare a dilute smoke solution by adding one part commercial smoke flavouring to nine parts water. Either soak the seeds in this solution overnight (or until they swell), or water the pot or flat once with this solution. Smoke flavouring is found in the spice and flavouring or barbecue section of the grocery store. It comes in small brown bottles of liquid, called "liquid smoke" or "hickory seasoning". Look for the "all natural" type that lists only water and natural smoke concentrate as ingredients. "Wright's" is the brand we find out here in California.

Smoke treatment is still experimental, and you may have to try different dilutions. Let us know your results! Another method is to burn a layer of straw or pine needles on top of the flat.

PLASTIC BAG METHOD

Best for very slow to germinate seeds, very tiny dust-like seeds that can't be allowed to dry out, and very slow-growing seedlings. Small clean pots are filled with damp, sterile, soilless mix. Milled sphagnum or a light dusting with powdered charcoal discourage fungi & algae. Seeds are sown and the whole pot is sealed in a plastic bag and placed out of direct sunlight. This creates a mini greenhouse and the soil will not dry out and the seeds are protected from mice, etc. Pots can be left for years with no care other than regular checking for seedlings. As soon as seedlings appear, begin hardening off. Bagged pots may be kept under fluorescent lights without overheating. Don't forget to label!

Hardiness Zones

The rating of plants according to the hardiness zones in which they will grow has improved greatly over the last twenty years. They are fine for common, widely grown plants, but for most plants, our knowledge of their ecological limits is incomplete. Too often, the assignment of zones is merely educated guesswork.

Many plants will do well far outside their usual ecological range, or seeds collected at the species northernmost limit will be much hardier than usual. Also, a heavy mulch, or planting by a south-facing wall, or other simple tricks can greatly extend a species range in cultivation. So don't rely too much on the zone numbers, or let them prevent you from experimenting!

The Limits of the average Annual Minimum Temperatures for each Zone
Zone 1: Below -50°F
Zone 2: -50° to -40°F
Zone 3: -40° to -30°F
Zone 4: -30° to -20°F
Zone 5: -20° to -10°F
Zone 6: -10° to 0°F
Zone 7: 0° to 10°F
Zone 8: 10° to 20°F
Zone 9: 20° to 30°F
Zone 10: 30° to 40°F

SEED LONGEVITY

Many people have the mistaken idea that seed must be very fresh to grow, and the occasional failure of a packet is blamed on 'old seed'. Properly stored, seed of most species is viable many more years than most people would expect, some giving high germination for decades. Though some species are short-lived, many others are dormant when freshly harvested, and actually need an aging period of weeks, months, or even years before reaching full germination.

Therefore the longevity of various seeds are included. And all of seeds are stored in glass jars in a low humidity atmosphere, ideal for preserving viability. Should you find you need to store some seed till next season, keep the packets in a well closed glass jar in a *cool dry place.*

Seed *half-life* is frequently given in the catalog. This is the average number of years it takes for seed of that species to drop to half of maximum germination, when stored under ordinary conditions.

"In several cases, maximum germinability was not attained until after some years of storage." —D. A. Priestly, in describing a longevity study of 36 species of common seeds. It was found that 14 of the 36 species retained half their viability for *ten years or more of open storage.* —J.L.H., 11/87.

About Difficult Seeds

Over millions of years, wild plants have evolved germination strategies which ensure their survival, but which may not be convenient for the home gardener who wants a quick and even stand of plants from a packet of seed. Many seeds sprout irregularly, so that if the first flush of seedlings is killed by adverse weather, insect predation, etc., more will come along to take their place. In adaptation to various environments, some seeds need periods of cold, warmth, darkness or light, fire, etc.

Some have seedcoats of varying hardness or impermeability, and others contain chemical germination inhibitors which must be leached from the seed before it can sprout. Some species disperse themselves over wide areas by being eaten by animals, the seed sprouting far from the mother plant, the seedcoat softened by digestive juices. Many seeds have internal clocks, and give much higher germination at certain times of the year, regardless of the treatment given.

All seeds wait for the correct time and conditions before sprouting, and the gardener must mimic those conditions to ensure successful germination.

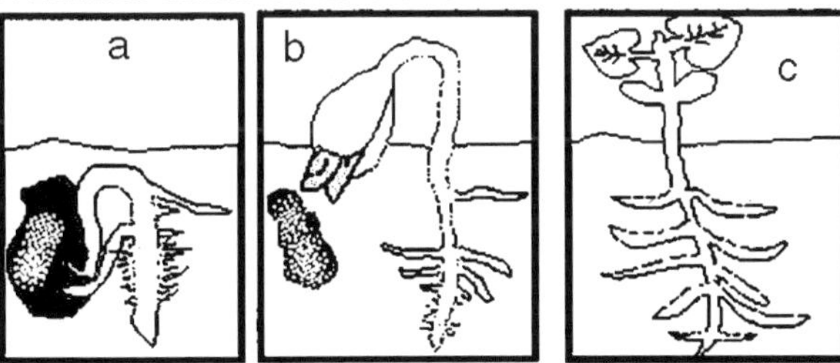

Fig. (a) First the young root appears. (b) The skin splits and the shoot straightens, pulling the cotyledons with it. (c) The main root gets bigger and grows side roots, and the first leaves appear

We are continually testing seed for germination, and conduct research into improving methods of handling difficult species. For slow seeds, which take months or years, making a standard test impractical, we may use 2,3,5-triphenyltetrazolium chloride to test for hydrogenase enzyme activity, which quickly indicates whether there is a living embryo in the seed. It can take years of research to determine the best ways to germinate a specific species.

Oryzopsis seed has been studied for over 50 years, yet we still do not fully understand its requirements. Even then, seed collected from one population may germinate readily, yet the same species gathered in a cold-winter area may need cold treatment.

We offer many seeds which are easy, and sprout quickly and evenly. But with some you must be prepared to experiment, be patient, and use your initiative and intuition. Remember that with some rare species, you are venturing into unknown territory. Most corporate seed companies will not carry difficult seeds, only selling, easy, mass-produced varieties.

GERMINATION OF WHEAT GRAIN

A wheat grain, which may be taken as an example, is not a seed, but a kind of nut with a single seed within it. The seed grows in such a way as to completely fill up the interior of the nut, and become practically united with its inside wall. The embryo occupies only a small part of the grain, the rest being taken up by the floury endosperm of the seed. The embryo is easily seen at the base of a soaked grain on the side opposite the furrow.

When removed it has the appearance. The part of it which lies close up to the endosperm is a flattened somewhat fleshy shield-shaped structure called the scutellum attached to the front of the scutellum is the plumule, consisting of a bud formed of an extremely short stem, upon which are sheath-like leaves enclosing each other. The embryo generally possesses five roots, one primary and two pairs of secondary. They are all completely enclosed by a sheath which continuous wjtn the scutellum, and are consequently not visible from outside; their position, however, is marked by projecting bosses. The sheath round the roots is termed the coleorhiza, and when germination takes place it expands and bursts the coats of the grain, the roots about the same time breaking through the enclosing coleo-rhiza.

When a wheat grain is sown in the ground it remains there, but the plumule grows upwards, its first leaf, the coleoptile, appearing above the soil as a single pale tube-like structure; from a slit in the tip of the latter the first flat green blade soon appears, and is followed by a succession of single green leaves, the younger ones growing from within the older ones in regular order. With a sharp knife or razor cut through from back to front, so as to divide the grain into two longitudinal halves, and note the floury endosperm and the shape and parts of the divided embryo. Place a folded sheet of damp blotting paper on a plate, sow some soaked wheat grains on it, and cover with a tumbler.

The grains will germinate. Watch their development up to the time the first green leaf appears, taking out the embryo and examining it at different stages of its growth. There is difference of opinion as to which part of the embryo is to be considered the cotyledon.

Soine authorities regard the scutellum as the cotyledon, while others give this name to the coleoptile or first sheathing leaf which comes above ground, and which has no green blade. Others, again, consider that the first sheathing leaf is an extension of the scutellum, and the two combined is therefore the cotyledon. In any case, there is only one cotyledon present, and wheat therefore belongs to the class of monocotyledonous plants. During the growth of the embryo of a wheat grain, it will be noticed that the endosperm becomes soft and decreases in quantity as the roots and plumule expand and develop; the endosperm is the food upon which the young plant depends during the early staggs of its life, the scutellum acting as a structure for changing, absorbing, and transferring this reserve-food to the growing parts which need it. Remove the embryos from well-soaked grains, and grow them without the endosperm on damp blotting-paper.

Allow ordinary uninjured grains to grow with them. Both the embryos in the grains and those removed from the grains develop, but there is a great difference in the results after a few days. The store of reserve-food on which the young plant depends for its early development is sufficient to enable it to form a root, stem, and several leaves, as is evident when seeds are allowed to germinate upon damp flannel or blotting-paper, from which nothing but water is absorbed. No food-materials or manures are needed for this primary development, and seeds germinate and the seedlings grow for a considerable time as well in poor soil or sand as in good rich ground.

As soon as the reserve store is exhausted hunger becomes apparent, and unless the plants are then supplied with suitable nutriment from the soil and air, and are also placed under conditions favourable for growth, weakness and death are likely to occur. Among the larger seeds, such as beans and peas, where there is an abundant store of reserve-food, the young seedlings begin to manufacture food for themselves from materials absorbed from the soil and air, long before their reserve is exhausted.

In small seeds the reserve is sometimes almost consumed before the roots and leaves are sufficiently developed to carry on their work properly, in which cases a more or less temporary starvation and check to growth ensues. Especially does this happen when seeds are sown too deeply, for a large amount of food is then used in the production of a stem long enough to lift the leaves up into the air.

SEED OF WHITE MUSTARD

The seed of white mustard (Brassica alba Boiss.) contains an embryo which like that of the bean consists of a radicle, plumule and two cotyledons; the latter, which are folded together, are relatively thinner than those of the bean and deeply notched. The radicle is bent round and lies in the fold of the cotyledons, between which is the very small, almost invisible plumule. On germination the cotyledons, instead of remaining within the seed-coat and below the ground as in the case of the broad bean, escape from the enclosing coats altogether and grow up out of the ground, enlarging at the same time, and becoming green like ordinary leaves. They are the first ' smooth' leaves of the seedling mustard plant. After a short time the plumule grows up from between the cotyledons and forms a stem upon which are gradually unfolded the ordinary divided ' rough' leaves. Soak white mustard seeds, and examine their structure, noting especially the way in which the embryos are packet in them.

Allow some to germinate and grow for a week or more on damp flannel, and examine them in various stages of development, noting the notched cotyledons with small. The term hypogean is applied to cotyledons which remain below ground, those coming above being epigean, the relative amount of growth in the hypocotyl and epicotyl determining their position. If the hypocotyl grows vigorously during or after germination the cotyledons are forced above ground; when only the epicotyl grows the plumule is lifted up above the soil, but the cotyledons remain below where the seed is placed. In the broad bean the hypocotyl is very short, and the point where it ends and the root begins is not clearly denned. In a mustard seedling, however, the point of separation between the root and stem is somewhat swollen and readily distinguished.

All plants whose embryos, like those of the bean and mustard, possess two cotyledons, are known as Dicotyledons they form a very large, well-marked class of the flowering or seed-bearing plants. The seeds contain within their coats nothing but an embryo plant, which depends for the development of its root and shoot upon the substances stored up in some part of its body, its cotyledons chiefly.

This is true even in the case of seeds like those of white mustard, in which the cotyledons of the embryo are comparatively thin. There are, however, a number of plants, such as the ash, mangel and potato, which, although belonging to the Dicotyledons, have seeds in which there are stores of food inside the seed-coat, but free from the embryo and its cotyledons. Such separate reserve-food is stored in that part of the seed known as the endosperm or 'albumen,1 and seeds in which it is present are called endospermous or albuminous seeds.

Those like the bean, pea, and vetch, mustard, and turnip, which have no separate reserve-food, are known as exendospermous or exalbuminous seeds. Take out a seed from the fruit of the ash tree in autumn; carefully cut thin shavings from the flat side of the seed, stafting about the middle of the seed and cutting towards the narrow end. Note the white embryo with its well-marked radicle, hypocotyl and two flat cotyledons lying within the semi-transparent endosperm. Some of the most commonly occurring endospermous seeds will be found to have embryos within them which are not dicotyledonous, and whose structure is in many respects. A good example is met with in the onion.

ONION SEED

The seed is black, somewhat oval in outline, with one side convex, the other almost flat. Each contains within it endosperm and an embryo which lies curled up inside in the form seen. When germination commences, the curved part imbedded in the middle of the endosperm grows and forces the end of the embryo out of the seed.

From this exposed end, which is the radicle, a straight, slender, primary root develops. The part of the young seedling which extends from the root into the interior of the seed, grows very rapidly at first, at the same time assuming a sharply bent outline.

It comes above ground in the form of a close loop but on further growth the end within the seed is pulled out of the soil, and grows up in the air. The tip within the seed changes and absorbs the endosperm, and usually remains there until all the nutrient material has been transferred from it to the various centres of growth in the young plant. After the food-reserve is exhausted the tip withers and becomes free from the seed coat.

In loose soils the latter is pulled above ground before the endosperm is exhausted, and remains on the end of the tip for some time. In other cases where the soil is damper and of a stiffer nature the seed-coat remains below ground altogether. The curved part of the embryo which comes above ground is a leaf. It is the cotyledon of the embryo, and is in reality a thin hollow leaf like those of the full grown onion plant: within it is the plumule, which consists of a series of hollow conical leaves arranged one inside the other. After one leaf emerges others soon follow, the younger ones coming out in regular order through slits in the sides of those immediately older than themselves. Soak fresh onion seeds in water for a few hours. With a razor cut through some parallel to their flat sides in order to show the embryo within.

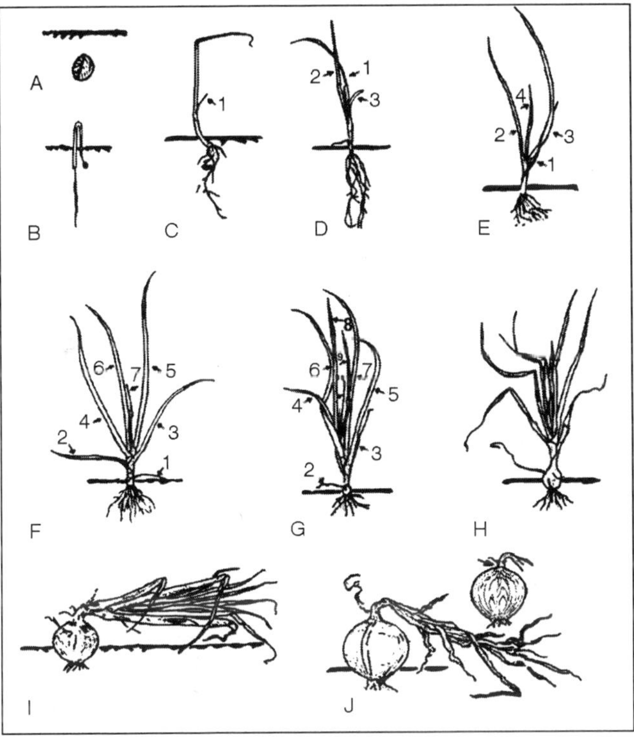

Fig. Germination of Onion

Sow others in damp blotting-paper; allow them to germinate and the seedlings to develop; make observations of them at different stages of growth. Watch the germination of seeds sown in boxes or pots containing ordinary garden soil. Plants whose embryos possess only one cotyledon are known as Monocotyledons, and form the second large class of seed-bearing plants. Few of the representatives of this class with which we are ordinarily familiar have true seeds large enough for examination.

The onion is probably one of the best commonly occurring examples which may be considered typical of the monocotyledons and easily obtainable. To this class, however, belong all the grasses, but their seeds and embryos are so different in many respects from those of the onion that it is necessary to examine one of them in detail.

The Common Bean of Seeds

A broad bean is one of the largest seeds met with in ordinary farm or garden practice, and as its parts are all sufficiently large to be observed without the special aid of anything more than an ordinary pocket lens, it is especially fitted for study. When a nearly ripe pod of a broad bean plant is opened, each seed within it is found attached to.

The inside by means and it is through this stalk that all the nourishment passes from the parent to enable the young seed to develop. A1 first the pod exists in a rudimentary form in the centre of a flower and its parts and contents are very small; they are nevertheless readily seen with a pocket lens. After the fading of the flower, the pod and seeds within it grow larger and larger at the expense of food supplied by the rest of the plant, and ultimately when ripe the funicles wither and dry up, and the seeds become detached from the parent which has produced them. When dry and ripe each bean seed is hard, with an uneven surface, but its internal construction cannot be clearly examined in this condition.

On soaking in water for twelve hours, however, it becomes softer, and the parts can then be easily investigated. The outside, which is a pale buff colour, is smooth, and has at one end a narrow elongated black scar called the hilum of the seed. It is known popularly as the ' eye' of the bean, and marks the place where the broad end of the funicle separated from the seed when it ripened in the pod.

Quite close to one end of the hilum is a very minute hole known as the micropyh, easily seen with a lens, and through which water oozes out usually accompanied by bubbles of air when soaked beans are squeezed between the finger and thumb. This opening communicates with the interior of the seed, and is the only one it possesses. The rest of the seed after the testa is removed, is of oval flattened shape similar to the complete bean, and is divisible into two large fleshy halves called cotyledons which, however, are not completely separate from each other, but connected at the side with a conical projecting body, one end of which is found to fit into a hollow cavity in the seed-coat exactly opposite the micropyle; the other end is bent and turned inwards between the fleshy cotyledons.

Small Curved Structure

The extent and shape of this small curved structure is most easily observed when one of the cotyledons is removed completely; it remains attached to the other. Soak some broad beans in water and keep them in a warm place all night. Examine them next day and make drawings of the various parts seen both before and after stripping off the testa. Observe the relative position of the parts of the embryo in reference to each other and to the seed-coat.

Examine and compare the structure of the following seeds after soaking in the same way:—Pea, scarlet runner beans, vetches, and red clover. The bean seed contains nothing more than what has already been described; the nature and relationship of its component parts only become intelligible when the seed is placed in the ground or maintained under certain conditions, and allowed to grow.

When growth commences the lower end of the small curved structure elongates and breaks its way through the coat of the seed at a point very close to the micropyle, but not, as often erroneously stated, through the micropyle itself. It soon assumes the form and is recognised as a root of a young bean plant The upper bent half, which lies between the cotyledons, also pushes its way out of the same opening in the seed-coat and develops into a stem, from the tip of which leaves are gradually unfolded.

It is thus seen that the seed of a broad bean is a packet containing a bean plant in a rudimentary condition. This plantlet is callai an embryo, and the portion of it which becomes root and stem is its primary axis.

The part of the primary axis which is below the point where the cotyledons are attached consists of a very small piece of stem, the hypocotyl at the end of which is the radicle or primary root. Where the stem ends and the root begins cannot be determined in the bean seedling without the aid of the microscope and examination of the internal structure of the axis of the plant. The curved end of the primary axis above the cotyledons is the plumule of the embryo, and consists of a very short piece of stem, the tpicotyl on the top of which is a bud.

From the latter is derived the ordinary stem which comes above ground with its green leaves and flowers. In the early stages of the growth of the embryo from the seed the hypocotyl grows very little. The upper part of the stem bearing the plumule comes out of the seed bent and it maintain this curved shape for some time after emerging.

By this behaviour the delicate leaves of the plumule are protected from injury during their progress upwards when a seed is sown in earth or sand. Fold up some soaked beans in two thicknesses of white flannel made damp, and place them on a plate. Cover them with another plate placed upside down, and leave them in a warm room. Examine them twice a day, leaving them, exposed to the fresh air for a few minutes each time, and keeping the flannel damp, not wet. When they sprout notice the place where the radicle has come out of the seed coat. Let some grow till the radicle and plumule are well out of the seed, and compare the various parts of the sprouted seeds with unsprouted ones.

PRIMARY AND SECONDARY BOOTS OF SEEDLING

It was noticed, when dealing with the bean seedling, that its two ends always grow in opposite directions; the plantlet can be considered as an elongated axis, one end of which bears the leaves and invariably comes above ground, while the opposite end never bears leaves, and persistently follows the plumb-line downwards.

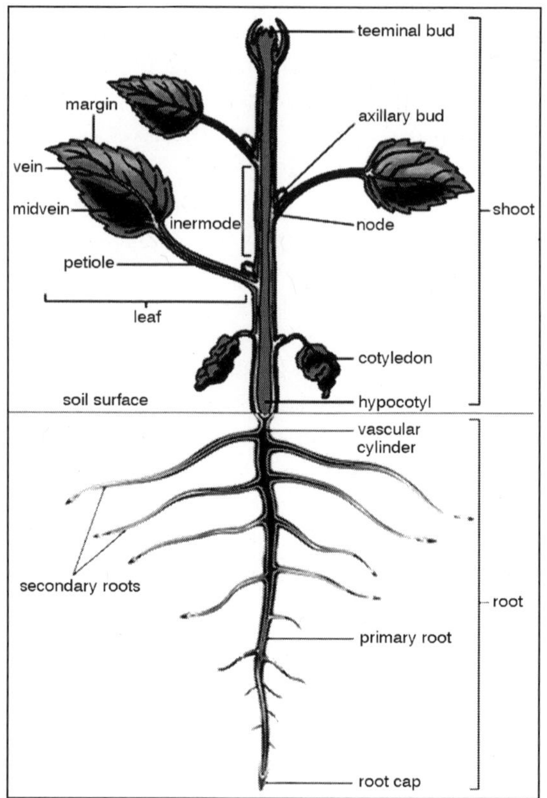

Fig. Dicotyledonous Plants

The descending part is known as the root. As will be pointed out later, all roots do not behave in this manner, and it is to be specially noted that many of the underground parts of plants are not roots; the exceptions, however, may be left for future consideration. The first or primary root which the bean plant possesses is merely an extension of the radicle of the embryo which exists nrithin the seed.

Soon after making its exit from the seed, it takes a downward course, and elongates by growth taking place near its tip. Another point to be noted is that the lateral roots do not arise as up-growths on the surface of the primary root, but come from within it, and are described as endogenous. The slits which they make in the substance of the primary root, and through which they emerge, can be readily seen in a bean seedling.

The secondary lateral roots are connected with its central more solid core; the three lowest ones, although they have just begun to grow, have not yet penetrated the outer layer of the root, and would not be visible on the outside of the latter. This mode of origin is characteristic of lateral roots generally wherever they are met with. Many dicotyledonous plants have roots similar to those of the bean plant.

When, as in this case, the primary one continues to grow, keeping distinctly larger than the lateral ones, it is called a tap root. Very good examples are met with among cultivated plants in the carrot, mangel, red clover, and mustard in shep-herd's-purse, poppy, and many other weeds, as well as in most broad-leaved trees. A number of plants have swollen fleshy roots in which food materials are stored for future use; they are described as tuberous roots, and must be distinguished from tubers, which are fleshy underground stems. To designate the different forms of thickened roots various special terms are in use.

The typical carrot root is conical; that of the turnip napiform. The root of the radish is spoken of as fusiform. In some instances the primary root is soon rivalled in size by its branches; it may even cease growth altogether. Such plants, on being pulled out of the ground, exhibit a bunch of slender roots, the chief of which are all much the same diameter and length; roots of this character are described as fibrous and are well exemplified in common groundsel and grasses.

GERMINATION CONDITIONS

The optimum conditions for different stages of germination and seedling growth are not identical and may even vary for different seeds within the same seed lot. The aim of much seed testing research has therefore been to determine a combination of conditions which will give the most regular, rapid and complete germination for the majority of the same species. Substrate. Soil is rarely used as a substrate for germination tests because each sample will vary greatly in physical, chemical and biological properties. While the germination results might be more comparable with field conditions, the lack of reproducibility and difficulty in comparing tests of different seed lots render it unsuitable. Artificial media are much more easily standardised. Most laboratory tests on small seeded species are made on paper. Other materials used include sand, granulated peat moss and expanded mica (Vermiculite and Terralite).

The main requirements for the substrate (Justice 1972) are:

- Non-toxic to the germinating seedlings
- Free of fungi and other micro-organisms
- Of porous texture to enable adequate aeration and moisture for the germinating seeds.

The choice of media on which the seeds are placed for germination depends on the equipment, the species, the working conditions and the experience of the operator. It is prescribed in the ISTA Rules for a number of forest trees.

Filter paper, heavy box paper, or any other absorbent paper may be linked with a filter paper or cotton wick dipped into water, or placed on top of a layer of sand or vermiculite. Cellulose paper is becoming more commonly used as a germination medium because it is easier to handle than sand, but still permits penetration of the radicle, thus providing for better evaluation of abnormal germination. Also moisture content does not tend to become stratified within the paper medium as it does with sand (Belcher 1974).

The best paper substrates are germination blotters, paper towelling, laboratory filter paper and creped cellulose paper wadding (Bonner 1974). Any paper substrates should be checked for presence of toxic chemicals. The 1976 ISTA rules give detailed specifications for paper and towels, including weight, bursting strength, capillary rise and acidity.

Seeds without a specific light requirement may be placed on top of paper or between folded paper. Folded paper increases the surface contact area between the seeds and the moisture supply. Large seeds may be rolled in paper towels which are then placed in an upright position; this allows the roots to grow downwards and avoids tangling of the roots (MacKay 1972). Rolled or folded paper packs may be kept moist without the necessity of a wick dipping into water.

Rolled seeds may be stored on glass plates above, but not in contact with, a water bath and covered by a polythene sheet (Knudsen 1982). Any rolls which show signs of drying out may be sprinkled with water. Use of rolled paper is a very rapid and convenient method but may cause curling of the radicles. This is of no concern if the germinated seeds are to be discarded after testing. For operational germination of large quantities of seed, however, the slightly slower folded paper method is preferable. This is used with success as standard practice for Pinus and Eucalyptus spp. in Thailand.

Sand is not suitable for very small seeds because they are difficult to locate, but it is widely used for larger seeds. Sand can be sterilised and fungi develop on it less freely than on paper. It also provides a good contact between the seed and moisture because the seeds can be pressed into the medium. A commonsense rule is to cover the seed with a thickness of sand not less than the length of the seed on its longest axis recommends a thickness of 1 – 2 cm according to seed size and prescribes that sand particle sizes should be between 0.05 and 0.8 mm and pH between 6.0 and 7.5. Sand as a germination substrate is preferred for tree species that have a longer germinative period, for instance, Rosa spp., Pinus caribaea, P. elliottii, and P. palustris, or with larger seeds, for example, P. pinea and Quercus spp. (Magini, 1962). A sand-perlite medium permits leaching of the water which can be important when testing repellent-coated seed (Belcher 1967), but may be a disadvantage when testing seeds which are sensitive to drying.

Moisture and Aeration. The moisture level of the substrate has been suggested as one of the major causes of variation in seed research results. The ISTA Rules specify that a sand substrate be moistened depending on its characteristics and the size of the seed to be tested and suggest that a moisture level of 50 – 60 per cent of the water-holding capacity of the sand is suitable for several groups of agricultural seeds. A paper substrate must not be so wet that a film of water forms around the seed. Belcher (1974), working with Pinus spp., concluded that most species have considerable tolerance to diverse moisture levels and the major variations in germination were due to dry conditions. He found some species to be sensitive to dry conditions and others to very wet conditions.

As a generalisation, the substrate must be moist enough at all times to supply the necessary moisture, but excessive moisture will restrict aeration. All tests must be examined daily to ensure that the moisture content of the substrate is near optimum. The water should be reasonably free of impurities. Temperature control. In laboratory seed testing the important temperature is that at the level of the seed. The temperature will vary for different species and appropriate values are listed for a large number of tree species in the International Rules for Seed Testing.

Temperature is one of the most critical factors in the laboratory germination of seeds and must be regularly checked. When alternating temperatures are required the test is usually held at the low temperature for 16 hours and at the high temperature for 8 hours each day. Temperature of 20° and 30°C alternating are commonly prescribed for tree species. Although natural fluctuations between day and night temperatures are less in lowland moist tropical forests than in other forest types, alternating temperatures may still affect germination of tropical species. In one experiment of Terminalia ivorensis in Nigeria, alternating temperatures of 34°C and 24°C gave 93 per cent germination in 41 days compared with 27 per cent at a constant temperature of 30°C, both with continuous light (Okoro 1976).

Light. Light is required for germination in many tree seeds. Fluorescent light is as effective as natural daylight and is preferred in testing, because the wavelength and intensity can be standardised, within limits. Cool white fluorescent lamps are recommended because of their light quality and low heat emission: Light should be evenly distributed over the tests, with an intensity within the range 750 – 1250 lux. Seeds should be subjected to light for only part of the test period, 8 hours in every 24 hours is usual but a longer or shorter period may be beneficial to seeds of some species.

Controlling fungi in germination tests. Laboratory practices that will minimize the spread of fungi include proper spacing of seeds, control of temperature, removing decayed seeds, proper aeration, and keeping the substrate only just wet enough to permit germination. Sterilisation of laboratory equipment and the periodic disinfection of germination cabinets and other apparatus will assist. As a general rule, seed is not disinfected because seeds which decay are usually of poor quality. Magini (1962) states that a number of experiments carried out with different methods of disinfecting the seed just before or during germination have shown no reliable improvement in germination percentage. A small quantity of fungicide added to the water used to moisten paper rolls has, however, been found beneficial in testing conifer seeds in Denmark (Knudsen 1982). In Australia spraying of Karathane fungicide at a concentration of 0.8 g in one litre of distilled water has also proved effective when testing eucalypt seed on moist filter paper over vermiculite.

GERMINATION TESTING

Of all the quality measurements of seed lots, none is more important than the potential germination of the seeds. The main aim of a laboratory germination test is to estimate the maximum number of seeds which can germinate in optimum conditions. The use of standardised ideal conditions in the laboratory such as those prescribed by ISTA ensures that results obtained from a given seed lot in one laboratory should be identical with those obtained from any other laboratory in the same or in other countries. A common standard for assessing germinative potential is very valuable for seed moving in international trade.

On the other hand, it is clear that results obtained under ideal controlled conditions in the laboratory are not directly applicable in the field nursery, where only limited control over environmental conditions is possible. Each nurseryman should apply his own correction factor, derived from experience over the years, to convert the germination potential of a seed lot, as determined from laboratory tests, to the actual field germination he can expect under his own local conditions.

At the other extreme, a nurseryman may prefer to carry out his own germination test in his own nursery in advance of large scale sowing. The results of the test should be directly applicable to the later sowings of the same seed lot in the same nursery, but would not be applicable to other nurseries. However, there is sometimes insufficient time for germination tests in advance of the main sowing and managers of large operational nurseries may be reluctant to involve themselves in small scale research.

Between the two extremes lies the case of the small silvicultural research station which lacks the laboratory facilities to comply with ISTA prescriptions for testing, but which carries out germination tests in the nursery on bulk seed supplies before distributing them to afforestation projects throughout the country. Here again the local forester or nurseryman must apply his own correction factor to convert actual germination figures from the research nursery to expected germination figures in his large operational nursery.

The following account is largely based on that of Turnbull (1975 d). It is in conformity with ISTA rules (1976), but some of the principles are equally applicable if lack of equipment or trained staff makes it necessary to use simpler methods. Germination is defined as the emergence and development from the seed embryo of those essential structures which are indicative of the seed's capacity to produce a normal plant under favourable conditions (Justice 1972, ISTA 1976). Germination is expressed as the percentage of pure seeds which produces normal seedlings or as the number of seeds germinating per unit weight of the sample.

In the laboratory the environmental conditions, including moisture, temperature, aeration and light, must not only be specific enough to initiate germination but also favourable for the development of the seedlings to a stage where interpretation of normal and abnormal types may be made. With a few exceptions, all germination tests should be made with pure seeds separated by the purity test.

The pure seed must be well mixed and counted at random into replicates. The seeds should then be spaced uniformly on the test substrate. Normally one test consists of 400 seeds in 4 replicates of 100 seeds each, but if 100 seeds overcrowd the test substrate, the replicates may be broken down into a larger number of smaller replicates of 50 or 25 seeds each (Bonner 1974). A general recommendation is to leave 1.5 to 5 times the normal seed width or diameter between seeds, in order to discourage the spread of fungal moulds (Bonner 1974, Justice 1972).

The exceptions are very small-seeded species in which it is impossible (some species of Eucalyptus) or very difficult (Alnus, Betula, Populus,Salix) to separate the seeds from the accompanying inert matter or chaff. In these cases the test is made with the same number of replicates but the replicates are of equal weight rather than an equal number of seeds.

GERMINATION EQUIPMENT

The kind of germinator can be selected according to the type and amount of seeds being tested and it is acceptable providing it can give good control of prescribed temperature, moisture, and light conditions.

Germinators range in size from small or portable individual units through cabinets of various sizes and large germination tables of the Jacobsen Copenhagen type, to walk-in germination rooms. The major types recommended by ISTA are as follows:

Jacobsen and Rodewald apparatus. In Europe, a germinator called the Jacobsen apparatus or Copenhagen tank is commonly used. It consists of a bath containing water, the temperature of which may be controlled by thermostats. The seeds are distributed on top of paper and placed on metal or glass strips suspended about 5–7 cm above the bath; paper or cotton wicks lie beneath the substrate and pass through slots into the water below. The moisture content of the substrate may be adjusted by altering the water level, and so increasing or decreasing the distance between the seed bed and the water (Kamra 1968).

High humidity is maintained around the seed either by covering the entire tank with a transparent lid or placing an inverted plastic funnel with a hole in the conical end over each germination pad. The apparatus may be exposed to natural daylight but artificial light is usually desirable.

One of the disadvantages of the standard Copenhagen tank is the absence of direct temperature control at the seed bed. In the tropics it tends to overheat (Robbins 1982 b). Improved models have been developed, such as the one by Overaa (1962) which uses hollow stainless-steel strips, through which the water for heating and cooling is circulated. Other disadvantages are that the equipment requires a large amount of floor space for the number of tests accommodated. Also, the handling of covers and substrata appears to be more cumbersome than moving a lightweight tray with paper substrate (Justice 1972).

The Rodewald apparatus consists of a zinc box covered with glass in which the seeds are exposed to direct or diffuse light. The bottom of the apparatus contains water over which a layer of moist sand is placed on a tray. The seed bed consists of unglazed porcelain dishes placed in the moist sand, or the dishes may be placed directly in the water. The temperature of the water is controlled thermostatically and the sand is moistened through wicks reaching into the water bath. A sand substrate is not desirable for species requiring alternating temperatures because of its slow adjustment to a change in temperature.

Germination cabinet. Another common type of apparatus is the closed chamber for germinating seeds in darkness, diffused light, or direct light. This type of germinator often consists of a double walled cabinet, suitably insulated against temperature changes by an air jacket or layer of insulating material. It is equipped with appropriate tray slides to accomodate the type of germination trays preferred by individual laboratories. The modern germination chamber has provision for both heating and cooling. Usually, water is chilled and passed between the walls of the chamber or through cooling pipes placed along the inner walls of the chamber. Either the air or a reservoir of water in the bottom of the chamber may be heated. In such cabinets the temperature can be regulated at any desired setting between approximately 8°C and 40°C (ISTA 1976).

Inexpensive germination cabinets are sometimes constructed locally. Gupta and Kumar (1977) describe the low cost germinators in use in the Seed Testing Laboratory in Dehra Dun, India. The outer wall of the cabinet is constructed of teak, the size 195 × 70 × 40 cm, and the electric controls include a thermostat, hot air blower, a separate air circulator and a time switch to control illumination. Glass wool is used for insulation between the inner and outer walls. Satisfactory performance is obtained at temperatures varying from room temperature up to 45°C (± 1°C).

A germination cabinet should meet as far as possible the following requirements (Oomen and Koppe 1969):

- *Air humidity:* As high as possible but preferably never lower than 90 per cent, to prevent the seed substrate from drying up;
- *Air temperature:* Adjustable between 10°C and 35°C, the temperature all over the working section of the cabinet over a number of days to be uniform in each of the temperature regimes, within plus or minus 1°C;

- *Light:* Uniform illumination of the trays, intensity 750 to 1250 lux at the level of the seed;
- *Air movement:* As low as possible to prevent the seeds from drying up;
- *Fresh air supply:* Small, of the order of one air change per hour, which should be sufficient to remove the carbon dioxide produced by the germinating seeds;
- *Day and night cycling:* A rapid initial temperature drop to be achieved in 30 minutes during the day to night change, reaching the ultimate night temperature in one hour, a similar requirement to hold for the night to day change;
- *Condensation:* Not allowed.

However, when cabinets operate with alternating temperatures it is very difficult to maintain the high humidity and the substrates dry out; this necessitates frequent watering, which adversely affects the germination results and adds to the manhours spent on the test.

There are many modifications of the germination cabinet and for detailed descriptions of some of them see Justice (1972) and Oomen and Koppe. Room germinator. When large numbers of tests are made, whole rooms may be used as germinators, with temperature, humidity and light controlled.

The room germinator is a modification of the cabinet. It is constructed on the same principle as the cabinet, but is large enough to permit workers to enter it and place the tests along either side of a central passageway. Usually, fans are required to prevent stratification of temperature and special equipment is necessary to maintain a high relative humidity. Another modification is the combination room-cabinet germinator. The temperature of the entire room is kept as low as the lowest temperature required. Germination cabinets are placed in this room and heated, individually, by electricity to maintain the various temperatures required. Either constant or alternating temperatures may be obtained by this type of germinator.

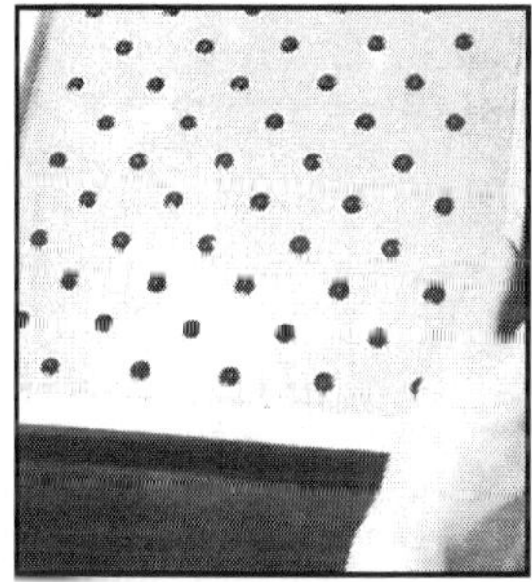

Fig. Counting Board with Seeds of Celtis Laevigata. Seeds are Spread on the Top Board to Place one Seed in each Hole. The Spring-held Top Board is then Moved to the Right Until its Holes Line up with another Set in the Bottom Board; the Seeds then Drop through. (USDA Forest Service)

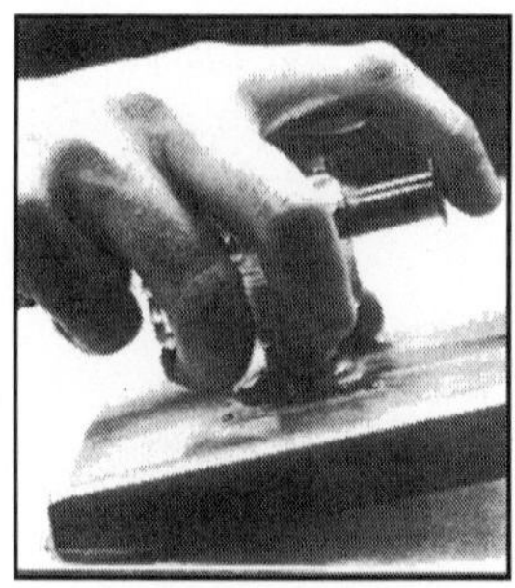

Fig. Counting Head on a Vacuum Seed Counter. A Pump Pulls a Vacuum through the Line and the Hollow Counting Head that has 50 or 100 Tiny Holes on its Bottom Surface. When one Seed per Hole is Attached, the Vacuum can be Released with the Finger Plunger to Drop the Seeds. (USDA Forest Service)

Fig. Seed Germination Equipment at the Division of Forest Research, CSIRO Canberra (A) Open Seed Germination Cabinet (B) Group of Cabinets. (Division of Forest Research, CSIRO Canberra)

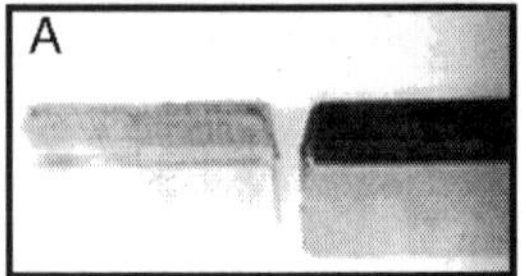

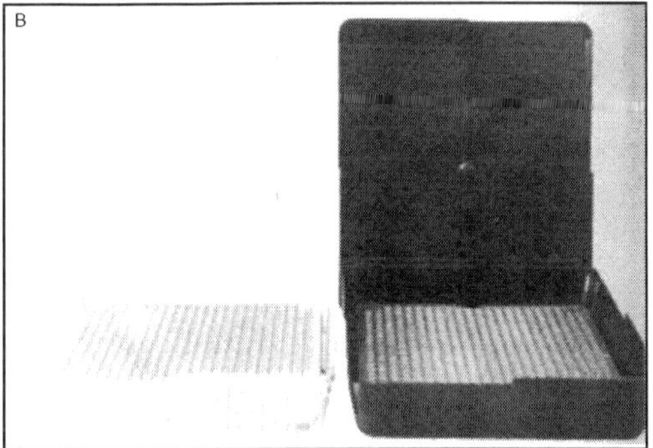

Fig. Clear and Black Germination Boxes Developed for Seed Testing in Canada, shown Closed (Top) and Open (Bottom).

Conviron G30 Germinator, with Over 95 per cent Relative Humidity and Programmable Temperature and Photoperiod within 24 hours, as Used at Petawawa National Forestry Institute, Canada. (Canadian Forestry Service)

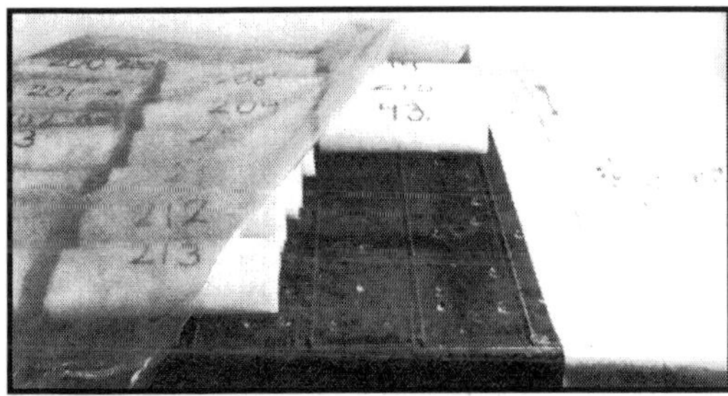

Fig. Copenhagen Tank and Rolled Filter Paper for Germination Tests. (DANIDA Forest Seed Centre)

Portable germination boxes. A simple and versatile germinator consists of a number of transparent plastic boxes with lids, which can be stacked one upon another.

Robbins (1984) describes the ideal container as being:

- Rectangular and stackable, for economony of space;
- Large enough for adequate spacing of at least one replicate of seeds (100, 50 or 25 seeds, depending on seed size);
- Sufficiently deep to allow for the required substrate depth and to permit development of the seedling for proper assessment;
- Provided with a wellfitting lid, to maintain a high moisture content of the substrate and surrounding air;
- Easy to sterilise by heat or chemical treatment;
- Transparent (at least the lid), to allow light if required for germination and subsequent development of the seedlings.

In Honduras the size of box used is 178 × 117 × 72 mm deep and each box contains one replicate of 100 pine seeds. This leaves a space of at least the width of one seed between each seed.

For larger-seeded species there are fewer seeds per replicate box. Any of the common substrates described later in this chapter, *e.g.*, filter paper, sand, may be used. By adding a suitable quantity of water to the substrate at the beginning of the test and then keeping the boxes covered with their lids at all times except during assessment and removal of seedlings, it is possible to maintain a high and constant moisture content in the substrate and the air inside the boxes, without the need for further additions of water or for control of humidity in the atmosphere outside the boxes. Where filter paper or blotting paper is used as the substrate, it can be kept permanently moist by placing it on a raised platform over a reservoir of water which is fed to the substrate by wicks.

This arrangement resembles, in miniature, the Jacobsen tank. The boxes may be placed inside an incubator where temperature and light can be controlled in accordance with *e.g.*, ISTA recommendations or, if no incubator is available, they can be stored in ambient room conditions of temperature and light. In either case the closed boxes should provide optimum humidity conditions for the germinating seeds.

Researchers at the Petawawa National Forestry Institute in Canada have developed a germination box, moulded out of light, unbreakable and heat-resistant polycarbonate plastic, which is large enough to contain 4 replicates of 100 seeds each of pines or similarly sized seeds (Wang and Ackerman 1983). The box is 28 cm long × 24 cm wide.

The base is 5 cm deep and the lid 1 cm deep, but the special universal locking mechanism makes it possible to fit two bases together; alternative combined depths of 6 and 10 cm are thus available for use according to the characteristics of the species under test. A perforated false bottom on eight legs 1 cm long supports the germination substrate and the space below can be used as a water reservoir if desired. The material is available in either clear or black plastic for germination in light or dark conditions; to ensure complete darkness it is necessary to seal the junction between base and lid with opaque tape. Four aeration holes on the side wall of the bottom component are optional in the clear plastic type.

Tests of this germination box gave results closely comparable with those from the much smaller, and thus less convenient, petri dishes. It was found that loss of moisture over a four week period was least in boxes which had the water reservoir filled at the start of the test and which had no aeration holes.

Both clear and black were equally good in this respect, but growth of mould was more severe in the (sealed) black than in the clear model. A substrate of blotting paper over Kimpak (cellulose paper) lost moisture more slowly than Kimpak alone. Selection of a germinator. Most forest seed laboratories with a moderate through-put of samples use either the cabinet-type germinators or germination tables. A report by Boeke *et al.* (1969)

recommends the exclusive use of cabinettype germinators for small seed testing laboratories. The advantages of using cabinets instead of germination tables of the Copenhagen type are a saving in space and, in well constructed cabinets, the more precise control of temperature and humidity.

It should be noted that a single-level Copenhagen tank requires five to eight times the floor space occupied by a cabinet of equal capacity. Where research is being carried out on germination conditions, or where testing covers a wide variety of seeds, the flexibility provided by several cabinets operating at different temperatures and/or light conditions can be very advantageous.

For nurseries or small research institutes concerned with carrying out simple tests of germination without maintaining strict control of temperature and light, the flexibility, storability and cheapness of plastic germination boxes have much to recommend them. They can also be used in combination with incubators in which light and temperature are controlled.

5

Seed Technology

INTRODUCTION

Seed technology is the science dealing with the methods of improving physical and genetical characteristics of seed. It involves such activities as variety development, evolution and release of varieties, seed production, seed processing, seed certification and storage. Thus seed technology is interdisciplinary science which includes broad range of subjects. It also includes seed quality control, seed marketing, seed distribution and research on above aspects of seed.

ROLE OF SEED TECHNOLOGY

The role of Improved Seed Technology is Given Below:

Improved Seed is Carrier of New Technology: Introduction of quality seed of new varieties wisely combined with other inputs significantly increases the yield level. In India cultivation of high yielding introduced dwarf wheat varieties helped to increases the production from 12 million tones making country self sufficient with in span of 10 years.

Improved Seed is Basic Tool for Secured Food Supply: Due to successful implementation of high yielding variety programme in India crop productivity is significantly increased as a result import is substantially reduced. The export is increasing in majority of crops in spite of rapid population increase.

Improved Seed is a Principal Means to Secure Crop Yields in Less Favourable Areas of Production:Good quality seed of suitable varieties helped in securing higher crop yields. In disaster region E.g M-35-1 drought resistant variety in Jawar served as a boon in drought prone area in rainy season.

Improved Seed is a Medium for Rapid Rehabilitation of Agriculture in case of Natural Disaster: In the wide spread droughts and floods, it is much economical if government had National Seed Reserve Stock (NSRS) at their disposal. NSRS plays two fold roles. i) It provide improved seed in emergency period in production areas, and ii) To supply seed for resowing in disaster region (Normally which is not available with farmer in time).

GOALS OR OBJECTS OF SEED TECHNOLOGY

Major goal of seed technology is to increase agricultural production through spread of good quality seed of high yielding varieties developed by the plant breeders, well before the planting season at reasonable cost. Following are important objectives of seed technology.

Rapid Multiplication

It is the quickest possible spread of new high yielding varieties and hybrids developed by the plant breeder.

Timely Supply

The improved seed of available well in time, so that the planting schedule is not disturbed and they are able to harvest the good crop.

Assured High Quality Seed

The expected divided in agriculture from use of improved seed is possible with use of good quality seed.

Reasonable Price

The cost of quality seed must be within reach of average farmer. The cost of the seed must be cheap and farmers should able to purchase it.

SEED DEVELOPMENT

One hundred fifty years ago the United States did not have a commercial seed industry; today we have the world's largest. Some view this as real progress, a form of genetic Manifest Destiny. A nation once a 'debtor' in plant genetics now supplies the world. In 1854, seeds were sourced in the U.S. by way of a small number of horticultural seed catalogs, farmer exchange, on-farm seed saving, and through the beneficence of the United States government.

Specifically, beginning in the 1850s, the U.S. Patent and Trade Office and congressional representatives saw to the collection, propagation and distribution of varieties to their constituents throughout the states and territories. The programme grew quickly so that, by 1861, the PTO had annual distribution of more than 2.4 million packages of seed (containing five packets of different varieties). The flow of seed reached its highest volume in 1897 with more than 1.1 billion packets of seed distributed.

The government's objectives in funding such a massive movement of seed stemmed from the recognition that feeding an expanding continent would require a diversification of foods. To the early colonies, the introduction of wheat, rye, oats, peas, cabbage and many other vegetable crops was as critical to food security as was the adoption of the corn, beans and squash. Immigrants were encourage to bring seed from the old country, founding fathers such as

Thomas Jefferson engaged in seed-exchange societies, and by 1819 the U.S. Treasury Department issued a directive to its overseas consultants and Navy officers to systematically collect plant materials.

The first commercial seed crop was not produced until 1866—cabbage seed produced on Long Island for the U.S. wholesale market. The industry flourished to some degree, but early seed trade professionals felt their growth was stymied by the U.S. government programs as well as the self-replicating nature of their product (that is, the factory contained within that product). In 1883, the American Seed Trade Association (ASTA) formed and immediately lobbied for the cessation of the government programs.

The organization developed powerful allies, such as Grover Cleveland's Secretary of Agriculture, J. Sterling Morton, who wrote that the government giveaway was "antagonistic to seed as a commodity-form and in direct competition with the private seed trade." But the programme was very popular with constituents, and the USDA's seed budget was kept intact – at one point counting for a full 10 per cent of the agency's overall annual expenditures.

In the early part of the 20th century, the first wave of hybrids began to provide seed companies with a potential increase in product profitability (as farmers would now need to return to the seed distributor for materials each year). However, most of the hybrid development was occurring at Land Grant Universities, and these universities refused to give the companies exclusive rights to the seed. Once again, the industry felt its growth hindered by federal programs and complained of unfair trade practices. Mounting data also indicated a slowing in yield increases from seed developed in government programs. The industry used this last point to strengthen its argument for the privatization of seed development in order to foster greater food security.

In 1924, after more than 40 years of lobbying, ASTA succeeded in convincing Congress to cut the USDA seed distribution programs. The USDA still supported breeding at the state agricultural schools, and for a time these programs continued to compete with seed companies by developing 'finished' commercial varieties. Associations such as the American Society of Agronomy and American Society of Horticulture Science eventually convinced the public programs that their appropriate role was in training plant breeders, performing fundamental research, and creating raw materials and technologies for private industry to capitalize on. The LGUs began to increasingly serve in this capacity, developing inbred parental lines and breeding stock that the seed trade would use to create proprietary varieties.

These changes in the public role, along with improvements in hybrid techniques, led to the growth of the seed trade following World War II. The trade was well represented during this period by regional companies. The conversion to monocropping and large-scale corporate agriculture had not yet moved into full swing. The Santa Clara Valley grew vegetables and fruit and not Internet startups, and Americans still planted their Victory Gardens. The seed trade reflected this diversity in food production.

In the 1960s, a few larger seed firms began to purchase smaller companies. But the consolidations of this period were minor compared to the frenzy that would come with a Supreme Court ruling on June 16, 1980, in the case of Diamond v. Chakrabarty. Prior to the Chakrabarty decision, a plant (or animal) could be owned, but the genetics could not. This case cleared the patenting of life forms on the bases of their genetic coding. The PTO granted more than 1,800 such patents following the ruling. Companies that had no historical seed interests—primarily chemical and pharmaceutical firms—began purchasing seed companies. In a few short years, there were billions of dollars in mergers and acquisitions—with little to no regulatory oversight—creating for the first time a majority ownership of plant genetics by a few multinational companies. No other natural resource (marine, timber, minerals) has ever shifted from public to private hands with such rapidity, such intensity of concentration, and so little oversight.

IMMEDIATE AND THE FUTURE

"There is a direct threat to our food system when we have a preponderance of genetic resources controlled by institutions whose only goal is profit," plant breeder Frank Morton expressed emphatically when asked for his perspective on the Monsanto acquisition. He went on to compare the present with the past, "When these services [breeding and production] were diffused amongst many individuals and groups with diverse motives, we had a much more diverse and healthy food system."

Diversity and competition have historically made for healthier economies as well as ecologies. Ecological and economic systems have another thing in common; as one group abandons a niche, it leaves room for others to move in and utilize it. Many of the seed company representatives and breeders I spoke with felt that the organic community can and should develop a less centralized seed system. Steve Peters, seed procurement manager for Seeds of Change, shared his firm's vision: "Part of our strategy is to go into neglected markets and respond to these needs. We want to offer true organic alternatives; this means we not only pay attention to the regional needs but also have a different approach to breeding – building new alliances in crop development. We're not chasing single-gene resistance but looking at horizontal resistance in traits like Downy mildew in spinach."

Adaptability has historically been an integral part of the organic movement– responding to customer inquiry and opinion, personal service (a face on the food), localized and decentralized – all traits that help make the organic sector healthy and promote its vigorous growth. "The organic seed world is where the organic produce market was twenty years ago," said veteran organic grower and seed producer Nash Huber.

"Produce quality wasn't always so great then, but we worked it out and now are better than the conventional systems. The potential in seed is that the customers – we farmers – will see that, in the long run, an organic seed trade will serve us." Farmers like Huber are investing in the organic seed trade. He works as an educator in the Organic Seed Alliance's WSARE organic seed production trainings, is leading a participatory plant breeding project with the Alliance, and produces seed for the organic seed trade as well as his own on-farm use. Carrots are the main cash crop on Huber's 400-acre farm, and he produces 80 per cent of his own seed to insure quality and availability. "I can't afford to have a company drop my source," he explained.

Seed companies and farmers are not alone in their rising to meet the need for changes in the seed world. In 2003, more than 70 university breeders, representatives from farmer-based NGOs, and policy specialists met at the "Summit for Seeds and Breeds for the 21st Century" in Washington, D.C. The breeders who represented a diverse set of crops, from schools that have traditionally served clients of large-scale Green Revolution style agriculture, called on each other to "reinvigorate public breeding" to meet the needs of organic and sustainable agriculture.

While the seed consolidation trend seems to have reached a particularly dark moment in the Monsanto-Seminis acquisition, it may also serve to fuel a momentum in the developing organic seed community. The issues are immediate and fraught with complexity; the answers are long-term and require commitment. Said C.R. Lawn of Fedco, "We need to keep working on creating a farmer breeding community. But this is long, slow work. You find and replace varieties one by one. You work slowly in the direction you want to go. Seed work is slow work.

SEED CLEANING TECHNOLOGY

Pre- conditioning refers to operations such as shelling, debearding, etc, that prepare seed lots for basic seed cleaning, and also to the removal of particles such as pieces of trash, stones, clods etc, larger in size than desirable crop seed, from threshed seed lots. Some pre-cleaners, in addition to removing larger sized particles, also remove particles that are lighter in weight and smaller in size than the crop seed. The necessity of both of these operations, other than shelling, is associated with advanced mechanised agriculture, no pre-cleaning is usually required on hand harvested and winnowed seed lots.

The main characteristics by which sound seeds may be distinguished from inert matter including sterile and empty seeds are size and shape, specific gravity, colour and surface texture.

The ease with which sound seeds can be differentiated depends on:

- The degree of difference which exists between the seeds and the matter to be separated from them
- The degree of uniformity among the seeds themselves.

Colour, size and shape are useful criteria for visual separation, while most seed cleaning machines make use of seed size and specific gravity. Screening and sieving methods separate by seed or particle thickness or diameter; the indented centrifugal cylinder by particle length; liquid flotation and blowing, fanning and winnowing methods by specific gravity; while frictional cleaning methods rely on differences in surface texture.

Modern cleaning machines often combine more than one method, so that the cleaning process is both effective and quick. However, the species and the amount of seed to be handled will determine whether cleaning is best carried out by hand, by improvised equipment or by specialist machinery. The following account of cleaning and grading methods is based on Turnbull.

BASIC SEED CLEANING

Basic seed cleaning refers to actual cleaning and grading of seeds. Unlike pre-cleaning/ Precleaning/ pre-conditioning, which may or may not be required, basic seed cleaning is an essential process in the seed cleaning operations. Many kinds of seeds can be completely cleaned and made into a finished product by, basic cleaning. In other instances, however, further separations to remove specific contaminants may be necessary, or desirable. Regardless of whether further specific separations are made or are not made, basic seed cleaning for every lot is invariably done.

Equipment for basic seed cleaning. Basic seed cleaning is done over an air screen machine, commonly referred to as an air screen cleaner. It is also basic equipment in seed processing plants.

Air Screen Machine

Principles of Cleaning: The separation of undesirable material and seeds from desirable seeds in an air screen machine is done on the basis of differences in seed size and weight. In some separations, seed shape could also be used. The air screen machine uses cleaned, but which were too large to drop through the second screen, are removed here.

As the crop seed drops off the fourth screen, they fall through the lower air separation. This removes the light seed and trash which was not removed by the upper air and the screens. For efficient cleaning, the lower air blast should be strong enough to blow out a few good seed.

SCREENING OR SIEVING

In most cases a number of sieves with different sized perforations are used and the cleaning is a process of gradually sifting out smaller and smaller particles. It is not only the size of the perforations which determines the quality and quantity of the seed cleaned; other important factors include the precision of the perforations, the angle at which the sieves operate, the amplitude and speed of movement of the sieves, and the correct cleaning and maintenance of the equipment. Sieves or screens may be made of flat perforated plate or wire mesh, and sometimes they may be three dimensional such as funnelshaped sieves. For small samples hand held sieves are adequate, but in larger scale cleaning a series of shaking screens is commonly used. In Brazil sieving with a mesh size of 32 per inch (approx. 12.5 per cm) was effective in separating seeds from chaff in Eucalyptus grandis. 84% of good seeds were retained and 89% of chaff was eliminated.

SORTING ACCORDING TO LENGTH

Cleaning with sieves relies on the separation of seeds, with diameter the critical factor. Fractionating according to length cannot be done with sieves, but this is possible with an indented cylinder. In addition to its use for separating good seed from the impurities, the equipment is used in agriculture for separating seed mixtures and can also be used for grading seeds.

The equipment consists of a slightly inclined horizontal rotating cylinder and a movable separating trough. The inside surface has small closely spaced hemispherical indentations. Small material is pressed into the indents by centrifugal force and can be removed. The larger material flows in the centre of the cylinder and is discharged by gravity. Depending on the type of impurities, the seed may be separated via the indentations or by passing down the cylinder.

BLOWING

Cleaning by blowing is a very important and widely used method. It is based on the principle that any object can float in an airstream of sufficient velocity. For separation in an airstream there are three possibilities: falling, floating or rising. The behaviour of the seed and other matter will depend on their weight, their resistance to the flow of air (volume and shape), and the velocity with which the air moves.

The operation of blowing is often referred to as winnowing or fanning. In its simplest form the uncleaned seed is thrown into the air on a windy day. The components separate out and the desired ones are retained. Indoors the airstream from a fan can be utilized.

Hand winnowing has been used successfully in Thailand to separate full seeds from empty seeds of Pinus kesiya (Bryndum 1975). The initial separation was followed by a second winnowing of the discarded fraction. Cutting tests showed that the original full seed percentage of 82% was raised to 98% in the improved fraction, while the discarded fraction (approximately 10% of the total volume) contained only 18% full seed or about 2% of the total full seed in the seed lot.

One operator took 8 minutes to winnow 1 kg. Loss of the very smallest and lightest full seed in the separation process is usually not serious since these seeds are likely to germinate slowly and to be low in vigour.

Simple winnowing machines can be easily constructed and Yim illustrates a machine made entirely of wooden components which is used in Korea.

Laboratory blowers can be classified as either the pneumatic type—a fan is situated near the air intake and pushes air through the system, or the aspirator type—a fan is situated near the air outlet and pulls air through the system, creating a partial vacuum. Small laboratory cleaners such as the 'Brabant' cleaner, and the 'Kamas' laboratory aspirator are available.

Another seed blower in common use is the South Dakota Blower. The principle of blowing is that a sample of seed, when suspended in an upward stream of air of a certain velocity, will split into a light fraction and a heavy fraction, the light fraction being carried upward and the heavy fraction falling down.

The two fractions are caught apart from each other. The heterogeneity of the heavy fraction can be further reduced by subjecting it to a second blowing at a higher wind velocity. In that way a light fraction, a medium fraction and a heavy fraction are obtained. The blowing apparatus in the South Dakota Blower consists essentially of a centrifugal blower, the outlet of which is connected to the bottom end of a vertical tube of a few centimeters' internal diameter and about 50 cm in length.

A more complex blower developed in Canada is described by Edwards. It consists of four plexiglas tubes of varying diameters which cause differential air velocity and, by choosing the appropriate combination of tubes, it is possible to use the equipment either to separate seed from chaff or full seeds from empty seeds. It works well with large seeds *e.g.* Abies amabilis but is not suitable for very small, light seeds such as Betula or Chamaecyparis.

In Zimbabwe a home-made cleaner was constructed by fitting a tapered aluminium sleeve with internal baffles onto a constant speed domestic fan (Seward 1980).

The narrow end of the sleeve projects over a compartmented receptacle; full seeds fall into the nearest compartment, while the lighter impurities and empty seeds are blown into the further compartments. A similar locally made seed winnowing chamber, which uses a stream of air from an electric fan, has been successfully used in the Solomon Islands to separate debris and empty seeds from dry seed batches of Swietenia macrophylla and Campnosperma brevipetiolata.

Many seed cleaning machines use a combination of winnowing and screening. A coarse upper screen removes larger material, a lower fine screen stops the seeds and lets through fine matter, and then the seed fraction passes through a transverse or nearly vertical airstream from a fan to remove chaff and empty seeds. The air-screen cleaner is the basic equipment of seed cleaning plants. The size of air-screen cleaners varies from the small two-screen model to a modern precision model, which uses several top and bottom screens and in one operation as many as three air separations.

LIQUID FLOTATION

Cleaning by flotation relies on the principle that the density of the seed of a given species is specific both for filled and unfilled seed.

There are two basic methods used:

1. Density method in which liquids with a density or specific gravity between that of the full and empty seed are used. The specific gravity of the liquids used is usually below 1.0 and such that the full seed sinks and the empty seed and light debris float.
2. Absorption method in which water is used and, although both full and empty seeds float initially, after some time the full seeds absorb water, become heavier and sink. The time of soaking can vary from a few minutes to several hours. This method is useful where there is a very small difference between the specific gravities of the full and empty seeds. The seed must be re-dried after being separated.

The flotation methods can separate out insect attacked, mechanically damaged, and immature seeds from filled mature seeds. The density method can only be applied if a liquid of suitable density is available which is not harmful to the seed. The application and the problems encountered are discussed by Simak (1973); both this and a method for separating viable from non-viable full seeds which was developed by the same author, involving an element of pregermination, are described.

FRICTION CLEANING

Most debris can be removed from the seed by air-screen combinations, but leaf fragments, resin particles and other objects of similar size and density to the seed are difficult to remove.

Friction cleaning relies on the principle that any object falling or sliding over a surface encounters a certain friction. The movement of the particle is proportional to its weight and to a coefficient of friction which depends on the nature of the particle's surface and the surface on which it moves. Separation of seed from debris is made on an inclined cloth or rubber belt on the basis that the angle necessary for the run-off of the seed differs from the angle necessary for run-off of the debris. A continuous upward moving belt removes seeds downwards by gravity and the lighter debris upwards by friction.

SPECIFIC GRAVITY SEPARATION

This method makes use of a combination of weight and surface characteristics of the particles to be separated. It is a method which is finding increasing use in separating and grading tree seeds.

The specific gravity (SG) separator employs a flotation principle. A mixture of seeds is fed onto the lower end of a sloping perforated table. Air, forced up through the porous deck surface and the bed of seeds by a fan, stratifies the seeds in layers according to density, with the lightest seeds and particles of inert matter at the top and the heaviest at the bottom. An oscillating movement of the table causes the seeds to move at different rates across the deck; the lightest seeds float down under gravity and are discharged at the lower end, whilst the heaviest ones are kicked up the slope by contact with the oscillating deck and are discharged at the upper end.

Deck coverings of linen, plastic, and wire mesh have been used to distribute the air uniformly beneath the seeds and provide the correct push to the heavier seed in contact with the deck. A closely woven covering such as linen gives the best results for small seeds. On modern gravity separators, it is possible to control the feed speed, the tilt of the table in two directions, the speed of oscillation and the force of air differentially at various points along the deck. The combination of these different controls makes it possible to adapt the machine to handle a wide variety of species and seed lots.

The specific gravity separator will separate particles of the same density but of different size, and particles of the same size but of different densities. It will not separate efficiently particles which differ both in density and size, *i.e.* separate a larger but less dense particle from a smaller but denser particle. It has been found practical for cleaning the chaff from some eucalypt seeds and for grading pine seeds. Purity% of uncleaned seed of E. grandis after extraction is about 10%. This was raised to 95% purity with 95% germination by SG separator treatment.

OTHER CLEANING METHODS

A number of other methods of seed cleaning have been used experimentally, but are not yet in widespread operational use. They include electronic and electrostatic separators, magnetic separators, electronic colour separators and shaking tables, which separate seeds by the angle at which they rebound when thrown against fixed walls.

Seeds of Ochroma can be successfully cleared of their floss by placing the uncleaned mass on a wire sieve of 0.3 mm mesh and setting fire to the floss. The fire flashes across and the seeds drop through the mesh. Good results are obtained when the seeds are allowed to drop through the mesh into a pan of water. The inflammable oil in the floss burns with intense heat and experience in Honduras has shown that the floss must be spread out thinly to avoid damaging the seed. This method has also been tried to clear Populusseeds of floss but may damage more than 50% of the seeds in this genus.

Seeds of Prosopis are often embedded in a gummy matrix within the pods. *One way of obtaining clean seeds is:*

- To remove one side of the pod mechanically with a knife
- To soak the pod contents in a 0.1 normal solution of hydrochloric acid for 24 hours

- To wash in water for one hour, then dry in the direct sunlight
- To beat or pound the dried mass to separate the clean seeds from the gummy coating. This method has been successful in India, yielding clean seeds with a germination rate of 65% in 12 days.

UPGRADING THE QUALITY OF CLEANED SEED

We have mentioned earlier that in certain instances it is necessary and in others rather desirable to further upgrade the seed quality by removing either specific contaminants, or by very precise size grading. The various processing operations conducted after basic cleaning to further improve seed quality are regarded as upgrading operating. The choice of upgrading operation, however, shall depend upon the type of contaminants and crop seed.

6

Evolution of Seed Structure and Seed Dormancy

INTRODUCTION

MD and MPD are present in B-type seeds, but can also be typical of some specialized species. MD is thought to be the ancestral dormancy type among seed plants and is the most primitive dormancy class. The dispersal of seeds with an underdeveloped embryo that need time to grow might have evolved as an ancient strategy to disperse germination over time. MD and MPD are not only typical for primitive angiosperms, but also for primitive gymnosperms like the Zamiaceae, Ginkgoaceae, Podocarpaceae and Taxaceae. Evolution of larger embryo size in MD seeds resulted in non-dormant (ND) seeds. Concurrently, the gain of physiological dormancy mechanism(s) led from seeds with MD to seeds with MPD, which upon gain in embryo size led to PD seeds.

PD is the most phylogenetically widespread dormancy class. This class of dormancy is distributed over the entire phylogenetic tree from gymnosperms and the basal angiosperms to the higher core eudicot Rosid clade. As ND is also distributed over the entire phylogenetic tree it has been proposed that gain and loss of PD occurred at several times and at several levels of seed evolution.

The most phylogenetically restricted and derived dormancy classes are PY and PY+PD. The occurrence of an impermeable seed or fruit coat combined with a ND embryo (PY) or a PD embryo (PY+PD) is probably the adaptation to specialised life strategies or habitats. PY and PY+PD are the only dormancy classes not found in gymnosperms, and PY is only evident in one Monocot family (Cannaceae).

A conceptual model for the evolution of PY and PY+PD within the Ancardiaceae has been proposed. Within this family the genus *Rhus* includes species with PY+PD- dormancy and species with only PY-dormancy. Anatomical evidence for PY of extinct *Rhus* spp. suggests that PY had evolved in the Ancardiaceae by 43 Ma in the Middle Eocene. The model begins in pre-

Eocene times with an ancestral species that has large, non-dormant (ND), recalcitrant seeds.

By the middle Eocene, a derived species with relatively small, orthodox seeds and a water-impermeable covering layer had evolved, i.e. a PY-dormant seed evolved. By the Oligocene, PD had been added to the seed dormancy mechanism, i.e. a PY-dormant seed evolved. Climatic drying (Eocene), followed by climatic cooling (Eocene-Oligocene transition), are proposed as primary selective agents in the development of PY.

An evolutionary connection between PY and recalcitrance is suggested by the relatively high concentration of these two states in the Rosid clade. Phylogenetic data and fossil evidence seem to support the PY —> PY+PD evolutionary sequence in Ancardiaceae, which also may have occured in Fabaceae.

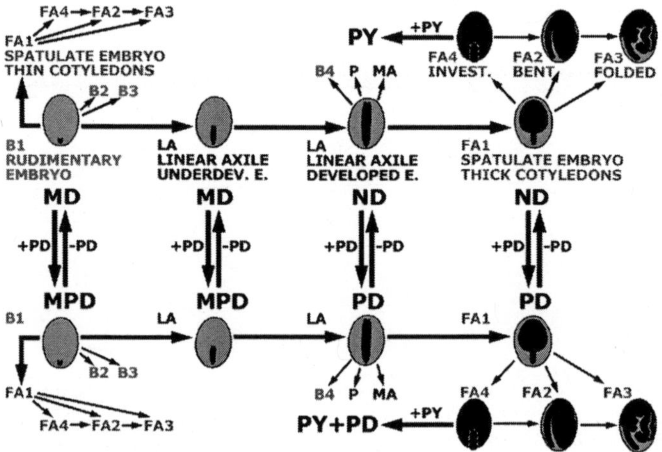

Summary of general trends in angiosperm seed evolution with respect to seed type and dormancy class.

Seeds with small embryo and abundant endosperm (B1 seed type) are primitive and evolution shows a general trend (indicated by arrows) towards mature seeds with little or no endosperm in which the embryo occupies most of the seed (through LA to FA seed types). Gain or loss of PD to each of these seed types (B1, LA and FA1) alters the dormancy class as indicated.

Similar changes can also occur in the other seed types. Addition of PY to FA4-type ND or PD seeds is indicated as a side branch occurring in specialized species. This addition can be reversed and PY can also be added to other seed types.

Embryo Dormancy and Coat Dormancy: The components of Physiological Dormancy

The hierarchical classification system by Baskin and Baskin accurately reflects the "whole-seed" ecologists view of the control of germination by dormancy. At the molecular level very little is known about MD, MPD, PY, PY+PD and deep PD. Recent work has provided insight into molecular mechanisms of non-deep PD.

Together with earlier physiological work it is clear that the intrinsic molecular mechanisms determining dormancy can have an embryo and/or a coat component. The terms embryo and coat dormancy are used to distinguish between these two mechanisms, but it should be emphasized that they are not used as part of a classification system for "whole-seeds" sensu Baskin and Baskin. Embryo dormancy and coat dormancy are components of PD; their sum and interaction determines the degree of "whole-seed" PD. Embryo dormancy is characterized by a block that inhibits extension growth and therefore excised embryos do not grow. Coat dormancy, is characterized by a block that is conferred by the covering layers. Non-dormant embryos excised from coat-dormant seeds will therefore extend and grow.

"Coat" is used in a loose sense and can be any embryo covering structure, e.g. testa, endosperm and/or pericarp and terms like testa dormancy and endosperm dormancy can be used to specify these. Mechanical resistance from combined testa and endosperm dormancy, which is greater than the embryo growth potential opposing it, appears to be the cause of non-deep PD in seed model systems like *A. thaliana* and Solanaceae species. When embryo and coat dormancy are considered as components of non-deep PD the influence of ABA and GA on dormancy and the control of germination can be summarised as follows:

Embryo Dormancy

Embryo dormancy (embryo component of non-deep PD): A dormant embryo is characterized by a high ABA/GA ratio, high ABA sensitivity and low GA sensitivity. Embryo dormancy release involves remodelling of hormone biosynthesis and degradation towards a low ABA/GA ratio, a decrease in ABA sensitivity and an increase in GA sensitivity.

Thus ABA dominates the embryo dormancy programme, and GA the embryo germination programme. A non-dormant embryo is characterized by increased growth potential, the ability for cell extension growth and the ability to induce the release of coat dormancy.

Coat Dormancy

Coat dormancy: The combination of an embryo with low growth potential and mechanical constraint from the seed covering layers can result in dormancy. GA can release this coat dormancy by increasing the embryo growth potential and/or by reducing the mechanical constraint. There are two forms which are distinguished based on whether the covering layer consists of dead (e.g. outer testa) or living (e.g. endosperm) tissue:

1. Testa dormancy: ABA during seed development determines the

subsequent GA requirement for germination since ABA influences the testa characteristics (e.g. thickness) and GA the embryo growth potential.

2. Endosperm dormancy: Endosperm weakening can be either part of the coat dormancy release or part of the germination programme. Since the endosperm is in most cases living tissue it can actively participate in regulating embryo constraint by influencing both the ABA/GA ratio and sensitivity to these hormones. GA acts by increasing the embryo growth potential and by promoting endosperm weakening which is achieved though ABA-independent and ABA-inhibited mechanisms.

Induction, maintenance and release of physiological dormancy by plant hormones and environmental signals.

There is considerable evidence that abscisic acid (ABA) is an important positive regulator of both the induction of dormancy and the maintenance of the dormant state in imbibed seeds following shedding.

- ABA-deficiency during seed development is associated with absence of primary dormancy in the mature seed, whereas over-expression of ABA biosynthesis genes can increase seed ABA content and enhance seed dormancy or delay germination.
- ABA produced by the seed itself during seed development can impose a lasting dormancy, whereas, maternal ABA or ABA application during seed development both fail to induce a lasting seed dormancy, but has other functions.
- In *Arabidopsis thaliana*, the members of the AtNCED gene family encode enzymes catalyzing the key regulatory step in ABA biosynthesis. ABA synthesis in the embryo and the endosperm both seem to contribute to the induction of seed dormancy.
- High ABA contents are present in the imbibed seeds of the strongly dormant *A. thaliana* ecotype Cape Verde Island (Cvi) and decrease as dormancy is lost. A recent transcriptome analysis with this ecotype strongly supports the view that increased ABA biosynthesis is associated with the dormant state.
- *De novo* ABA biosynthesis during imbibition of dormant, but not non-dormant, seeds has been demonstrated in the *A. thaliana* ecotype Cvi, as well as in other species including *Nicotiana plumbaginifolia, Helianthus annuus,* and *Hordeum vulgare*. This de novo ABA biosynthesis has been interpreted as a mechanism for dormancy maintenance.
- Work with the strongly dormant *A. thaliana* ecotype Cvi shows that dormancy may depend on an intrinsic balance of gibberellin (GA)- and ABA-biosynthesis and catabolism, which will determine the dominance of either of the hormones. While PD release of Cvi seeds

occurs effectively by after-ripening, stratification or inhibition of ABA biosynthesis, the addition of GA appears less effective. GA treatment of dormant Cvi seeds caused a transient increase in ABA levels, suggesting that in dormant seeds a feedback mechanism exists that maintains a high ABA/GA ratio. Thus, the net result of the dormant state is characterized by increased ABA biosynthesis and GA degradation. It appears to be the ABA/GA ratio, and not the absolute hormone contents, that controls germination.

Model for the Regulation of Dormancy and Germination by ABA and GA in Response to the Environment

According to this model ambient environmental factors (e.g. temperature) affect the ABA/GA balance and the sensitivity to these hormones. ABA synthesis and signalling (GA catabolism) dominates the dormant state, whereas, GA synthesis and signalling (ABA catabolism) dominates the transition to germination. The complex interplay between hormone synthesis, degradation and sensitivities in response to ambient environmental conditions can result in dormancy cycling. Change in the depth of dormancy alters the requirements for germination (sensitivity to the germination environment); when these overlap with changing ambient conditions, germination will proceed to completion.

ORIGIN AND EVOLUTION LEACHING

The origin of seed plants is a problem that still remains unsolved. However, more and more data tends to place this origin in the middle Devonian. The description in 2004 of the proto-seed *Runcaria heinzelinii* in the Givetian of Belgium is an indication of that ancient origin of seed-plants. As with modern ferns, most land plants before this time reproduced by sending spores into the air, that would land and become whole new plants.

The first "true" seeds are described from the upper Devonian, which is probably the theater of their true first evolutionary radiation. The seed plants progressively became one of the major elements of nearly all ecosystems.

ECONOMIC IMPORTANCE

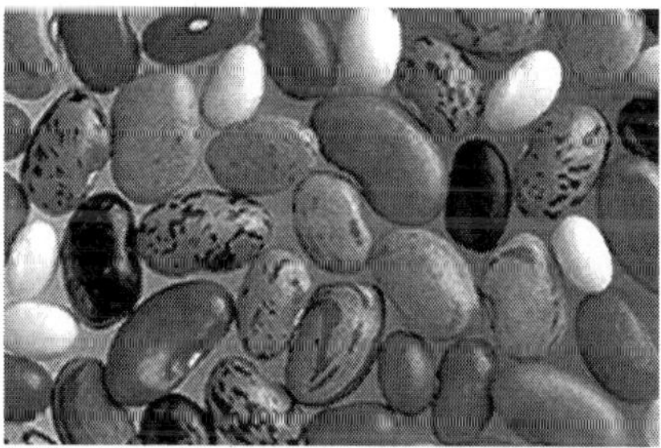

Fig. A variety of bean seeds.

Edible Seeds

Many seeds are edible and the majority of human calories comes from seeds, especially from cereals, legumes and nuts. Seeds also provide most cooking oils, many beverages and spices and some important food additives. In different seeds the seed embryo or the endosperm dominates and provides most of the nutrients.

The storage proteins of the embryo and endosperm differ in their amino acid content and physical properties. The gluten of wheat, important in providing the elastic property to bread dough is strictly an endosperm protein.

Seeds are used to propagate many crops such as cereals, legumes, forest trees, turfgrasses and pasture grasses.

Seeds are also eaten by animals, and are fed to livestock. Many seeds are used as birdseed.

Poison and Food Safety

While some seeds are considered by some as healthy to eat, other seeds may be harmful or poisonous, Plants and seeds often contain chemical compounds to discourage herbivores and seed predators. In some cases, these compounds simply taste bad (such as in mustard), but other compounds are toxic, or breakdown into toxic compounds within the digestive system.

Children, being smaller than adults, are more susceptible to poisoning or death by plants and seeds. One should be satisfied with reliable food safety information before choosing to eat any particular seeds.

An infamously deadly poison, ricin, comes from seeds of the castor bean. Reported lethal doses are anywhere from two to eight seeds, though only a few deaths have been reported when castor beans have been ingested by animals. In addition, seeds containing amygdalin; apple, apricot, bitter almond, peach, plum, cherry, quince, and others, when consumed in significant amounts, may result in cyanide toxicity. Other seeds than contain poisons include annona, cotton, custard apple, datura, uncooked durian, golden chain, horse-chestnut, larkspur, locoweed, lychee, nectarine, rambutan, rosary pea, sour sop, sugar apple, wisteria, and yew. Another seed poison is strychnine.

The seeds of many legumes, including the common bean (*Phaseolus vulgaris*) contain proteins called lectins which can cause gastric distress if the beans are eaten without cooking. The common bean and many others, including the soybean, also contain trypsin inhibitors which interfere with the action of the digestive enzyme trypsin. Normal cooking processes degrade lectins and trypsin inhibitors to harmless forms.

ALFALFA SEED STRUCTURE

Germination is the resumption of growth by the embryo within the seed leading to the development of a new plant. The germination process is complete as soon as the radicle ruptures the seed coat. The process is influenced by available soil moisture, soil temperature, and the nature of the soil surrounding the seed (i.e., salty vs. normal soil) or residual herbicides that may be present in the soil.

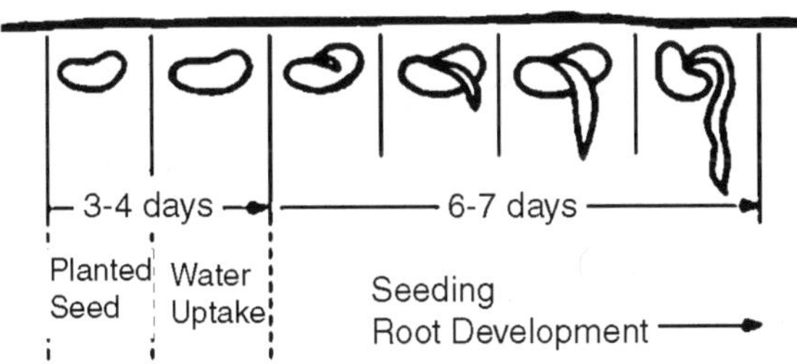

Fig. Germination and seedling root growth

SEED STRUCTURE

ALFALFA GERMINATION PROCESS

Seedling growth is the developmental period of the young plant from the time germination is completed until it can manufacture enough food through photosynthesis to sustain growth. The seedling root is the first structure to emerge from the seed during germination. It penetrates the soil very rapidly, forming a slender, usually unbranched taproot, which may penetrate 5 to 6 feet into the soil during the first growing season.

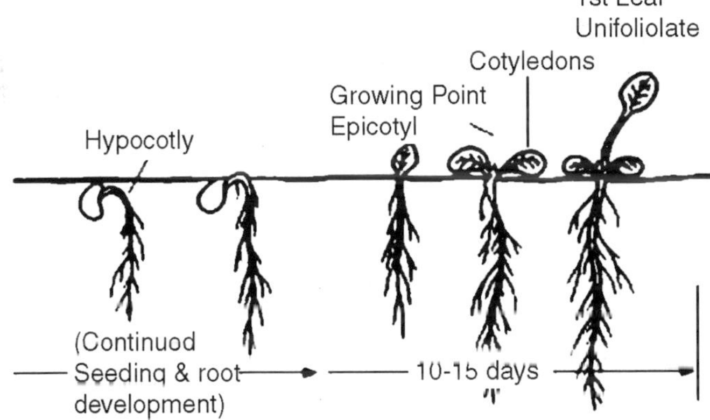

Fig. Seedling emergence through unifoliolate leaf stage

Once the seedling root is anchored firmly in the soil, the seedling axis below the cotyledons elongates in an arch (hypocotyl arch) pulling the cotyledons upward to the soil surface. Seed germination and seedling emergence occur in about three to seven days. Fewer days may be required for seedling emergence under ideal soil moisture and temperature conditions. As the hypocotyl arch emerges from the soil, growth stops on the side exposed to light and continues on the underside until the seedling is in an upright position. This raises the cotyledons above the soil surface where they expand. The growing point (epicotyl) of the seedling is now exposed. The first true leaf is unifoliolate and emerges from a bud at the first stem node above the cotyledons. Seedling growth is complete. Under good growing conditions, the seedling is developed fully 10 to 15 days after planting.

ALFALFA SEEDLING GROWTH

A fully developed alfalfa seedling does not assure plant establishment. It must continue to develop deeper roots and grow more leaves to survive and become an established stand. Vegetative growth continues through cell division and expansion in the epicotyl or growing point of the young plant.

The second leaf of the alfalfa plant is usually trifoliolate (three leaflets) and originates from the second primary stem node. All subsequent leaves are trifoliolates except in new multifoliolate varieties that have 5, 7, or 9 leaflets per leaf. The unifoliolate leaf and all subsequent trifoliolate/multifoliolate leaves are useful characters to distinguish alfalfa from weed seedlings in stand evaluation.

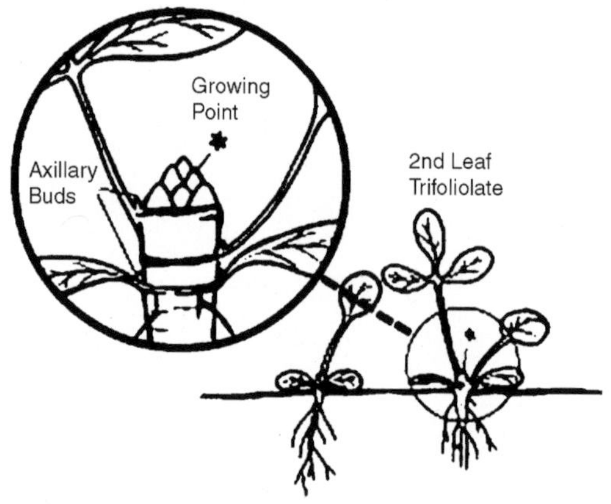

Fig. Unifoliolate through fourth trifoliolate leaf stage

Axillary buds develop in the axils of all leaves. After three or more trifoliolate/multifoliolate leaves have appeared on the primary stem, new secondary stem growth may occur from any of the axillary buds, but frequently only one axillary bud stem (usually the unifoliolate) develops early, especially if seedlings are shaded in a companion crop or under competition from weed growth or other alfalfa plants.

ALFALFA FIRST TRIFOLIOLATE LEAF AND BUDS

The primary and secondary stem(s) of the young plant increase in length by cell division and internode elongation from the first stem node upwards. The second and subsequent leaves are trifoliolate/multifoliolate and develop alternately at each stem node as growth continues. Once the first true trifoliolate leaf develops, further growth and development is best described by the number of trifoliolate/multifoliolate leaves that develop on the main shoot as the plant continues to grow. Any axillary buds in axils of leaves can develop new stem tissue and generally do in less competitive conditions.

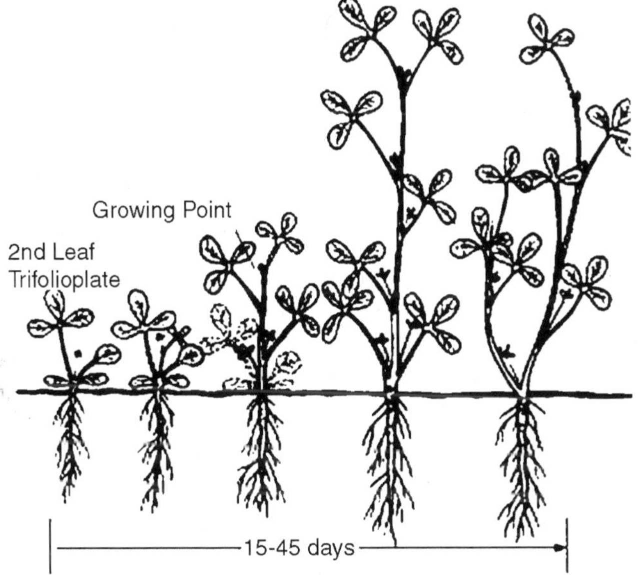

Fig. First-year vegetative growth and development

ALFALFA FIRST-YEAR VEGETATIVE DEVELOPMENT

Growth and development of new shoots from axillary buds gives the young plant a branched appearance, especially if light is adequate and the stand is not too thick. Vigorously growing alfalfa plants quite often have three and sometimes four secondary shoots in addition to the primary or central stem, which forms the characteristic first-year crown.

The branches from the unifoliolate leaf bud appears first, followed by branches from the cotyledonary and first trifoliolate leaf buds. Fewer branches may form the crown with late seedings, under severe companion crop competition, or with high seeding rates. Under these conditions the cotyledonary, unifoliolate, and first trifoliolate leaf buds may remain dormant or may give rise to a branch rhizome or underground stem in the fall.

The young alfalfa plant, at about 8 to 10 weeks of age, undergoes a growth phase known as contractile growth. This process in alfalfa and sweetclover involves a change in the shape of cells in the hypocotyl or seedling axis below the cotyledons and upper portion of the primary root from long and narrow to short and wide as a result of carbohydrate or food storage. This shift pulls the lower stem nodes beneath the soil surface.

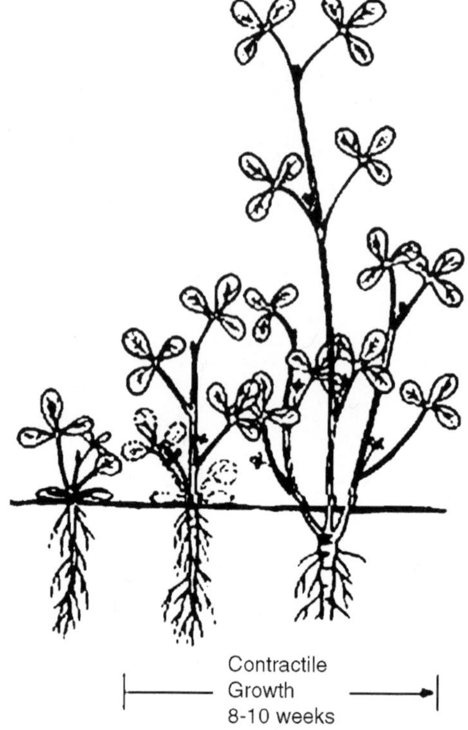

Fig. Crown development showing contractile growth

Most winter hardy alfalfa varieties have several nodes pulled well below the soil surface (1 to 3 inches) in the seeding year. Contractile growth greatly aids winter survival of alfalfa by providing soil insulation for the perennial over-wintering crown structures.

CROWN DEVELOPMENT

The lower internodes or nodes of the primary or secondary stems may produce adventitious or crown buds. Crown buds or dominant axillary buds produce the vegetative regrowth following harvest the seedling year, or these buds may give rise to branch rhizomes in the fall. The branch rhizome is the structure from which new growth will initiate in the spring unless winter injury occurs.

The branch rhizome must be developed adequately in the fall by the first killing frost or the seeding will fail, even under the most ideal growing conditions. In most years, alfalfa seeded later than about August 15 under good moisture (irrigation) conditions in North Dakota will not have adequate time to initiate branch rhizomes before the first killing frost. A dryland seeding after the first week of August is not recommended because of the increased risk of stand loss. A crown of a perennial can be described best as the young overwintering stem tissue. The crown in the fall of the first year may be as small as the lower portion of the main stem and dormant cotyledonary node buds, or as large as the lower portion of the main stem (i.e., cotyledonary, unifoliolate, and first trifoliolate leaf nodes), the secondary stems from these nodes, and the branch rhizomes that develop.

Spring growth will occur from the rudimentary leaves on the branch rhizome if winter injury hasn't "burnt" them off. If winter injury has occurred, spring growth will initiate from adventitious buds on the branch rhizome or from a dormant crown bud. But, spring growth is much slower in this case. All subsequent growth in the first-harvest year occurs from adventitious or dormant crown buds.

The crown of the alfalfa plant increases in size during the second year. Branch rhizomes develop on last year's branch rhizome. These grow outward and upward increasing the circumference of the plant. After two to four years, the typical multibranched crown of alfalfa develops.

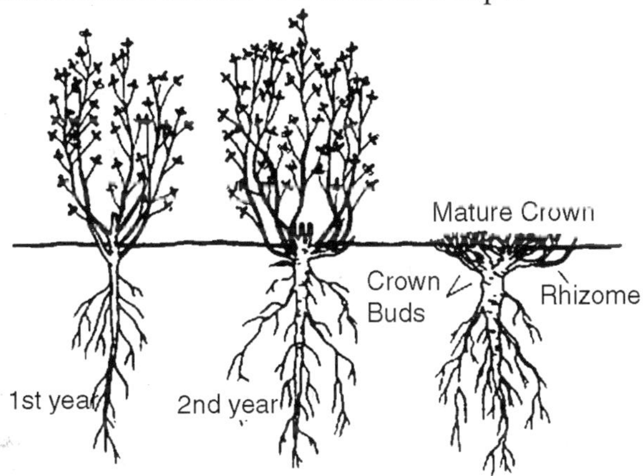

Fig. Fully developed alfalfa crown

ALFALFA WINTERING STRUCTURES

The crown and associated root tissue are the storage center for carbohydrates (food) produced through the process of photosynthesis. The carbohydrates are used to develop cold resistance or winter hardiness, for plant respiration during winter, and to provide the energy source for the initiation of the new growth in the spring and following each harvest. First-year management influences subsequent growth. Alfalfa management is discussed in R-571, Alfalfa Management in North Dakota.

Over the next several lectures we will follow the early growth of young trees and other vascular plants from seed through germination and growth in the first year. We will draw examples from many species, but will illustrate some phases of growth by contrasting two tree species native to British Columbia: Douglas-fir representing gymnosperms, and red alder representing angiosperms.

Evolution of Seeds

The evolution of seeds was a critical factor in facilitating the proliferation of land plants into a wide range of environments. Seeds protect and nourish embryos, and embryos can survive a wide range of environmental conditions. Seeds can also persist in a dormant state, enabling plant species to survive and regenerate following catastrophic events or periods of weather that are not conducive for growth.

Seeds can remain viable for long periods of time, centuries, even millenia. An extreme example is provided by frozen 10,000 year old lupine seeds found in the Yukon. These seeds germinated within 48 hours of thawing!

Gymnosperms

Gymnosperm seeds are, by definition, naked, that is they lack a surrounding fruit. The outermost layer of the seed is the seed coat. It originates from the mother tree and is diploid. The seed coat has three layers: the outer layer; the thicker, tough 'stony' or middle layer; and the inner layer. The colour of the seed coat, which can be quite variable, is due to tannins that accumulate during seed development. Some species of conifers, including *Abies* spp. (true firs), *Tsuga* spp. (hemlocks) and *Thuja plicata* (western redcedar) have resin vesicles in the middle or outer layers of the seed coat.

These resin vesicles may play a role in:

1. Seed coat dormancy,
2. Protecting the embryo from excessive dehydration,
3. Deterring seed herbivory.

These vesicles can also create problems for the extraction and cleaning of conifer seed. Many conifer seeds have a seed wing to aid in dispersal.

Immediately inside the cell wall is the nucellus, a papery layer surrounding the megaspore cell wall. Inside the megaspore cell wall is the megagametophyte (the female gametophyte), the haploid nutritional tissue found in gymnosperm seeds. Early in the development of cones, megagametophytes produce egg cells which are fertilized by male gametes to produce zygotes. Zygotes develop into embryos. The megagametophyte then plays its second functional role, surrounding the embryo, protecting and nourishing it.

The megagametophyte in Douglas-fir is 60% lipids, 16% proteins, and 2% sugars, making it a high-energy and nutrient tissue both for the embryos it contains and for a plethora of seed predators including small mammals, birds and insects. In a dry seed (<10% moisture), the megagametophyte is creamy yellow and does not entirely fill the seed coat. When a seed is imbibed (soaked in water until the moisture content is 30-40%), the megagametophyte is white and shiny.

The embryo is found in the corrosion cavity, a pit in the centre of the megagametophyte that is fully filled by the embryo in mature seeds. It consists of the cotyledons, shoot apical meristem, root apical meristem, root cap and suspensor. The cotyledons and shoot apical meristem point towards the wider chalazal end of the seed.

The radicle (embryonic root including the apical meristem and root cap) and suspensor are at the more pointed micropylar end of the seed towards the micropyle (the opening in the integument through which the pollen entered the ovule). The number of cotyledons varies from 2 in western redcedar to 12 or more in *Pinus ponderosa* (ponderosa pine).

The suspensor is found at the base of the root cap and plays a role early in embryo development by pushing the embryo into the megagametophyte.

Angiosperms

Angiosperm seeds are found inside fruit. The many types of fruit will be discussed in detail in lab. Angiosperm seeds generally vary more in structure than conifer seeds, but share some features with conifer seeds and have different parts playing the same functional roles for others.

Like conifer seeds, angiosperm seeds are surrounded by a seed coat protecting the embryo. Inside the seed coat (which is usually formed by the integument, but is sometimes fused with the fruit), nutritive reserves for the mature embryo may be stored either in endosperm or in large cotyledons.

Endosperm is formed through the fertilization of the two polar nuclei by a male gamete, producing triploid cells, at the same time that the egg cell is fertilized by a second male gamete (more on this after the Christmas break). All angiosperm seed have endosperm early in embryo development. In mature seeds of some species, there may be a substantial amount of triploid endosperm surrounding the embryo (e.g., *Arbutus menziesii* (arbutus)).

Commonly in Eudicots, however, embryos largely absorb the endosperm and the nutritional reserves are stored in the two (of course!) large cotyledons that fill the seed cavity (e.g., *Alnus* spp. (alders), *Acer* spp. (maples), and *Quercus* spp. (oaks)). Some angiosperm seeds already contain a primordial shoot with leaves, called a plumule, above (distal to) the cotyledons prior to germination.

The seeds of monocots have several noteworthy features. They typically have endosperm at maturity, and the single cotyledon often plays a role in absorbing food digested from the endosperm. Grass embryos have a large cotyledon called a scutellum attached to one side, as well as a plumule and a radicle. The plumule is protected by a structure called the coleoptile, while the radicle is protected by the coleorhiza.

Angiosperm seed also have suspensors, but these have a different function than those of conifers. In angiosperms, suspensors are involved in supporting the early development of the embryo both through providing nutrition and growth regulators. The suspensor cells die in angiosperms relatively early in embryo development.

STRUCTURAL FEATURES

Seeds are the dispersal and propagation units of the Spermatophyta (seed plants): Gymnosperms (conifers and related clades) and Angiosperms (flowering plants). A comparison of these to major groups is presented on the "Seed evolution" webpage. Several seed–related issues differ between gymnosperms (750 species) and angiosperms (250000 species). If not otherwise stated, the information given in this website refers to typical Angiosperm seeds.

Seeds are mature, fertilized ovules. Ovules are structures of seed plants containing the female gametophyte with the egg cell, all being surrounded by the nucellus and 1–2 integuments. In angiosperms the double fertilization results in formation of the diploid embryo and the triploid endosperm.

EMBRYO

Young sporophyt, dipoid (2n), result of fertilization. The mature embryo consists of cotyledons (seed leaves), hypocotyl (stem-like embryonic axis below the cotyledons), radicle (embryonic root). Seed and embryo types were defined by Martin (1946). These and the resulting evolutionary trends are found on the "seed evolution" webpage.

ENDOSPERM

Food storage tissue, triploid (3n), result of double fertilization, 2/3 of the genome is of maternal origin.

TESTA (SEED COAT)

Outer protective layer of the seed, developed from the integuments of the ovule, diploid maternal tissue. Fruits are mature, ripened ovaries containing seeds. The pericarp ("fruit coat") is diploid maternal tissue.

PERISPERM

Diploid maternal food storage tissue originates from the nucellus. Only in some species, *e.g.* Beta vulgaris, Piper nigrum, Coffea arabica, many Caryophyllales.

ENDOSPERMIC SEEDS

The endosperm is present in the mature seed and serves as food storage organ. Testa and endosperm are the two covering layers of the embryo. The amount of endosperm in mature seeds is highly species-dependent and varies from an abundant endosperm layer (Nicotiana tabaccum) to a single layer (Arabidopsis thaliana).

NON-ENDOSPERMIC SEEDS

The cotyledons serve as sole food storage organs as in the case of pea

(Pisum sativum). During embryo development the cotyledons absorb the food reserves from the endosperm. The endosperm is almost degraded in the mature seed and the embryo is enclosed by the testa.

In mature seeds of Lepidium sativum (garden cress) the embryo is surrounded by 1–2 cell layers of endosperm. FA2-type seed.

We found that Lepidium seeds exhibit, as tobacco, a two-step germination process with distinct testa rupture and endosperm rupture. We found that Lepidium seeds provide an excellent model system for the experimental investigation of endosperm weakening. Lepidium and Arabidopsis are closely related species and exhibit, except for the size, highly similar seed structure.

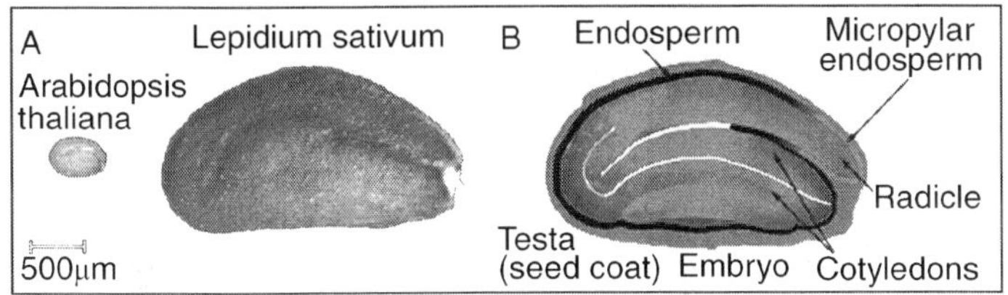

ENDOSPERMIC SEED STRUCTURE

Monocot species like wheat and other cereal species have caryopses (cereal grains) as propagation units. Caryopses are single-seeded fruits in which the testa (seed coat) is fused with the thin pericarp (fruit coat). Cereal grains have highly developed embryos and in cereal grains the triploid endosperm consists of the starchy endosperm (dead storage tissue) and the aleurone layer (living cells).

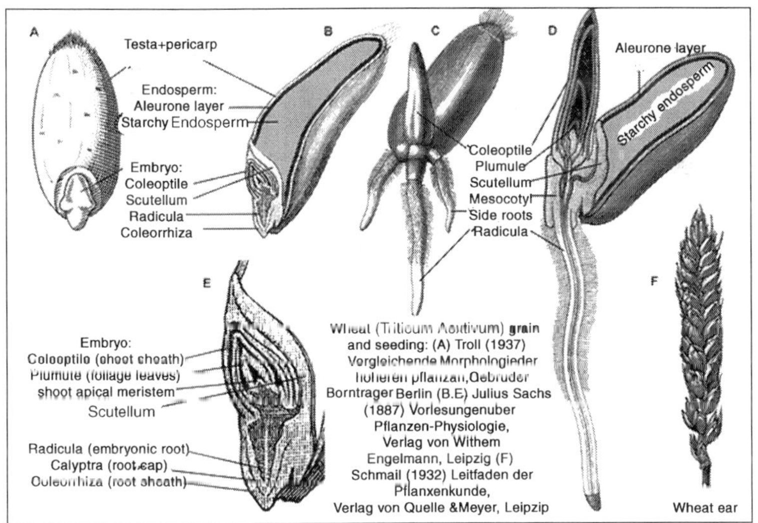

Fig. Structure and Germination of a Cereal Grain (Caryopsis): Triticum Aestivum-wheat

Organs of the cereal embryo are the coleoptile (shoot sheath), the scutellum, the radicula and the coleorrhiza (root sheath). The scutellum is homologous to the leaf blade of the cotyledon and the coleoptile to the leaf sheath of the cotyledon. After germination, the scutellum functions in absorbing the nutrients from the starchy endosperm and delivering them to the growing seedling. After germination, the coleoptile protects the shoot meristem and plumule when growing through the soil. The coleorrhiza has a similar function for the radicle prior to germination. The coleorrhiza is the first structure that grows through the pericarp, it is then ruptured by the radicle.

Germination and seedling establishment of cereal grains is hypogeous. After the primary root emerges, the coleoptile is pushed upward by elongation of the mesocotyl. The coleoptile elongates and reaches the soil surface. Here it ceases to elongate and the first leaves of the plumule emerge through an opening at its tip. Mobilization of the food storage in the starchy endosperm is a post-germination event and requires an embryo signal (gibberellin) which induces the production and secretion of hydrolytic enzymes from the aleurone layer.

SEED DORMANCY

Seed of most temperate and boreal species is dormant when it disperses, that is it won't germinate even if conditions of temperature, moisture and light are favorable. Why? To avoid active growth until winter has passed.

An actively growing plant cannot be adequately frost hardy. Dormant seeds cease growth and maintain a very low metabolic rate until specific environmental signals are received or other requirements have been met.

Embryo Dormancy

Dormancy in many species is under the physiological control of the embryo. Embryo dormancy results from the levels of plant growth regulators (plant hormones). Dormant embryos have growth-inhibitors present and lack growth promoters. Breaking embryo dormancy is dependent on decreasing growth inhibitor concentrations and increasing growth promoters above some threshold.

Embryo dormancy in many species can be overcome through exposure to low temperatures to meet the chilling requirement of the seed. To accomplish this, seeds are imbibed (soaked in water), then placed in a fridge, usually on damp paper or moss. This is called stratification.

In Douglas-fir, holding seeds at 4°C for 4 weeks will meet the chilling requirement. Many species require a longer stratification period – up to 100 days. Seeds of some species with embryo dormancy may germinate eventually without chilling after being soaked in water, but will germinate more slowly than stratified seed, and the percent of seeds that germinate will likely be lower. Embryo dormancy can also be overcome in some cases by artificially applying growth regulators. Some species require alternating day and night temperatures, or cycles of wetting and drying following low temperatures. Some species have both embryo dormancy and other types of seed dormancy.

Seed Coat Dormancy

Some desert and tropical species have chemical growth inhibitors in the seed coat that must be

leached out by water before seed will germinate, increasing the probability that seed will germinate during adequately moist periods. Other species require a heat treatment to germinate. Germination of buckbrush (*Ceonothus* spp.) is promoted by pouring boiling water over seeds.

Other Types of Seed Dormancy

Embryo immaturity – Seeds of some species are immature at the time fruit or seed are shed and require 'after-ripening'. The seed of whitebark pine are typically immature when harvested by Clark's nutcrackers and continue to mature after caching.

Seed coat impermeability – Seed coats can be impermeable to water or oxygen, which physically imposes dormancy through preventing growth. This is common for leguminous species, among others. These seed may need scarification (scratching or abrading of the seed coat, either in nature or artificially with sandpaper), infection of seed coat by fungi or bacteria, or treatment with concentrated sulphuric acid (simulating passing the seed through the digestive tract of an animal).

Light requirement – Some species (e.g, some spruce and pine) require exposure to light to germinate. Other species require exposure to light of a specific quality. Red light promotes germination in these species, while far red light inhibits it. It is the ratio of far red to red light that is critical. As light passes through vegetation, this ratio increases, and germination in light-sensitive species will decrease. This means that seeds will be less likely to germinate when they are below competing vegetation. Once the competing vegetation is gone, the amount of red to far red light will increase, and the seed will germinate.

Serotiny – Most interior lodgepole pine (*Pinus contorta* ssp. *latifolia*) has cones that are serotinous. The cone scales are 'glued' closed by resin that has a melting point around 40°C. These cones must be exposed to either a fire or to particularly warm temperatures (e.g., on a dark surface in the full sun) to open. The cones of juvenile interior lodgepole pine, and Shore pine (*P. contorta* ssp. *contorta*) do not have this adaptation – why?

Species Without Seed Dormancy

Red alder seeds lack dormancy, but since they are shed late in the fall, environmental conditions are not favorable for germination until spring. Other examples of seeds with no dormancy include sugar maple and willows, which are viable for only a few days following dispersal.

Many agricultural crop species have no seed dormancy, but their wild progenitors do – humans have selected against seed dormancy in these species as it complicates farming.

GERMINATION

Germination is the resumption of growth by embryos when exposed to favorable conditions. It requires several steps:

- Seeds imbibe and swell; the seed coat ruptures;
- Cell division occurs in the embryo and the radicle elongates and extends out of the seed;
- Further elongation of the radicle pushes either the cotyledons or just the epicotyl hormones produced in the meristems and possibly elsewhere initiate enzymatic activity, hydrolysis of stored nutrients, and movement of these nutrients to regions of growth.

The process of germination is complete when enough photosynthetic tissue is exposed to support further growth.

The radicle develops into the tap root of the seedling (except in the case of species with fibrous root systems). It is positively gravitropic (grows in the direction of gravitational pull) while the shoot is negatively gravitropic (grows against gravity). The root cap consists of loose, slimy cells that protect the root apical meristem as the root elongates through the soil. The radicle develops root hairs for absorption of water and dissolved nutrients soon after germination.

In species with epigeous germination (e.g., conifers, red alder, maples), the cotyledons are pushed out of the ground, turn green and become photosynthetically active. In species with hypogeous germination (e.g., oaks and other species with large seeds), the cotyledons remain in the ground and only the epicotyl emerges. The energy reserves in the cotyledons supply the energy for early seedling growth, but the cotyledons never become photosynthetically active in this case. Rather, the first true leaves are the first photosynthetic organs with hypogeous germination.

Symbiotic relationships between seedling roots and other organisms are often initiated soon after germination. Lateral roots can become infected by mycorrhizal fungi soon after germination. In addition to becoming mycorhizal, red alder germinants becomes symbiotic with *Frankia* bacteria. Nodules form on the roots, and in these nodules N_2 gas is fixed into NH_4^+, which can then be converted into amino acids.

This allows red alder to 'fertilize itself'. Up to 320 kg N can be fixed per ha per year! Other species that fix nitrogen include *Ceonothus* spp. (deer brush) and *Shepherdia canadensis* (soopallalie).

SEED DRESSING AND PELLETING

USES

In seed dressing or seed pelleting treatments, the surface of the seeds is coated with some inert material, the sticker, to which chemicals of various kinds may be added.

Some of the advantages claimed for pelleting and listed by Magini are:

1. The incorporation of fertilizer material in the pellet furnishes the young germinated seedlings with needed nourishment.
2. Plant growth regulators or stimulants may promote rooting or hasten emergence of the seedlings.
3. Fungicides and insecticides are more effective when in direct contact with the seeds.
4. Seeds may be protected against rodents by adding unpalatable, repellent or poisonous substances.
5. Small seeds become larger and heavier and this improves the scatter pattern in aerial seeding.

Pelleting can also be used to incorporate a brightly coloured substance with the seed, which renders it more visible against the soil and thus facilitates uniform sowing rates. Coating with an anti-desiccant material such as alginate can preserve the seed against drying out, until adequate rain has fallen to permit safe germination.

Investigations have shown that some fungicides used in seed dressings may have an adverse influence on germination. The adverse effect was found to increase with temperature and this calls for extra caution in tropical countries. The cost of seed dressings is seldom justified when plants are raised in a nursery. Fertilizers, fungicides or insecticides can be applied more conveniently to the nursery soil than to the seeds, while good nursery hygiene coupled with presence of nursery staff through the day and periodic trapping or poison-baiting should be effective against birds and rodents. Pelleting is, however, sometimes used to improve the uniformity of seed for precision sowing in the nursery.

The main use of pelleting is for direct seeding, including aerial seeding. Protective measures to assist individual seeds after sowing are impracticable and pelleting is the only possible means of achieving some degree of protection. Most emphasis is now given to seed protection through incorporation of fungicides, insecticides and repellents; inclusion of fertilizers is seldom practised. One example of successful use was in aerial seeding of pines in the southern USA. The coating formulation consisted of endrin and arasan as the protectants, with a latex sticker to act as a bond. The seedling yields in field studies comparing coated with untreated seed were 55 to 1 for Pinus palustris and 12 to 1 for P. taeda. Seed coating of eucalypts is practised for aerial seeding of cutover mountain forests in southern Australia, where an area of 8,000 to 12,000 ha annually has been regenerated by this method. Aerial seeding has been used very little in the tropics but some success was obtained in Indonesian trials in Central and Eastern Java on areas dominated by Imperata grass, using Leucaena leucocephala, Calliandra calothyrsus and Acacia auriculiformis.

MATERIALS AND METHODS

Latex, methylcellulose and hydrol emulsions are used as stickers. Flaked aluminium powder is sometimes incorporated, because its shining quality is effective in scaring off birds, or to hasten drying and prevent clumping of the treated seed. A variety of chemical protectants is available. Thiram or arasan, endrin and anthroquinone are among those commonly used. Thiram has been found effective against damping off. Red lead was long used as a dressing to protect seeds against birds and rodents but it is now considered to be valueless in this role. It has reduced germination and growth in Larix and may sometimes be toxic to the operator in hand-sowing techniques. A more suitable dye which is equally effective as colouring and which has no apparent adverse effect on subsequent germination is a Waxoline dye such as Lithofar red or the later Waxoline red A. It can also be used as a marker for special certified seed batches in storage.

Magini describes one method of coating seeds through use of a small cement mixer. The seed is placed in the mixer (a batch of about 12 kg at a time is suitable) and moistened with a latex sticker solution of one part latex to nine parts of water at the rate of one eighth to one quarter of a litre per kg of seed.

Enough treating powder is then added to dry the sticker, usually in a volume relationship of four parts powder to one part sticker. The thickness of the seed coating is dependent on the amount of sticker in relation to the amount of seed. The total mixing time should not exceed four minutes, since prolonged agitation damages seeds or chips off the pelleted coat.

SEED MORPHOLOGICAL DORMANCY

Seeds with underdeveloped embryos at the time of dispersal or collection will not germinate until the embryos have had time to mature. In a few species maturation is possible during dry storage *e.g.* Gingko biloba. More commonly, a period of moist warm pretreatment is necessary before the embryos develop sufficiently for germination to take place.

The recommended treatment is the same as that for mechanical dormancy i.e:

1. Soak seeds in several times their volume of cold water at approximately 3–5 °C for 48 hours.
2. Drain off the water and mix the seeds with two to four times their volume of a moistened, water-retaining medium such as sand, sand/peat mixture or vermiculite.

3. Store at a warm temperature. A constant 20–25°C or alternating 20° and 30° C is suitable for many species.
4. Open containers weekly, mix seeds and, if surfaces show signs of drying out, remoisten with water spray.

OVERCOMING PHYSIOLOGICAL DORMANCY—COLD STRATIFICATION

Far more common than morphological dormancy among temperate species are the cases where seeds are fully developed at dispersal or collection but are inhibited from immediate germination for physiological reasons. The pretreatment most effective in removing this physiological dormancy is that which approximates to the conditions of overwintering in nature *i.e.* a moist cold treatment, or cold stratification.

Cold stratification will not only overcome physiological dormancy but can also reduce the sensitivity of both dormant and non-dormant seeds to their optimum requirements for light and temperature, resulting in increased rate and uniformity of germination over a variety of conditions. If properly carried out, cold stratification does no damage to non-dormant seeds which are undamaged and have not deteriorated from excessive physiological ageing. It can therefore be safely used when different degrees of dormancy are expected in the same seed lot.

Stratification (sensu stricto) refers to the method of placing seeds in layers alternating with layers of a moisture retaining medium, such as sand, peat or vermiculite, and keeping them at a cool temperature for a period, which is commonly between 20 and 60 days but varies considerably from species to species. The combination of high moisture and low temperature appears to trigger off biochemical changes which transform complex food substances into simpler forms utilized by the embryo when it renews growth at germination. The use of the word "stratification" has recently been extended to include all forms of cold, moist treatment whether or not the seeds are placed in layers. The present section gives a brief description of stratification in the original sense (outdoor and indoor), while the following section on "Other moist prechilling methods" describes non-layered methods of cold, moist treatment.

The three main requirements for success in both stratification and moist prechilling are a renewable source of moisture for the seeds, low temperature and adequate aeration. Only imbibed seeds will benefit fully from cold, moist treatment, while good aeration is needed to supply oxygen for respiration and to dissipate heat and CO_2. Low temperature not only favours biochemical changes in the seed but reduces micro organism activity and the risk of overheating and premature germination of the after-ripened seeds.

In temperate climates low enough temperatures to accomplish prechilling are experienced by overwinter stratification in outdoor pits.

The method used in the UK is described by Aldhous. Its main features are:

1. The pit should be located in a cool, shady and well-drained site. The bottom 10 cm should be filled with either sand or gravel for drainage. Pit contents should remain moist during stratification, but should never become waterlogged.
2. 60–80 cm is convenient for depth and width, the length can be adjusted to the volume of seed to be stratified.
3. The pit should be lined at the bottom and sides with a frame carrying 6 mm mesh wire netting as a protection against mice. After the pit is filled, it should be covered by a lid of the same mesh netting.
4. The seed to be stratified should be mixed with 4 times its weight of sand and the pit filled with the seed/sand mixture to within 15 cm of the surface. The top 15 cm should be filled with pure sand.
5. The start of stratification should be timed in relation to the anticipated sowing date and the optimum period of stratification expected for the species. Seeds should be inspected periodically, starting a few weeks before the sowing date. They should be sown when most seeds are beginning to split and the tips of the radicles are visible but before the radicles have elongated. Delayed sowing leads to broken or injured radicles; hardwoods can sustain such damage without too serious ill effects, but it can cause severe losses in conifers.

Where cold-room facilities are available, stratification may be carried out indoors, where closer control of moisture and temperature can be achieved than with the pit method. A temperature of +1° to +5°C is commonly recommended. In the USA one method in common use is to place 10- to 25-pound (4.5 – 12 kg) lots of seeds in loosely woven bags which are flattened into disks no more than 3 inches (7.5 cm) thick and alternated with layers of moist medium. Boxes, trays, cans or drums make suitable containers but must have perforated bottoms to facilitate drainage and aeration. Soaking seeds overnight in water at room temperature is common practice with southern pines and hardwoods. Containers should be covered loosely to prevent the seed and medium drying unevenly. The seeds should be inspected periodically to prevent heating, poor aeration and excessive drying and to detect the first stages of germination.

Seeds should be sown soon after removal from stratification. Stratified but ungerminated seeds of some genera, *e.g.* Prunus, may undergo a secondary dormancy if subjected to extreme drying or to temperatures over 20°C. A new cold stratification is then needed to overcome the secondary dormancy (Suszka 1978 b).

SEEDCOAT DORMANCY

Seeds of some species have hard, cutinised seedcoats that completely prevent the imbibition of water and sometimes also the exchange of gases.

Without imbibition and gas exchange renewal of embryo growth and germination are impossible. Physical seedcoat dormancy of this kind occurs most frequently in species adapted to alternating dry and wet seasons, including several leguminous genera such as Acacia,Prosopis, Ceratonia, Robinia, Albizzia, Cassia. In some species, *e.g.* Tectona grandis and Pterocarpus angolensis, the hard layer is formed by the pericarp or fruit. The unit of dispersal, storage and sowing in Tectona is the fruit, often referred to loosely as the "seed", but for practical purposes the origin of the impermeable layer does not affect the choice of pretreatment.

Pretreatment to overcome physical seedcoat dormancy are designed to soften, puncture, wear away or split the seedcoat in order to render it permeable, without damaging the embryo and endosperm within. They include physical and biological methods, dry heating and soaking in water or chemical solutions. Any treatment which destroys or reduces seedcoat impermeability is commonly known as scarification (Bonner 1984 a). Destruction of impermeability at a single point in the seedcoat is normally sufficient to allow imbibition and gas exchange.

Chemical seedcoat dormancy, caused by the presence of chemicals located in the seedcoat but inhibiting germination of the embryo, can usually be overcome by some form of liquid treatment which leaches out the chemicals. The wet treatment methods described in this chapter should therefore be effective in breaking chemical, as well as physical seedcoat dormancy, and a combination of the two. Dry methods on the other hand, which are effective in overcoming physical dormancy, are unlikely to have any effect on chemical dormancy.

PHYSICAL METHODS

One of the simplest and most direct physical methods is to cut, drill or file a small hole in the coat of each seed before sowing. This method has been found appropriate in the Philippines for large leguminous seeds in the genera Afzelia, Albizzia, Intsia andSindora and in Honduras for Acacia, Prosopis, Enterolobium and other legumes (Robbins 1982 b). Intsia seed is nicked at each end and a third time in the area of the hilum and micropyle; the last area is the most important. In Tanzania the hard and bristly pericarp of Pterocarpus angolensis is chipped at one edge by knife or cracked by beating with clubs. In the Philippines the hard seedcoat of Eusideroxylon is cracked by hammer. In Calophyllum complete removal of the seedcoat has produced better germination than nicking.

Sand-paper may also be used to reduce seedcoat thickness by abrasion. In trials in Pakistan sandpaper proved the most effective treatment in increasing and accelerating germination in a number of hard-coated species. As an example Leucaena seeds had zero germination in the untreated control and after 24 hours' soaking in cold water. This was increased to 42% in 26 days by 1 minute soaking in concentrated H_2SO_4, to 60% in 13 days by 2 minutes in boiling water, and to 100% in 3 days by sandpaper treatment. In laboratory trials in Sweden, sandpaper scarification followed by a three hour cold water soak, was the most effective treatment for Acacia farnesiana, yielding 88% germination in 7 days and 100% in 21 days, compared with 63%, 23% and 3% in 21 days, respectively for soaking in concentrated sulphuric acid, absolute alcohol and hot water.

In six other hard-seeded species from Iraq tested at the same time, it was less effective than one or other of the soaking treatments. Manual treatment of individual seeds is slow but, with experienced workers, is safe and effective. It is best suited to the larger and more refractory seeds *e.g.* Delonix regia. The effect of physical scarification may be reinforced by soaking in cold water before sowing.

For treatment of large quantities of seed, mechanical scarification is more suitable than the manual method. Seed may be tumbled or churned in a concrete mixer with sharp gravel or sand, or in a special drum lined with abrasive material such as sandpaper, cement or crushed glass or incorporating rotating abrasive disks. If gravel or sand is used, it should be sieved to ensure that it can be readily separated from the seed by use of an appropriate mesh screen. The method is not suitable for seeds with abundant resin or pulp which may clog the machine. Care must be taken to avoid overtreatment causing damage that can reduce or destroy germinability of the seeds. Examination of the seedcoat surface, if necessary by hand lens, or testing the capacity of the seed to take up water, demonstrated by swelling, can be used to estimate effectiveness of scarification.

In India mechanical scarification is used successfully to overcome seedcoat dormancy in Acacia catechu, A. nilotica subsp. indica, Albizzia falcataria, A. lebbek, Cassia fistula, C. javanica, C. nodosa, Delonix regia, Dichrostachys cinerea, Santalum album, Terminalia arjuna and T. tomentosa (Pattanath 1982).

Horizontal shaking of seeds of Acacia farnesiana, bottled with crushed glass, for five hours gave a germination of 20.7% plus additional imbibed seeds 10.7% in 43 days, compared with 0.7% + 0.7% for the control in the same period. But this result was much inferior to that obtained with the same species by manual scarification with sandpaper.

SOAKING IN WATER

A number of treatments involve soaking seeds in water or other liquids. These wet treatments may combine the effects of softening hard seedcoats and leaching out chemical inhibitors.

Some seeds which have little resistance to germination may respond well to soaking for 24 hours in water at ambient temperature. This may be the result of a more rapid imbibition than can be achieved in a moistened nursery bed. It is recommended for some species after manual, mechanical or acid scarification. Matias *et al.* found that 48 hours' soaking ofPinus caribaea seed in water at room temperature gave more uniform germination than in the untreated seed.

In India soaking in water, for periods varying from two to 48 hours according to species, accelerates germination in Acacia mearnsii, A. melanoxylon, A. nilotica subsp. kraussiana, Adenanthera microsperma, Albizzia amara, A. procera, Grevillea robusta and Trewia nudiflora.

A more effective treatment, especially in hot climates, is alternate wetting and drying of seed. This has frequently been used for Tectona fruits. In Thailand, after scarification, alternate soaking and drying was applied, 4 times soaking and 3 times drying, each for 30 – 45 minutes. Elsewhere, if preliminary scarification is not applied, common practice is to spread the "seed" out on a hard surface in the sun in a layer about 5 cm deep and to soak it thoroughly; then to turn it over from time to time, allowing it to dry out and bake in the sun for a day or two.

This process of alternate soaking, drying and baking is repeated a number of times, usually between five and ten cycles, until signs of germination appear. Each cycle of wetting and drying may be one day for soaking and 3 – 5 days for drying and baking. As soon as germination starts, the "seed" should be sown in the nursery (Laurie 1974). In Tanzania the preliminary soaking is for 72 hours in full sacks in a stream or in large drums. It is then sown on the surface at the rate of 5 kg/m^2 and, after two days in the sun, is covered with about 2 – 5 cm depth of soil and watered daily.

Tectona grandis is one tropical species in which there is evidence that chemical dormancy caused by inhibitors in the pericarp may be more important than physical dormancy. Fairlamb and Davidson found that an aqueous extract, obtained by soaking T. grandis fruits for 4 days and used to moisten filter paper, inhibited germination of cress seeds. Germination was 11% in 144 hours in the extract, compared with 76% in tank water and 96% in distilled water. On the other hand Pattanath could find no evidence for physical dormancy in T. grandis. She found that, after soaking the fruits in water for 24 hours, water had soaked right through to the locules where the seeds were attached. There is great variation in the degree of dormancy between provenances of teak.

Some require no pretreatment, some require alternate wetting and drying as described above, others respond to soaking for four hours in Sach's nutrient solution, which may indicate a nutrient imbalance in the seeds (Gupta *et al.* 1975).

Chemical dormancy apparently occurs also in Terminalia ivorensis. Alternate daily soaking and drying of the fruits of this species for seven days was effective in producing 50 – 70% germination, whereas germination of untreated seed is usually poor. In a later experiment the same author found that an aqueous extract from Terminalia ivorensis fruits had a slight inhibitory effect on germination of lettuce seeds but a more serious effect in the number of malformations in the germinated seedlings. He identified Coumarin as one of the more important inhibitants involved. Soaking in running water for one to two weeks has been used successfully to remove inhibitors in Atriplex spp.

Hot water treatment has given good results with a number of leguminous seeds. The seeds are usually placed into boiling water which is immediately removed from the heat source and left to cool gradually, the seeds remaining in the water for about 12 hours. They imbibe and swell as the water cools. The proper relationship of the volume of water to volume of seeds can be determined by experiment. It may vary considerably according to species, ratios of 2 – 3 times, 4 – 5 times and 5 –10 times as much water as seed have been suggested. Some species respond better to an initial temperature well below boiling *e.g.* Albizzia falcataria (Valencia 1973).

An initial temperature of 90° C, cooled to ambient 20° C, has given good results with Parkinsonia aculeata and Ziziphus spina-christi. Of various initial water temperatures and periods of soaking and cooling tested forLeucaena leucocephala in the Philippines, one minute soaking from an initial temperature of 80° C gave the best result 90% germination. The period of soaking and cooling appeared to have little effect, *e.g.* an initial 80° C and 6 hours' soak and cool gave 89% germination. The initial water temperature had a big effect; germination was only about 30% after an initial temperature of 100° C and only about 25% after an initial 40 °C.

Bowen and Eusebio (1981b) found that in Acacia mangium seeds in Sabah there was a close correlation between the initial temperature of the water and the subsequent germination. Germination increased progressively from 5% after immersion in water at 30° C to 91% after immersion at 100° C. Prescribed treatment for this species is now to immerse seeds in 5 × their volume of water at 100° C, stir in the cooling water for 30 seconds, pour off hot water and soak overnight in 20 × their volume at room temperature. For Albizzia falcataria a lower initial temperature of 80°–90° C is recommended. Some species of Acacia require more severe treatment. Larsen found that, whereas the standard treatment with boiling water improved germination of Acacia sieberiana, an exceptionally resistant species, from 2% to 10%, continuous boiling of the seed in water for 60 minutes achieved 60% germination within two weeks.

Prescriptions for hot water treatment must be applied meticulously if seedcoat dormancy is to be removed without killing the seeds through excessive heating. It is often easier to ensure this at a central research laboratory than in a number of scattered field nurseries. In Sabah it was found possible to redry seeds of Acacia mangium and Albizzia falcataria after the initial 30 seconds' hot water treatment, package and despatch them to field nurseries and, after 3 days' storage, to give a simple cold water soak before sowing. Germination was just as good (c. 80%) as when seeds were sown immediately after treatment. Drying was done in a current of dry air at 45°C for periods of 6 – 18 hours.

Hot water treatment is relatively easy and safe to apply, as well as being effective with some species. It is not well adapted to large lots because of the difficulty in handling and sowing the swollen seeds.

ACID TREATMENT

The chemical most commonly used to break seedcoat dormancy is concentrated sulfuric acid. For some species it is more effective than hot water treatment. Seed which has been kept for a long period in store may require a longer period in the acid than fresh seed, which could be severely damaged by the same length of treatment. Great care is needed in the handling of sulfuric acid and this method is not suitable for use by unskilled workers.

Materials and equipment required are as follows: commercial grade (specific gravity 1.84, 95 per cent pure) sulfuric acid; acid-resistant containers (thick plastic preferred); wire containers and screens for handling, draining, and washing the seeds; an abundant supply of running water; a safe place to drain the dilute acid resulting from rinsing the seeds; and facilities for drying the seeds after rinsing.

Safety precautions are a must! All workmen must understand and obey safety precautions in the use of acid. Seeds, containers, implements, and the acid itself must be handled with great care to avoid injury. Water must not be splashed into the acid, as a violent reaction will occur. All workmen should wear suitable safety clothing, gloves, and goggles or other eye protection.

Toughness of the seedcoat varies between lots and even between individual trees in most species. The optimum period of immersion in acid for each lot may be determined by treating a small sample for different periods and then soaking the lots in water at room temperature for 1 to 5 days (depending on species).

The treatment period that yields a high percentage of swollen seeds (by water uptake) without visible injury is the right one. Oversoaking may pit the seed and even expose the endosperm. Insufficient soaking leaves the seedcoats of most species glossy; coats of correctly treated seeds are dull, but not deeply pitted. If tests reveal only small differences between lots, then all may be lumped together for treatment, unless there are other reasons for keeping them separate (such as seed source distinctions). Large differences between lots should occasion separate treatment.

The steps in acid treatment are as follows.

1. Allow seeds to come to air temperature. If they have been removed from cold storage, do not open the container until temperature equilibrium is reached. Moisture will form on cold seeds exposed to warm moist air, and this moisture can react with acid to raise temperature to the danger point
2. Thoroughly mix seeds to be treated as one lot.
3. Immerse seeds in the acid for the required period, making sure that

all are covered. Treatment should be carried out at 65° to 80° F (18°–27°C), preferably on the upper end of the range (Heit 1967a). Lower temperatures require longer soaking times than do higher temperatures. Careful stirring will reduce the length of treatment necessary.

4. Remove seeds from the acid and wash them promptly and thoroughly in cool, running water for 5 to 10 minutes to remove all traces of acid. Water should be applied copiously at the start, and the seeds stirred carefully during rinsing.
5. Spread the seeds in a thin layer for drying, unless wet sowing is preferred.

Fig. A Rinsing Trough for Washing Acid from Treated Seeds.

Fig. One Apparatus for Treating Large Lots of Seed with Acid.

Fig. Outdoor Stratification of *Juglans Nigra* in Sand.

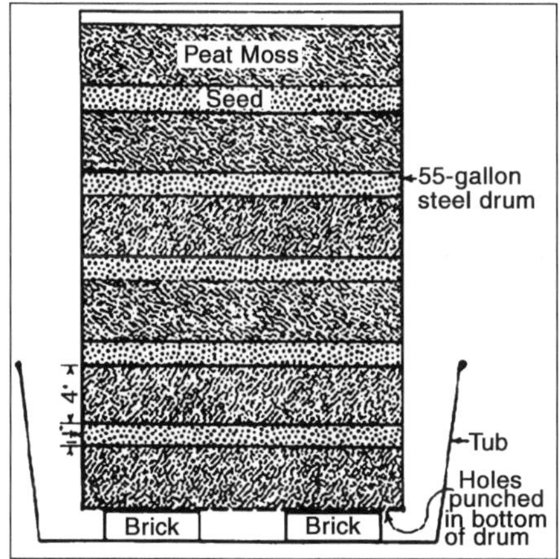

Fig. Desired Arrangement for Stratification in a Large Barrel.

Fig. Loblolly Pine (*Pinus Taeda*) Seed Prepared for Stratification in a Plastic Bag.

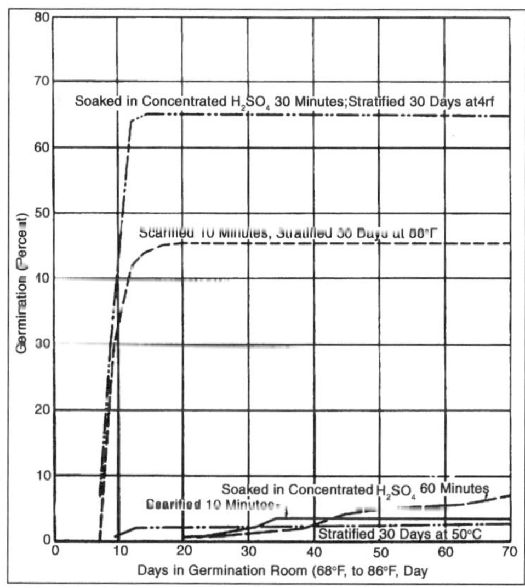

Fig. Effects of Several Treatments to Overcome Double Dormancy in Eastern Redbud.

Fifty-pound (20 kg) lots can be treated in screen-wire cylinders (reinforced with heavier wire) that can be lowered into the acid. In this way, most of the acid is retained for reuse. After a short draining period, the seed should be washed. Extra care must be taken in large-scale treatments to avoid excessive temperatures that can damage seeds.

There are several advantages of the acid treatment. It is effective for many species and requires little or no special equipment. Cost is reasonable. Most of the acid can be recovered and reused (unless the acid is poured on a pile of seed). Treated seeds can be held from a week to a month or more before sowing, without appreciable deterioration. Since the process leaves the seeds dry, firm and unswollen, they can be sown with mechanical seeders as well as by hand.

There are also disadvantages. Length of treatment must be carefully determined, and temperature must be carefully controlled, especially in large lots, to prevent serious injury to the seeds. Workmen also face a safety hazard.

If a large quantity of one seed lot is to be treated in several batches, it is desirable to standardise treatment as far as possible *e.g.* by using fresh acid of full strength and exactly the same volume (two volumes of acid to one volume of seed) for every batch.

In addition to protective clothing, a concentrated solution of sodium or potassium bicarbonate should be kept handy as an antidote against accidental splashes. Sulfuric acid treatment has been effective for several temperate and subtropical species, *e.g.* Gleditsia triacanthos (1 hour) and Ceratonia siliqua (2 hours) (Kisou *et al.* 1983). Examples of tropical species which respond well are Intsia palembanica (60 minutes of soaking), Parkia javanica (15 minutes), Dialium maingayi (5 minutes), Acacia albida (20 minutes), Acacia nilotica (60–80 minutes) and Acacia senegal (40 minutes), Acacia planifrons (2 hours) and Prosopis tamarugo (7 minutes). Hulled seed of Pterocarpus angolensis treated with acid gave 60% germination between 4 and 19 days after treatment, compared with 16% between 11 and 37 days in the control. In the Sudan it was found that seeds of Albizzia lebbek, Cassia fistula and Prosopis chilensis could be stored successfully for a further 3 – 4 months after treatment in H_2SO_4 or hot water. Storage after treatment is also practicable for several species of Acacia. Other chemicals have been tested from time to time to overcome seedcoat dormancy, but none have been adopted for largescale use in the same way as hot water or sulphuric acid. They include ethyl and methyl alcohol, xylene, ether, acetone, chloroform, hydrochloric acid, nitric acid and sodium hydroxide.

BIOLOGICAL METHODS

In nature animals and micro-organisms are an important factor in the breakdown of seedcoat impermeability. It is difficult to make use of these organisms as a controlled pretreatment of seed, but in a few cases successful results have been obtained.

Seeds of Acacia senegal and Ceratonia siliqua that have passed through the digestive tracts of goats germinate readily when placed in favourable conditions, because of the action of the strong digestive chemicals. Feeding the pods to penned goats and collecting the seeds from the droppings is a convenient pretreatment for these species. Seeds of some species are said to be regurgitated after partial rumination *e.g.* Gmelina arborea. Troup states that seeds of Acacia nilotica are ejected after rumination by sheep and goats but pass right through the digestive tract in cattle. In either case germination is improved by the digestive action.

Termites are an important agent for breaking down seedcoat dormancy in many parts of the tropics. In Thailand teak fruits were spread on the ground in a 5 cm thick layer immediately after collection and covered with cardboard. After about 5 weeks the termites had removed the exocarp and subsequent germination, after alternate wetting and drying, was significantly improved in comparison with fruits sown with intact exocarps. Termites have been used in a similar way to break down the tough winged and bristly fruit ofPterocarpus angolensis. Periodic inspection is essential to ensure that the process is not carried too far.

Partial fermentation, which is damaging to many seeds, can be beneficial in overcoming seedcoat dormancy in some. In the southern Sudan fruits of Tectona grandis are allowed to lie on the ground through the rainy season for partial fermentation. They are then collected, stratified in a pit with layers of (a) Seed (b) Organic matter (c) Soil and watered daily for 10 days.

DRY HEAT AND FIRE

Solar radiation is not used alone to promote germination but is an important component of the alternate soaking and drying treatment described above under "Soaking in water".

In the seasonal wet and dry tropics, fire is a powerful natural factor in the removal of seedcoat dormancy. A fierce fire will kill the seeds but a light to moderate fire, such as those associated with controlled early burning, will reduce seedcoat impermeability and stimulate germination. Fire has been used in a number of countries to stimulate germination of Tectona. The fruits may be spread thickly on the ground and covered with grass which is burnt off, or they may be lightly scorched by a flame gun. Adjusting the heat of the fire to achieve the maximum effect on the pericarp without damaging the seed embryo requires experience. Similar treatment is used for Aleurites moluccana in the Philippines.

The nuts are spread evenly on the ground and covered with a 3 cm thick layer of dry Imperata grass which is set on fire. As soon as the grass is burned, the seeds are placed in cold water. The quick change of temperature causes the nuts to crack and they are ready for sowing. An alternative is to sow the nuts at correct spacing with only half their diameter in the soil. A layer ofImperata grass is spread over the seed bed and set on fire. After burning, the seedbed is sprinkled with water and the nuts are pushed 2 cm deep into the soil and watered thoroughly.

In Sabah Bowen and Eusebio found that ten minutes' exposure of Acacia mangium seeds to dry heat at 100° C was nearly as effective in overcoming dormancy as the immersion in water at 100° C. Germination was 83% compared with 92% in the hot water treatment.

PLANT DORMANCY

In plant physiology, dormancy is a period of arrested plant growth. It is a survival strategy exhibited by many plant species, which enables them to survive in climates where part of the year is unsuitable for growth, such as winter and dry seasons.

Plant species that exhibit dormancy have a biological clock that tells them to slow activity and to prepare soft tissues for a period of freezing temperatures or water shortage. After a normal growing season, dormancy can be brought on by decreasing temperatures, shortened day length, or a reduction in rainfall.

DORMANT SEEDS

When a mature seed is placed under favorable conditions and fails to germinate, it is said to be dormant. There are two basic types of seed dormancy. The first is called seed coat dormancy or external dormancy, and is caused by the presence of a hard seed covering or seed coat that prevents water and oxygen from reaching and activating the embryo. The second type of seed dormancy is called embryo dormancy or internal dormancy, and is caused by a condition of the embryo which prevents germination. The oldest seed that has been germinated into a viable plant was an approximately 1,300-yr-old lotus fruit, recovered from a dry lakebed in northeastern China.

WHAT IS SEED DORMANCY?

Seed dormancy is defined as a state in which seeds are prevented from germinating even under environmental conditions normally favorable for germination. These conditions are a complex combination of water, light, temperature, gasses, mechanical restrictions, seed coats, and hormone structures.

In the sprouting industry, dormancy is commonly referred to as a seed producing plants ability to withstand winter freezes. The more dormant or "Winter Hardy", a seed producing plant is, the further north (in northern climates) it can grow and still survive the winter to come up the following season.

Below are the different ratings:

1 Very Dormant
2 Dormant
3 Moderately dormant
4-6 Categories of semi dormant and intermediate dormancy

7 Moderately non-dormant
8 Non-Dormant (moapa 69)
9 Very Non-Dormant (cuf 101)

Seed dormancy is nature's way of setting a time clock that allows seeds to initiate germination when conditions are normally favorable for germination and survival of the seedlings. For example, dogwoods produce mature seeds in the fall, but conditions are not suitable for seedling survival at that time. Thus, dogwoods have developed a mechanism that keeps the seeds dormant until spring when conditions are favorable for germination, as well as, seedling growth and survival.

Viable seeds that do not germinate are said to be dormant. Dormancy can be regulated by the environment or by the seed itself. If a seed is not exposed to sufficient moisture, proper temperature, oxygen, and for some species light, the seed will not germinate. In this case, the seed's dormancy is due to unfavorable environmental conditions. On the other hand, some seeds may not germinate because of some inhibitory factor of the seed itself. This particular kind of dormancy consists of two general types:

- Seed coat or external dormancy
- Internal (endogenous) dormancy

TECHNIQUES TO BREAK DORMANCY

SEED SCARIFICATION

Seed coat (external dormancy) results from a seed's hard seed coat that is impervious to water and gases. The seed will not germinate until the seed coat is altered physically. Any process of breaking, scratching, or mechanically altering the seed coat to make it permeable to water and gases is known as scarification. In nature, this often occurs by fall seeding. Freezing temperatures or microbial activities modify the seed coat during the winter. Scarification can also occur as seeds pass through the digestive tract of various animals.

Scarification also can be forced, rather than waiting for nature to alter the seed coats. Commercial growers scarify seeds by soaking them in concentrated sulfuric acid. Seeds are placed in a glass container and covered with sulfuric acid. The seeds are gently stirred and allowed to soak for 10 minutes to several hours, depending on the species. When the seed coat has been modified (thinned), the seeds are removed, washed, and sown. Sulfuric acid can, however, be very dangerous for an inexperienced individual and should be used with extreme caution!

Vinegar is safer (but less effective treatment) and can be used for species that do not have an extremely hard seed coat; the technique is the same as with sulfuric acid.For mechanical scarification, seed coats can also be filed with a metal file, rubbed with sandpaper, nicked with a knife, or cracked gently with a hammer to weaken the seed coat. Another method is hot water scarification. Bring water to a boil (212°F), remove the pot from the stove, and place the seeds into the water. Allow the seeds to soak until the water cools to room temperature. Remove the seeds from the water and sow.Following scarification, the seeds should be dull in appearance, but not deeply pitted or cracked as to damage the embryo. Scarified seeds do not store well and should be planted as soon as possible after treatment.

SEED STRATIFICATION

The second type of imposed dormancy found in seeds is internal dormancy regulated by the inner seed tissues. This dormancy prevents seed of many species from germinating when environmental conditions are not favorable for survival of the seedlings. There are several different degrees or types of internal dormancy. One type of internal dormancy is "shallow" and simply disappears with dry storage. Many vegetable seeds display this type of dormancy. No special treatments are necessary to overcome this kind of dormancy.

However, another type of internal dormancy requires special treatments to overcome. Seeds having this type of dormancy will not germinate until subjected to a particular duration of moist-prechilling and/or moist-warm periods.Cold stratification (moist-prechilling) involves mixing seeds with an equal volume of a moist medium (sand or peat, for example) in a closed container and storing them in a refrigerator (approximately 40°F). Periodically, check to see that the medium is moist but not wet. The length of time it takes to break dormancy varies with particular species. This type of dormancy may be satisfied naturally if seeds are sown outdoors in the fall. Warm stratification is similar except temperatures are maintained at 68°F to 86°F, depending on the species.

Seeds of some species exhibit what is known as double dormancy. This is a combination of seed coat (external dormancy) and internal dormancy. To achieve germination with seeds having double dormancy, the seeds must first be scarified and then stratified for the appropriate length of time. If the treatments are administered in reverse order, the seeds will not germinate. After these treatments, sow the seeds under the proper environmental conditions for germination.

It is a condition of a seed in which it is temporarily inactive. It does not die but does not grow also. The dormancy of seeds varies widely from seed to seed.

Some seeds remain viable only for few days (40 to 50 days). While some seeds remain dormant for few years. It is this dormancy which is to be preserved during storage. Whether the dormancy is less or more care has to be taken during storage specially for seeds which have less dormancy.

INTENSITY OF DORMANCY

This refers to the level of breaking dormancy by artificial means, i.e., based on the germination percentage after heat treatment for four days at 50 degrees centigrade, and the rice varieties classified based on germination percentage as follows

STRONGLY DORMANT

Varieties in which 50% of dormancy is broken after 4 days of heat treatment.

MODERATELY DORMANT

Varieties in which 50 to 79% of dormancy is broken after 4 days of heat treatment.

WEAKLY DORMANT

Varieties in which above 80% of dormancy is broken after 4 days of heat treatment.

SEED STORAGE

NEED FOR SEED STORAGE

Seed produced in one season is used in the next season. Till that time it has to be stored.

Extra unused or unsold seed has necessarily to be stored.

In research plant breeders have to store seeds of variety of crops for 5-10 years to maintain germplasm.

TECHNIQUE OF STORAGE

Technique of Storage Involves following Steps

To control environmental factors temperature & humidity. High temperature and high humidity reduce the viability of seed. The temperature should be less than 30o C and humidity 50-60 %. The seed should be stored in cool dry places.

Moisture Content of Seeds

The seed should be dried properly in the sun or by mechanical driers as to bring down moisture content within range of 8 to 9 per cent.

Cleanliness

Store room/godown should be properly cleaned, it should be painted with while colour and sprayed with Malathion 50EC insecticide.

Care During Storage

The seed should be inspected every after 15 days and manipulated accordingly. If there is any incidence of storegrain pests fumigation should be done with Aluminium phosphide

Per Tonne

Ethylene dibromide 32 gm per cubic meter space. After fumigation there should be good ventilation. The godown should be disinfected every after 3-4 weeks with sprays of Malathion 50 EC. The godown should always be kept clean.

SEED TREATMENT

Seed treatment refers to the application of fungicide, insecticide or a combination of both to seeds, such as to disinfect and disinfest them from seed - borne or soil-borne pathogenic organisms and crop pests both in field and in storage. It also refers to the subjecting of seeds to solar energy exposure, immersion in conditioned water etc.

BENEFITS OF SEED TREATMENT

Seed treatment improves the germination of seed.
Good establishment of healthy and vigorous plants.
Increase the production.
Seed does not rot after sowing or plant does not die.
It helps to save the expenses of disease control.
It helps to control the seed borne diseases
Protection of seeds in storage.

DRY STORAGE

For species where dormancy is naturally of short duration, it is often sufficient to store the samples in a dry place for a short period.

PRE-CHILLING

The replicates for germination are placed in contact with the moist substratum and kept at low temperature for an initial period. Agricultural and vegetable seeds are kept at a temperature between 5 and 10 degrees Centigrade for an initial period of upto 7 days. In some cases it may be necessary to extend the pre-chilling period or to re-chilling.

PRE-HEATING

The replicates for germination should be heated at a temperature not exceeding 40 degree C, with free air circulation, for a period of upto 7 days before they are placed under the prescribed germination conditions.

LIGHT

The test should be illuminated during atleast 8 hours in every twenty four hours cycle and during the high temperature period when the seeds are germinated at alternating temperatures. The light intensity should be approximately 750 - 1250 lux from cool white lamps. Illumination is recommended especially for certain tropical and sub-tropical grasses.

SEED TREATMENT BY APPLICATION OF CHEMICALS

Several insecticides and fungicides are used in seed treatment to protect the seeds from pests and diseases.

7

Design and Engineering of Seed Storage Facilities

STORAGE CONTAINERS

Some form of container is necessary for most seed storage, to facilitate access to, and handling of, individual seed lots while keeping them separate, to make the best possible use of storage space, to provide protection against animal and insect pests and, for some seeds, to prevent passage of moisture and gases between the enclosed and the outside atmosphere.

Many types of container have been used for tree seeds; they may be conveniently divided into:

- Materials freely permeable to moisture and gases
- Materials completely impermeable, when sealed, to moisture and gases
- Materials resistant, but not completely impermeable, to moisture.

MATERIALS FREELY PERMEABLE TO MOISTURE AND GASES

These include hessian or burlap sacks, cotton bags and containers of paper, cardboard and fibreboard. Hessian and cotton have the advantage that seed triers can be inserted through the cloth mesh to withdraw samples for testing without the need to open the mouth of the container. The resilience of the cloth will close the hole and avoid subsequent loss of seed, which is not possible with containers based on paper or paperboard. Hessian and cotton are also robust materials which can be used more than once.

None of these materials is entirely proof against insect and rodent pests, and all are freely permeable to water vapour and gases. For orthodox seeds in uncontrolled conditions they are therefore suitable only for rather short storage periods; these can be extended in the case of hardcoated seeds or where ambient conditions are cool and dry. If seeds are stored in large containers after drying to the correct MC, the outer seeds themselves provide some barrier to the passage of moisture. Viability of the inner seeds may thus be preserved for a period even though there is some deterioration from increased MC in the outer layers. If a seed store has facilities for controlling both temperature and relative humidity, then permeable containers can be safely used for orthodox seeds for several years, provided that pests can be excluded.

For moist storage of recalcitrant seeds, open or freely permeable containers such as hessian sacks should be used in order to allow free exchange of air and so avoid the overheating which can occur if moist, rapidly respiring seeds are enclosed without adequate ventilation. Periodic spraying of the sacks may be necessary to maintain the high MC which is appropriate for this type of seeds.

MATERIALS COMPLETELY IMPERMEABLE, WHEN SEALED, TO MOISTURE AND GASES

After drying of orthodox seeds to the correct MC, the MC may be maintained in storage by dehumidifying the whole storage space. Another very efficient way, commonly used in storing forest seeds, is to place the seed in sealed moistureproof containers. This avoids the need for expensive dehumidification equipment. For long-term storage the most effective method is a combination of moistureproof containers with controlled low temperatures provided by refrigeration. An added advantage of most materials in this type is that they also exclude oxygen and so reduce still further the rate of respiration. Impermeable sealed containers are not suitable for storing recalcitrant seeds nor are they suitable for orthodox seeds at high MC, which deteriorate more rapidly in sealed than in open storage. Some seeds absorb moisture quickly, so it is important that they be sealed inside the container as soon as possible after drying is complete, preferably within the drying room itself.

Moistureproof containers include tin or aluminium cans and drums, glass jars of the Mason or Kilner types, plastic vials and laminated aluminium foil packages. Rigid and unbreakable metal cans provide maximum protection against mechanical damage to the seeds and are equally suitable for storage and subsequent shipment.

Containers are only as moistureproof as their sealing. For rigid containers screw-top or clamp-down gasketed lids should be used, if periodic opening for seed extraction and subsequent resealing are anticipated; aluminium foil should be heat-sealed. The effectiveness of sealing is particularly important in long-term storage. Three types of container are considered suitable for hermetically sealed long-term storage of agricultural seeds: glass jars or vials; metal cans; and laminated foil packets. They should be equally appropriate for orthodox forest seeds. But the report by IBPGR recommended sealed metal cans as the most reliable and convenient. It noted that the seals on screw-cap jars are not always perfect and that further experience of the lasting qualities of laminated foil packets is needed before they can be recommended for general use in storage which will often last for several decades.

MATERIALS RESISTANT, BUT NOT COMPLETELY IMPERMEABLE, TO MOISTURE

They include polyethylene and other plastic films and aluminium foil. These materials are resistant to the passage of moisture but, over a long period of time, there will be a slow passage of water vapour tending to equilibrate the RH inside with that outside the container. Some of the figures quoted by Justice and Bass for transmittal of water vapour appear surprisingly high, *e.g.* 0.13 g per 100 square inches (645 cm^2) per 24 hours for low density polyethylene film 10 mil (250 microns) thick and about ten times this figure for low density film 1 mil (25 microns) thick.

However, the standard conditions for testing these materials are 0% RH on one side and 90–100% on the other. The RH gradient during storage is never so severe as this and hence the rate of passage of water vapour is much less rapid in practice. In one test using 6 mil (150 microns) high density polyethylene the rate of passage over two years from an outside RH of 95–100% at 20°/30°C was four times that from an outside RH of 50% at 10°C. The thicker the film, the greater the resistance to passage of water vapour and, for a given thickness, high density polyethylene is more resistant than low density.

Although polyethylene is not suitable for long-term storage of orthodox seeds for genetic conservation, it is very suitable for short-or medium-term storage and has given excellent results for up to 5 years' storage of Pinus caribaea and P. oocarpa seeds in Honduras, with no significant change in MC. For Honduran conditions a thickness of at least 4–5 mil (100–125 microns) is recommended; thinner polythene can permit a significant passage of water vapour in time and is also subject to mechanical damage in handling.

Harrington considered 3 mil (75 microns) high density or 5 mil (125 microns) regular suitable for temperate conditions and 7 mil (175 microns) high density or 10 mil (250 microns) regular as adequate for even severe tropical conditions. Proper sealing of bags is essential and can be done by a combination of heat and pressure. In the past hot irons were used, but sealing can now be done more efficiently and conveniently by commercial heat sealers, of which a number of different models is now on the market.

Different materials, each alone slowly permeable to water vapour, may become completely impermeable when laminated together. Various combinations of laminated polyethylene, aluminium foil and kraft paper proved completely impermeable to water vapour over a two year period, even when there was a high differential between the inside and outside RH.

USE OF DESICCANTS IN CONTAINERS

If orthodox seeds are dried to the correct MC and stored in sealed impermeable containers, the MC should remain constant for years. If, however, the seeds are stored in moisture resistant but not completely impermeable material such as polythene bags, or if it is necessary to open and reclose the containers periodically to extract seeds, there will be a slow build up of moisture in time. A convenient way to prevent this is to enclose some desiccant such as silica gel in the containers. The capacity of silica gel to adsorb moisture depends on the relative humidity of the ambient air, as shown in the following table: Moisture content of silica gel in equilibrium with various relative humidities

% RH	% H_2O adsorbed	% RH	% H_2O adsorbed
0	0.0	55	31.5
5	2.5	60	33.0
10	5.0	65	34.0
15	7.5	70	35.0
20	10.0	75	36.0
25	12.5	80	37.0
30	15.0	85	38.0
35	18.0	90	39.0
40	22.0	95	39.5
45	26.0	100	40.0
50			29.0

A convenient method is to use silica gel treated with cobalt chloride, which changes colour from blue to pink about 45% RH; the corresponding equilibrium MC for many orthodox species would be 7–9%. Dried silica gel is enclosed with the seeds and, whenever the granules turn pink, the silica gel is removed and reactivated by drying in an oven at 175° C and cooling in a sealed container before reuse. A weight of silica gel equal to one tenth the weight of seeds is recommended. Care should be taken not to include too much silica gel which could lead to overdrying of the seeds. Even with silica gel of one tenth the weight of seeds, the MC of seeds enclosed at 6% would be lowered to below 5% during the initial phase of storage. More frequent reactivation of silica gel would preserve an equilibrium of RH and seed MC at lower levels than the 45% and 7–9% mentioned above but would forego the convenience of the colour indicator. Another use for desiccants is where the MC of seeds is known to be higher than the optimum for sealed storage, for example because only air-drying is possible. As mentioned, the enclosure with the seed of approximately an equal weight of silica gel in sealed containers should reduce the MC of the seed to a suitable level and maintain it. As an example

1 kg (oven-dry weight) of seed at initial 19% MC (dry weight basis) contains		190 g H_2O
1 kg (oven-dry weight) of seed at 6% MC (dry weight basis) contains		60 g H_2O
Therefore moisture to be removed	=	130 g H_2O
RH in equilibrium with 6% MC	=	25 % RH
At RH 25%, 1 kg silica gel adsorbs		125 g H_2O

Therefore a weight of silica gel equal to the weight of seed will reduce the initial 19% MC to just over 6% MC for storage.

CHOICE AND USE OF CONTAINER

The following factors, which should be considered when choosing the best storage container for a given use, are based on those listed by Stein *et al*: When seed requires further drying in storage, do not use a tight-closing container because enclosing excess moisture is harmful to the seed. Use a tight-closing container if gain in seed moisture content can be damaging and relative humidity in the storage facility is high.

Containers and seed can quickly gather unwanted condensation when brought out of cool or subfreezing storage. Warming to room temperature is recommended before opening a container brought out of such storage. 4 to 10 mil (100 – 250 microns) polyethylene bags will greatly restrict exchange of moisture, but still allow exchange of oxygen and carbon dioxide with air outside. Such exchange may be beneficial or harmful, depending on species.

A container that is easy to open and close is desirable when quantities of seed are likely to be added or removed repeatedly. In order to minimize temperature and relative humidity fluctuations, open only when necessary. Alternatively, store seed in small containers, so that the entire contents can be stored or emptied at one time. For orthodox seeds, fill containers completely to ensure minimum exchange of moisture between the seed and the entrapped air and, more importantly, to limit the amount of oxygen enclosed.

When exchange of moisture through the container walls must be eliminated or restricted, the container must be made impermeable or of moisture resistant material. The longer the storage period and the higher the differential between external RH and RH within the container, the more impermeable the material must be. When seeds are fragile and easily damaged, a rigid-walled container should be used. Moisture-proof plastic bags are often used as liners for rigid containers. Choose a container shape and stacking arrangement which facilitates uniform temperature and aeration throughout the storage facility. Some containers may be made of substances that are harmful to tree and shrub seeds (Barton 1954). Unproven containers should be tested for toxicity.

Fig. Airtight Containers used for Storing Seed

Fig. Interior View of Cold Storage Room at Humlebaek

Fig. Examples of Different Types of Container used for Storage or Shipment

Static electricity can build up slowly on some materials such as PVC. This makes them difficult to clean between exhaustion of one seed batch and insertion of the next.

It should be stressed that no one container or packaging material can be the best for all sizes, conditions and objectives that occur in seed packaging. The relative merits of the various containers will have to be weighed against their disadvantages and costs before deciding the final choice.

REQUIREMENT OF STORING SEEDS

Seeds must be stored in a way which maintains their viability for long periods. Seeds left at ambient temperatures and relative humidities will lose their viability quickly whilst seeds stored in conditions of low moisture content and temperature will retain their viability for longer periods. Accessions held in a genebank are valuable and represent plants which are no longer available or which are endangered in their natural environment. These seeds must be conserved in the genebank for use in plant breeding in the future.

As soon as the seed has matured on the plant the slow process of deterioration begins. Therefore the sooner that seeds are placed into store the better. In practical terms this means that seeds collected in the field should be quickly returned to genebanks, processed and placed into store as soon as the cleaning, drying and packaging processes are complete

The Seed Storage for materials showing little morphological variation (genetically homogeneous) 3000 seeds are acceptable, but 4000 seeds are preferred, to represent each accession. For materials showing a large amount of morphological variation (heterogeneous) an accession should consist of at least 4000 seeds, but 12000 seeds are preferred. These sample sizes are recommended as number of seed but many genebanks would prefer to work with weight. An inter-conversion of numbers and weights is possible from the thousand seed weight of any accession.

PURPOSE OF SEED STORAGE

Seed storage is the maintenance of high seed germination and vigour form harvest until planting. Is important to get adequate plant stands in addition to healthy and vigourous plants. Every seed operation has or should have a purpose. The purpose of seed storage is to maintain the seed in good physical and physiological condition from the time they are harvested until the time they are planted. Seeds have to be stored, of course, because there is usually a period of time between harvest and planting. During this period, the seed have to be kept somewhere. While the time interval between harvest and planting is the basic reason for storing seed, there are other considerations, especially in the case of extended storage of seed.

Seed suppliers are not always able to market all the seed they produce during the following planting season. In many cases, the unsold seed are "carried over" in storage for marketing during the second planting season after harvest. Problems arise in connection with carryover storage of seed because some kinds, varieties, and lots of seed do not carryover very well.

Seeds are also deliberately stored for extended periods so as to eliminate the need to produce the seed every season. Foundation seed units and others have found this to be an economical, efficient procedure for seed of varieties for which there is limited demand. Some kinds of seed are stored for extended periods to improve the percentage and rapidity of germination by providing enough time for a "natural" release from dormancy.

Regardless of the specific reasons for storage of seed, the purpose remains the same maintenance of a satisfactory capacity for germination and emergence. The facilities and procedures used in storage, therefore, have to be directed towards the accomplishment of this purpose.

In the broadest sense the storage period for seed begins with attainment of physiological maturity and ends with resumption of active growth of the embryonic axis, *i.e.*, germination. Seeds are considered to be physiologically and morphologically mature when they reach maximum dry weight. At this stage dry-down or dehydration of the seed is well underway. Dry-down continues after physiological maturity until moisture content of the seed and fruit decreases to a level which permits effective and efficient harvest and threshing. This stage can be termed as harvest maturity. There usually is an interval of time between physiological maturity and harvestable maturity, and this interval represents the first segment of the storage period. Any delays in harvesting the seed after they reach harvest maturity prolongs the first segment of the storage period–often to the detriment of seed quality.

The second segment of the storage period extends from harvest to the beginning of conditioning. Seed in the combine, grain wagon, and bulk storage or drying bins are in storage and their quality is affected by the same factors that affect the quality of seed during the packaged seed segment of the storage period. The third segment of the storage period begins with the onset of conditioning and ends with packaging. The fourth segment of the storage period is the packaged seed phase which has already been mentioned. The packaged seed segment is followed by storage during distribution and marketing, and finally by storage on the farm before and during planting.

The control that a seedsman has over the various segments of the storage period for seed varies from a high degree of control from harvest to distribution, to much less control during the postmaturation-preharvest, distribution-marketing, and on-farm segments. Despite variable degrees of control over the various segments of the storage period, the seedsman's plans for storage must take into consideration all the segments. The things that can be done must be done if the quality of the seed is to be maintained.

METHODS

The conditions which prolong viability during storage have been well defined for seeds which are tolerant of desiccation. Storage conditions have been recommended by the IBPGR Advisory Committee on Seed Storage. For base collections, seeds of between 3-7% moisture content should be stored in sealed containers. Sub-zero temperatures are acceptable, but –18 °C or less is preferred. For active collections sealed storage of seeds dried to 7% moisture content or less is recommended at temperatures of less than 15 °C. Unsealed storage is not encouraged. In particular, it is not recommended in tropical areas.

CHECK THE NUMBER OF SEEDS IN THE ACCESSION

1. Weigh the seeds of each accession.
2. Convert the weight of seeds to seed number, using the thousand seed weight of each accession for an accurate conversion.
3. For accessions containing mixtures of genotypes, the sample size should be at least 4000 seeds. For genetically uniform accessions, the sample size should be at least 3000 seeds.
4. If the sample contains more than the required number of seeds, proceed to storage.
5. If the sample contains less than the required number of seeds, either proceed directly to regeneration or store the seeds temporarily in the genebank and regenerate at the earliest opportunity.

Notes and Examples

An approximate inter-conversion of seed number and weight can be done easily using the thousand seed weight.

Example for Sorghum (1000 seed weight – 17 g) 100 g contains:

$$\frac{1000}{17} \times 100 = 5750 \text{ seeds}$$

These are minimum sample sizes for the start of storage and if both space and seeds are available, more seeds should be held.

Regenerate as soon as possible. Seeds processed and stored under good conditions will not loose viability before regeneration.

Equipment

Coarse balance

DETERMINE WHERE THE SEEDS SHOULD BE LOCATED

1. Check the inventory data file of the genebank to find the next available space where a container can be located.
2. When seeds from the same regeneration cycle of the same accession are stored in several containers, keep all the containers of the accession together.
3. Make a list of where each accession will be placed.

Notes and Examples

The storage arrangements will vary among genebanks. The most important point is to know exactly where to locate each accession within the store.

Equipment

- Racks
- Trays, boxes or drawers
- Coldroom or freezer

PLACE IN THE SEED STORE

1. Place the container into the seed store in the listed location.

ENTER THE DATA INTO THE DATA BASE

1. Fill in the data on the location and date of storage of each accession and each container into the data file.
2. Record the date of the next monitoring test for germination in the data file. This date will be determined by the curator after considering the viability and moisture content of the seeds, storage conditions and the IBPGR recommendations.

Notes and Examples

A code can be used to locate an accession within the store. Each unit can be identified by a number or letter in the store. The code can indicate the number or the letter of the freezer, store, rack, basket or drawer, etc.

Example

Colour codes can also be a quick and easy way to locate accessions. A colour can be used for each rack, shelf or species. This both speeds up the work in the coldroom and makes it easy to spot errors. Owing to the very cold temperatures, the faster that one can locate accessions in the cold room the better.

A010201 could be used to indicate the location as:

Coldroom:	A
Rack:	01
Shelf:	02
Box:	01

CHOICE OF STORAGE METHOD

A number of different storage methods are available, as described below. The main factors affecting choice are the seed characteristics of the species in question, the period for which it is to be stored and the cost. If more than one method is suitable to maintain viability for the required period, the simplest and cheapest will normally be chosen.

STORAGE AT AMBIENT TEMPERATURE AND HUMIDITY

Seeds can be stored in piles, single layers, sacks or open containers, under shelter against rain, well ventilated and protected against rodents. Best results are obtained in cool, dry climates. In these conditions several species of Pinus, Eucalyptus Pseudotsuga and Tectona will store satisfactorily for at least six months, while leguminous trees with impermeable seedcoats and naturally low MC, *e.g.* Acacia Prosopis, Robinia will retain viability for years.

DRY STORAGE WITH CONTROL OF MC BUT NOT OF TEMPERATURE

Orthodox seeds will retain viability longer, when dried to a low moisture content (4 – 8%), as described on pp. 126–128, and then stored in a sealed container or in a room in which humidity is controlled, than when stored in equilibrium with ambient air humidity. Storage life is further prolonged if cool, but not controlled, temperature conditions can be provided, *e.g.* at high latitude or altitude and in a cellar or other room screened from direct sunlight.

Seed is sometimes stored in open containers in a room maintained at an RH of 15 – 20% by dehumidifying machinery. In forestry it is more common to rely on predrying of the seed to the correct MC, followed by storage in full, sealed containers. Provided that the containers are not opened too frequently and that the sealing is effective, the method will maintain a low MC for many years. It is cheaper than using a humidity-controlled room, especially during periods when only a little seed is in store, and it is not subject to hazards of mechanical breakdown.

This method is suitable for a range of species including many Pinus and Eucalyptus species, for which it should maintain viability for one or more years.

DRY STORAGE WITH CONTROL OF BOTH MC AND TEMPERATURE

This storage treatment is the standard practice for many orthodox species which have periodicity of seeding but which are planted annually in large-scale afforestation projects. For many species a combination of 4 – 8% MC and 0 to +5° C temperature will maintain viability for 5 years or more. Some cool-temperate genera benefit by storage at sub-freezing temperatures, *e.g.* -4°C or lower for Abies, -10°C forFagus, -5°C for Fagus, -18°C for Pinus strobus, Populus deltoides and others.

Pinus merkusii is an example of a tropical pine which responds well to storage at low temperature and MC. 80% germination was obtained after 3 years' storage at 2° C and 6 – 10% MC of seeds of the Zambales (Philippines) provenance, while seeds stored in equilibrium with room temperature and humidity showed a significant loss of germination after 3 – 4 months. P. caribaeaand P. oocarpa behave similarly in this respect. In addition to true seeds, this method is also suitable for certain types of fruits. For example in the Jari project of Brazil depulped, cleaned and dried stones of Gmelina arborea are successfully stored in sealed containers at 5° C and 6 – 10% MC. The fresh stones have a germination of 90% and after two years of storage germination is still 80%.

DRY STORAGE FOR LONG-TERM GENE CONSERVATION

The preferred storage treatment for long-term conservation of gene resources of orthodox agricultural seeds is -18° C temperature and 5% +-1% MC. This is likely to be equally appropriate for orthodox seeds of forest trees requiring storage for genetic conservation. The quantity of seeds requiring this standard of storage is small in comparison with the quantities used each year for operational afforestation, and the cost per kg of seed is higher.

For many countries it would therefore be desirable that forest and agricultural crop genetic resources should share a common long-term storage facility. A good example is that of the Banco Latino Americano de Semillas Forestales at CATIE, Turrialba which has its own seed store (55 m^3 capacity at 5°C) for short or medium term storage, but also has access to the long-term storage facilities (at -20°C) of the Regional Genetic Resources Unit, which are also at Turrialba.

Loss of viability in storage, in addition to reducing the number of plants which can be produced by a given seed lot, may result in a shift in the genetic constitution of the seed being stored. This could be particularly important in forest trees which are predominantly outbreeding, variable populations. First, loss of viability may occur more rapidly in some genotypes than others; if losses are high, say 50% of the total, the genotypes with short-lived seeds may be eliminated altogether. Yet they may have valuable traits for adaptation, growth or disease resistance as growing trees, and in any case they contribute to genetic variation in the species which it is the purpose of genetic conservation to preserve. Secondly, it is an accepted fact in agriculture that chromosome damage or change occurs and accumulates in the seed in storage, and that the risk of such heritable gene mutations depends not so much on the age of the seed as on changes in its viability (Roberts 1972, Barner 1975b). A seed lot which has suffered a serious loss of viability is likely to have experienced some gene mutations among the survivors, but there is very little direct evidence that heritable mutations are induced under good storage conditions which lead to only small losses in viability.

The high standard of storage conditions recommended by the IBPGR and referred to above, if combined with regular testing of seed and regeneration as soon as germination falls to 85% of the initial germination rate (Ellis *et al.* 1980) should minimize the risk of genetic change in storage. It is possible that still lower temperatures would increase longevity even more. Research on storage in liquid nitrogen has been pursued for some years and considerable progress has been made, but testing for several more years will be needed before the method can be recommended for general adoption in gene banks.

MOIST STORAGE WITHOUT CONTROL OF MC OR TEMPERATURE

Suitable for storage of recalcitrant seeds for a few months over winter. Seeds may be stored in heaps on the ground, in shallow pits in well-drained soils or in layers in well ventilated sheds, often covered or mixed with leaves, moist sand, peat or other porous materials. Seeds stored outdoors are kept moist by rain or snow, but those under shelter may need to be moistened periodically. The aim is to maintain moist and cool conditions, combined with good aeration to avoid overheating which may result from the relatively high rates of respiration associated with moist storage. This may be accomplished by regular turning of the heaps of seed or by inserting bundles of straw or twigs into them.

This method is suitable for short-term storage of large-seeded hardwood species in the temperate zone *e.g.* Quercus, Castanea, Aesculus. It is unlikely to be suitable for tropical recalcitrant species because ambient temperature is too high.

MOIST COLD STORAGE, WITH CONTROL OF TEMPERATURE

This method implies controlled low temperatures just above freezing or, less commonly, just below freezing (Magini 1962). Moisture can be controlled within approximate limits by adding moist media *e.g.* sand, peat or a mixture of both to the seed, in the proportions of one part media to 1 part seed by volume, and remoistening periodically, or more accurately (but more rarely) by controlling the relative humidity in the cold store. The latter type of control is often too expensive. Respiration rate is reduced and storage life prolonged by the low temperature, but seed should not be stored in sealed gas-proof containers which would limit oxygen supply. Closed polyethylene bags of 4 – 10 mil (100–250 microns) thickness will allow exchange of oxygen and CO_2

with air outside, while severely restricting exchange of moisture. The method is suitable for the same temperate recalcitrant genera as listed in the previous section, and with temperatures of 0 – 5° C should extend viability up to 1 ½ to 2 years.

Sub-freezing temperatures have given improved results in a few cases but frequently injure seeds with high MC and should be used only after research has demonstrated their applicability to the species in question. Less is known about the application of this method to tropical species, but it merits much more investigation that it has so far received, for the dipterocarps and genera such as Araucaria, Agathis and Triplochiton. As mentioned earlier there is evidence that some species are killed at low but above-freezing temperatures and Gordon has proposed a division of recalcitrant seeds into those which can withstand temperatures below 10° C without loss of viability and those which cannot.

Tamari, summing up several years of research on dipterocarps in Malaysia, concluded that the best treatment for several species was,

- Dry at a temperature not above 35° C, to reduce MC to 35%
- Seal with a fungicide inside polythene bags
- Store at 15° C, or for 3 weeks at 15° C followed by further storage at 10° C.

This treatment has been successful in extending longevity from a week or two up to two months in Hopea helferi, but this is still a long way from providing safe storage between seed years, reported to vary from 3 – 6 years apart in many dipterocarps (Tang 1971). Storage at 3.5°C and MC over 32% over periods of at least 6 months has been successful on the recalcitrant Araucaria hunsteinii.

In some recalcitrant species newly germinated seeds may retain viability better under moist cool storage conditions than ungerminated seeds. Gordon reported that some pregerminated seedlots of Quercus spp. showed no significant change in the number of living seedlings after one year's storage in 500 gauge (125 microns) polythene bags lightly sealed at 3° C, whereas a large proportion of the seeds which were viable but ungerminated when placed in the same bags died during the same period.

OTHER METHODS

Other storage methods have been used in the past, but are not yet of wide application.

They include:

a. Storage of recalcitrant seeds in running (not stagnant) water.
b. Storage under partial vacuum.
c. Storage in gases other than air *e.g.* nitrogen or CO_2.
d. Coating individual large seeds with paraffin or latex to prevent moisture exchange. This method may also be used to maintain moisture content during shipment.

FACTORS AFFECTING LONGEVITY IN STORAGE

SEED CONDITION

Even in ideal storage conditions seed will soon lose viability if it is defective from the start. Factors to be considered are:

Seed Maturity

Fully ripened seeds retain viability longer than seeds collected when immature. Certain biochemical compounds, essential for preserving viability, may not be formed until the final stages of seed ripening. These include dormancy-inducing compounds in certain species, and dormancy is sometimes associated with seed longevity. In a few species *e.g.*, Gingko biloba,Fraxinus excelsior, seed embryos are underdeveloped when the seed is shed. Maturation of these embryos is necessary before sowing but need not be done before storage. In Fraxinus excelsior drying of freshly collected samaras to 9–10 per cent MC, followed by storage in sealed containers at -3°C, gives satisfactory results provided that successive moist warm and moist cold treatments are applied after storage.

Parental and Annual Effects

In seed harvest, quantity and quality often go together. The percentage of sound seeds in a high-yielding mother tree is usually higher than in one with a poor crop.

Similarly, a given mother tree will have a higher percentage of sound seeds in a good crop year than in a poor one. Collection from high-cropping mother trees in a seed year is likely to yield seeds with the best longevity in storage. On the other hand, high-cropping "wolf" trees should be avoided because of their potentially undesirable wood properties, even though they may produce long-storing seeds.

Freedom from Mechanical Damage

Seeds damaged mechanically in extraction, cleaning, dewinging etc. rapidly lose viability. The danger is greatest for species which have thin or soft seedcoats. Excessive heat during extraction or drying also damages seed.

Care should be taken to use the minimum times, lowest temperatures and minimum machine speeds necessary during the preparation of seed for storage. In some species, damage during dewinging may be reduced by partly restoring the moisture content between extraction from cones and dewinging, since moist seeds suffer less mechanical damage than dry ones.

Freedom from Physiological Deterioration

Poor handling in the forest, during transit or during processing causes physiological deterioration of seeds even if mechanical and fungal damage are absent. Adequate ventilation of orthodox seeds is necessary to avoid rapid respiration and overheating, while recalcitrant seeds must be protected against excessive drying.

Freedom from Fungi and Insects

For species stored at low temperature and low moisture content, the storage conditions themselves should prevent the development of fungi and insects. It is necessary, however, to avoid collection of crops showing a high incidence of fungal or insect attack and to carry out all operations of collection, transport, processing etc. as quickly as possible to ensure seed is not already damaged before it goes into storage.

Attack by fungi and insects is most rapid on the forest floor, so collection from the ground should be carried out as soon after fruit fall as possible. Fungicidal treatment cannot be generally recommended since it can be harmful to seeds; many fungicides are only effective when dissolved in water and are inappropriate for dry storage. Insects are usually killed if seeds are dried at temperatures above 40° – 42°C. For seeds which cannot be dried, other measures may be needed. For example seeds of Quercus are fumigated with serafume or other chemicals or heated in warm water for control of weevils, while methyl bromide or carbon bisulphide are also commonly used to kill insects.

Initial viability. Seed lots with high initial viability and germinative capacity have a higher longevity in storage than those with low initial viability. Germination tests, preceded if necessary by appropriate pretreatment to overcome dormancy, should be carried out on a sample of each seed lot before storage, in order to determine how long the seed is likely to retain viability in storage. Longevity of the viable seeds is correlated with the percentage which germinate in the initial test.

As an example, samples of two seed lots of the same species, from which 80 per cent germination of fresh seeds is normally expected, might give results of 90 per cent and 50 per cent initial germination. Not only would storage of the second seed lot involve wasting space in storing dead seed, but even the 50 per cent of initially viable seeds are likely to lose their viability more quickly in storage than would the 90 per cent viable seeds in the first seed lot.

Deterioration in initial viability may not be serious if the seeds are to be sown within a few weeks or months, but only good quality seed should be stored for long periods. In long-term storage of agricultural seeds for genetic conservation, it is recommended that no seeds should be accepted for storage which have an initial viability of less than 85 per cent of that considered typical for the species or variety in question.

STORAGE CAPACITY

The weight of seeds to be kept in store can be estimated in the manner and will depend on the annual planting area, the maximum number of years' seed supply to be stored at any one time because of seeding periodicity, and the number of seeds per kg, for each species. Weight of seed in kg can be converted to net volume in litres (or in g to cm^3) by a factor related to average specific gravity. An average factor of 2.0 is appropriate for many forest species and corresponds to an apparent specific gravity of 0.5 (true specific gravity would be slightly higher because of the air-spaces between the seeds).

For conversion from net volume to gross storage space, allowing for shelving, ventilation, air spaces within and between containers, access and fittings within the cold room, a factor of about X8 is commonly used; this is with fixed shelving. Use of mobile shelving may double the quantity of seed which may be stored in a given space; in this case a factor of about X4 is appropriate.

Thus 500 kg of seed of S.G. 0.5 would need a gross storage space of 500 × 2 × 8 = 8000 litres or 8 m^3 if fixed shelving were used and 4 m^3 with mobile shelving. Where relatively few seed lots and large quantities of each lot are being stored, it is possible to use standard sizes of container, each filled to the brim, and for shelving space to be adapted to fit container size exactly. Under these conditions considerable savings in storage space can be effected. Thus in the Danish seed store at Humlebaek, which uses fixed shelving, a factor of only 3.12 has been calculated.

DESIGN AND EQUIPMENT

It is essential that designs and equipment be adapted to local conditions and local resources. The best installation in the world is of little use if it cannot be maintained, so it is essential to investigate the local provision for servicing and spares before committing oneself to any particular item. The reliability of mains electricity services and the need for a voltage controller and standby generator are of primary importance. Ready availability of a spare compressor may also be necessary.

The correct siting of a seed store may reduce the need for much expensive equipment. For example a tropical country with variable climate and topography might solve many problems by moving its store from a hot humid coastal site to the dry rain-shadow side of a mountain at 2000 m. In such a case a well-ventilated room might provide perfectly suitable conditions for several years' storage for relatively "easy" species such as pines and eucalypts and could be supplemented by one or more deep freeze chests for small quantities of more "difficult" species requiring sub-freezing temperature.

STORAGE CONDITIONS AND AGEING OF SEEDS

In common with all other living things, seeds are subject to ageing and, eventually, to death. In the case of orthodox seeds, the process of ageing and deterioration is so greatly affected by the conditions of storage that the "age" of seeds, expressed solely in terms of the period elapsed since ripening and harvesting, is an inadequate measure of the degree to which they have "aged" in the sense of losing viability and progressing towards the irreversible deterioration of death. The term "physiological age" is commonly used to describe the degree of deterioration of seeds measured by their reduced capacity for germination.

Nomographs for the effects of temperature and MC on physiological ageing of seeds have been constructed for several agricultural crops.

As an example, the nomograph for barley indicates that the same degree of deterioration (from initial 95% to final 50% germination) would occur in about 16 days in seeds stored at 25° C and 21% MC, but in about 100 years in seeds stored at 8° C and 8% MC. Both seed lots would have the identical physiological age, though stored for very different periods of time. Similar effects can be expected in orthodox seeds of forest trees. A number of physiological changes in cell tissues may be associated with physiological ageing in seeds.

They include:

- Loss of food reserves caused by respiration, *e.g.* decrease in proteins and non-reducing sugars, accompanied by increase in reducing sugars and free fatty acids
- Accumulation of toxic or growth-inhibiting by-products of respiration
- Loss of activity of enzyme systems
- Loss of ability of dried proteid molecules to recombine to form active protoplasmic molecules on subsequent rehydration
- Deterioration of semi-permeable cell membranes
- Lipid peroxidation, leading to production of free radicals which react with, and damage, other components in the cell
- Alterations to DNA in cell nucleus, causing genetic mutations as well as physiological damage.

It is still uncertain to what extent these various effects are the causes or only the symptoms of deterioration, but it has been suggested that the production of free radicals is the first effect of ageing and that damage to the several systems in the cells is the subsequent result of the release of free radicals.

Whatever the exact mechanism of seed deterioration, there is a consensus that, in orthodox seeds, loss of seed viability is largely governed by the rate of respiration. Any measures which reduce the rate of respiration without otherwise damaging the seed are likely to be effective in extending longevity during storage.

These are the control of oxygen, the control of moisture content and the control of temperature. Inrecalcitrant seeds the safe minimum levels of oxygen, moisture content and temperature, and hence of respiration, are all considerably higher than those for orthodox seeds but, provided levels are maintained above the safe minima for each species, it appears that longevity can be extended by keeping them as close to the minima as possible in order to avoid an excessively high respiration rate.

STORAGE ATMOSPHERE

The most obvious method of reducing the rate of aerobic respiration is to exclude oxygen from the atmosphere surrounding the seeds. This can be done by replacing oxygen by other gases such as CO_2 or nitrogen, or by using a partial or complete vacuum. In an example with lettuce cited by Roberts, seeds were stored in sealed containers at 6% MC and 18° C. After 3 years, seed stored in an atmosphere of pure oxygen had 8% viability, those in air 57%, those in nitrogen or argon or CO_2 78% and those in a vacuum 77%. The value of excluding oxygen during storage of dry orthodox seeds has also been demonstrated in Pinus radiata. Best results were obtained with a storage atmosphere of nitrogen, followed by CO_2, while vacuum and air both gave poorer results.

At the highest temperature used, 35°C, at which deterioration in viability was most rapid, the loss in final germination after 50 weeks' storage in sealed containers at 8% MC was 8% in nitrogen, 14% in CO_2, 21% in vacuum and 29% in air. The same ranking was obtained by comparing the speed of germination and the vigour of germinated seedlings (measured as dry weight 49 days after sowing). Although increases in seed longevity of this magnitude have been achieved experimentally, some of the methods are expensive to apply and the effects on seed life are less dramatic than the effects of differences in temperature and humidity. Exclusion of oxygen will prevent aerobic, but not anaerobic, respiration, whereas reduced MC and temperature will decrease the level of both. While systematic predictions have been made of seed longevity under a range of temperature and MC for several agricultural crops, similar quantitative predictions of the effect of oxygen levels on longevity are lacking.

One simple method which is recommended is to fill sealed containers as nearly full as possible. If there is only a small amount of air inside the container as compared with the volume occupied by the seeds, oxygen will be consumed and CO_2 produced. The resulting high CO_2/O_2 ratio is probably favourable for seed longevity in orthodox seeds.

Whereas the complete exclusion of oxygen from the storage atmosphere appears beneficial to most dry orthodox seeds, there is evidence that some oxygen is necessary for recalcitrant seeds. Seeds of Araucaria hunsteinii, which had an initial germination of 56%, had all died within a month if stored in pure nitrogen, in two months if stored in 1% oxygen, and in three months if stored in 5% oxygen, while germination was still 18% after four months' storage in 10% oxygen. Storage in a polythene bag of 25 micron thickness, which was ventilated (21% oxygen) periodically when opened for extraction of samples, gave results very similar to those in 10% oxygen. King and Roberts record a general consensus that adequate ventilation (*i.e.* adequate oxygen), is necessary for the successful storage of recalcitrant seeds at relatively high MC, as well as for storage of imbibed seeds of orthodox species.

SEED MOISTURE CONTENT

The relationships of seed moisture content on a wet weight or fresh weight basis to seed MC on a dry weight basis, and of the equilibrium moisture content of seeds to the relative humidity of the surrounding atmosphere, are important in seed processing. They are equally important in seed storage. In the first case manipulation of RH can effectively change MC of seeds to the optimum for storage, in the second case MC can be maintained at or near that optimum by maintaining a suitable RH in the atmosphere around and between the seeds.

Effect of MC

In orthodox seeds, moisture content is probably the most important single factor in determining seed longevity. Reduction in MC causes a reduction in respiration and thus slows down ageing of the seed and prolongs viability. Harrington, cited by Barner, has related MC to various processes within and around the seed as follows:

Table. Seed Moisture Content%

(wet weight)	
Above 45 – 60 %	Germination begins
Above 18 – 20 %	The seed may heat (due to a rapid rate of respiration and energy release)
Above 12 – 14 %	Fungus growth can occur
Below 8 – 9 %	Insect activity much reduced
4 – 8 %	Sealed storage is safe.

Prevention of fungal activity is more easily achieved by controlling MC than by controlling temperature. If MC and RH are high enough, fungal activity is possible between -8° C and +80° C and it is easier to keep MC below 12–14% (or RH to the equilibrium of around 65%) than to maintain sub-zero temperatures.

Within the range of 4 to 14% MC, Harrington has suggested a rule of thumb applicable to many agricultural species—the life of the seed is doubled for every 1% decrease in MC, Schönborn (1965) found a relationship of a similar order when measuring the respiration rate, expressed in terms of CO_2 production, of Picea abies. At 20° C and 20% MC the seed gave off 80 ml. CO_2 per hour per kg of seed; at 20° C and 5% MC, the rate of CO_2 production was reduced to 0.11 ml/hr/kg, a reduction of nearly a thousand times for a difference of 15 percentage points in MC.

An MC of 4 – 8% is considered safe for most orthodox species; 5% ± 1% is recommended for long-term storage for genetic conservation. Oily seeds will usually tolerate drying to a somewhat lower moisture content (calculated on the basis of total fresh weight) than will non-oily seeds. Drying below 4% can lead to damage or more rapid loss of viability in some species, although certain species can be dried to a considerably lower MC. Betula papyrifera was successfully stored at 0.6% MC without injury, while Schönborn (1965) succeeded in drying small samples of Picea abies, Pinus sylvestris, Pseudotsuga menziesii and Larix decidua down to 0% MC without any observable drop in germination after 6 months, compared with germination at the normally applied 6 – 8% MC. Drying in this case was done, not by exposing the seeds to high temperatures but by leading a current of dry air through the seeds at 20 °C. The same treatment killed Pinus strobus and Abies alba, and earlier attempts by Barton to store seeds of several species of Pinus and Picea at 0% MC also failed. Below about 2% MC desiccation injury becomes a strong possibility in many species. Drying to very low MC is also more costly than drying to the usual 4 – 8% and is likely to be used only in exceptional cases.

Some orthodox forest trees store best at appreciably higher MC. For seeds of Abies spp., an MC of 12–13% is recommended for storage of one to three years, but for longer periods this should be reduced to 7–9%. Species which benefit from storage at higher than average MC also need particular care in the timing and speed of drying.

Fluctuation in the moisture content of seed in storage due to open storage without humidity control or to frequent opening and resealing of sealed containers results in deterioration in the germinability of the seeds. In fact a steady MC slightly above the optimum is usually less harmful than one which fluctuates between the optimum and a higher moisture content.

Some cases have been reported in which the usual trend of decreasing seed longevity associated with increasing MC is reversed at or near the moisture content of fully imbibed seeds. If the species in question needs exposure to light in order to germinate, it is possible to store fully imbibed but ungerminated seeds for some time in the dark. Fraxinus americana was stored at 22 °C at varying MCs with the following results.

MC%	Germination (%) after storage periods indicated			
	1 month	**2 months**	**3 months**	**4 months**
6.0	98	92	96	94
9.5	94	88	76	4
18.6	81	22	0	0
Fully imbibed (in dark)	96	95	98	96

It has been postulated that imbibed seeds can repair damage to cell membranes, enzymes and DNA in the cell nucleus caused by free radicals in a way which is not possible for seeds at lower MC. Prolonged imbibed storage may, however, be difficult in practice, because of the need to maintain constant high moisture for imbibition and adequate oxygen without allowing the seeds to germinate or encouraging the multiplication of fungi and bacteria.

Moisture content is also important in recalcitrant seeds, but in this case the critical MC is the minimum to which it is allowable to dry the seeds rather than the maximum content for prolonged storage. For many of the large temperate hardwood seeds, moisture contents in the range of 25 – 79% are appropriate.

Storage should be carried out at close to the minimum safe MC, since the higher the MC the higher the respiration rate and the more rapid the loss of viability. Higher respiration rates release higher amounts of energy and there is a risk of overheating and death of the seed, unless great care is taken to provide adequate aeration. High MC also increases fungal activity and the spread of rot. Wang quotes results from two seed lots of Acer saccharinum; germinability of one lot stored at 58% MC and 1 – 2 °C dropped from 94% to 12% after 6 months' storage, while that of the other, stored at 45% MC at the same temperature, was still 78% after 16 months. Loss of viability in this species may be sudden. Tylkowski found that seeds stored in sealed bottles at 50–52% MC and -1° to -3°C had over 90% germination after 18 months but this dropped to nearly zero after 24 months.

Less research has been done on tropical recalcitrant species, but there is some evidence *e.g.* in Triplochiton, that viability may be significantly prolonged if the minimum MC for the species can be determined and if particular care is taken over the period of drying down to that MC. Trials of Shorea platyclados in Malaysia indicated that a gradual reduction of MC to 20 – 27%, followed by sealing in charcoal, sawdust or vermiculite at 15° – 22° C allowed storage for at least one month, compared with the week or so of natural viability. On the basis of experiments carried out on Shorea parvifolia and Dipterocarpus humeratus, Maury-Lechon *et al.* recommended reduction of MC to between one quarter and one half of the initial MC in freshly collected fruits. Although tropical recalcitrant seeds cannot yet be stored for more than short periods, there is a growing body of useful research on the problem. King and Roberts contains a good summary of achievements and possible approaches.

STORAGE TEMPERATURE

Temperature, like moisture content, is negatively correlated with seed longevity; the lower the temperature the lower the rate of respiration and thus the longer the life-span of the seed in storage. Harrington suggested another rule of thumb for agricultural seeds—between 50° C and 0° C, every 5 °C lowering of storage temperature doubles the life of the seed. For orthodox seeds, which can be dried to a low moisture content, still greater longevity can be assured by storage at sub-freezing temperatures. For long-term storage for genetic conservation of agricultural seeds, a temperature of -18° C has been recommended as the "preferred" standard for most species and –10 °C as an "acceptable" standard for those species known to have an intrinsic high viability. Much lower temperatures have been used with success on an experimental basis, *e.g.* in liquid helium at –269 °C, but the high cost of maintaining such low temperatures for a long period would outweigh any (as yet unproven) advantages in increased longevity.

Choice of storage temperature varies considerably according to species and the period for which the seed is to be stored. The lower the temperature that has to be maintained in a cold store, the higher the cost, and provision of subfreezing temperatures may be unnecessary if seed is to be stored for only a year or two for afforestation projects. Holmes and Buszewicz noted that, in a number of experiments on conifers, the superiority of sub-freezing temperatures became evident only after prolonged storage over periods of about 5 years or more. Sub-freezing temperatures (–18 °C) appear to prolong viability in tropical orthodox species such as Eucalyptus deglupta and Flindersia brayleyana.

Some seeds keep well at room temperature, *e.g.* many leguminous and rosaceous genera, Eucalyptus, Tilia and many other hard-seeded or stone fruits. Most species, however, only keep well for longer periods at lower temperatures. In 3 – 5 years' storage of most conifers and Alnusand Betula, the crucial temperature seems to lie at a maximum of +4 °C. The temperature should therefore be kept at 1 to 4 °C. For longer periods of storage, say 5 – 15 years, the temperature should be –4 to –10 °C. For Abies, however, a temperature of –4 °C is used for short storage periods and -10° to -20° C for longer periods.

Temperature and moisture factors are so interrelated that it is very difficult to separate them. Seeds at a relatively high moisture level can be stored for considerably longer periods at near freezing temperatures than at higher temperatures, while higher storage temperatures (30 °C) are less harmful when the moisture content of the seed is low. In short, it can be said that the critical moisture content lies at a higher level when storage temperatures are low than when they are intermediate or high, *i.e.* to some extent, a low temperature can compensate for a high moisture content, and vice versa.

It is, however, necessary to avoid any risk of freezing damage caused by ice formation in seeds of high MC. Roberts has suggested that 20% MC may be the critical upper limit for storage at 0 °C, 15% for –20 °C and 13% for -196 °C. If seeds are dried to 4–8% MC as commonly recommended for orthodox species, there should be no danger of freezing damage, even at temperatures well below zero. The equilibrium moisture content of many seeds at a given RH varies with temperature. Barton and Crocker (1948) have shown that, at a range of RH from 35% to 76%, the amount of water contained by seeds increased progressively as temperature decreased from 30° C to 10 °C. Species included Pinus as well as several agricultural crops. At low (35%) RH moisture absorption by dry seeds was approximately the same at 5 °C as at 10 °C, but at higher (55% and 76%) RH the seeds absorbed less moisture at 5 °C than at 10 °C, thus reversing the trend above 10°C. Change of equilibrium moisture content with change of temperature can be of importance in open storage. In sealed containers the effect is minimal because the final EMC is dominated by the initial MC of the seeds and not the moisture of the enclosed air. As with moisture content, repeated fluctuations in temperature lead to loss of viability. As far as possible, temperature should be maintained at a uniform level.

The effect of temperature on seed longevity of temperate recalcitrant species is similar to that of orthodox species—within certain limits the lower the temperature the longer the period of viability. Some tropical species are killed by temperatures above freezing *e.g.* some dipterocarps at <14° C, cacao at <10 °C and mango at < 3–6 °C. Seeds of Hopea helferi stored at 15 °C with high moisture content in unsealed polythene bags retained 98% germination after 37 days and 80% after 60 days.

Germination was much reduced if temperature was dropped to 10 °C or raised to 25°–28 °C. Shorea ovalis is another species which does not withstand low temperatures; it stores best at 21 °C. Shorea talura, in contrast, stores well at 4 °C and 40% MC (wet weight basis); after six months germination was reduced from an initial 95% to 69%. Other species of dipterocarp have a much shorter seed life.

For most temperate recalcitrant species the lower the temperature down to 0 °C the longer the safe storage period. Below-freezing temperatures, on the other hand, often kill recalcitrant seeds which need to be stored at a high moisture content. Some success has been achieved in storing Quercus seeds in the USA by maintaining them at 35 – 45% moisture content and -1° to + 3 °C temperature. Temperature can be critical because less than –1 °C will usually kill the seeds, while temperatures above 2° or 3 °C cause excessive germination. In Europe more northerly species of Quercus can be stored at slightly lower temperature –1° to –3 °C and 38 – 45% MC.

LIGHT

Light, particularly ultra-violet light, is reported to be harmful to seed, but very few studies have been made. Use of opaque metal containers would be preferable to glass jars or bottles for species which are affected by light. But light appears to be much less important than either moisture content or temperature.

SEED SHIPMENT

The benefits of exemplary seed collection, processing and storage methods may be largely lost if care is not taken over shipment from seed store to nursery. It is seed viability at the time of sowing, rather than at the time of despatch from the seed store, which determines the number of healthy plants produced from a particular seed lot. It is therefore essential to provide shipment methods which will ensure the minimum loss of viability in the interval between storage and sowing. The selection of appropriate packing material will depend on the characteristics of the species, the quantity to be shipped, the length of time in transit, the mode of transport and the temperature and moisture conditions to which the shipment will be exposed (Baldwin 1955).

High and fluctuating temperatures and adverse humidity are the chief causes of viability losses during shipment (Stein *et al.* 1974). These factors are identical with those that cause deterioration in freshly collected fruits between the collecting site and the processing depot, as described on pp. 83–84. However, seeds between storage and sowing should start with advantage of having had optimum conditions of temperature and moisture content during the storage period. In fact, maintenance of storage conditions during transit would be ideal, but is often not possible.

Provided that the initial moisture content of the seeds is correct, it can be easily maintained during transit by the use of sealed containers. In some cases the seeds can be despatched in the same containers in which they were stored. In others it may be advisable to transfer them from a large container in storage to a smaller container for despatch. Individual nurseries may require only a small quantity of a given seed lot. In addition, small and light packages are often less subject to mechanical damage in transit than large, heavy ones. Magini (1962) recommends separate packages of 1 – 20 kg but not larger. A variety of moisture-proof or moisture-resistant material is available, as described earlier in this chapter under storage containers. Polyethylene of 4–8 mil (100–200 microns) has the advantage of restricting moisture passage while allowing exchange of oxygen and CO_2.

Sealed containers are highly suitable for orthodox species, of which the seeds must be kept dry during transit. The addition of a desiccant such as silica gel may be a useful additional insurance if there is any risk that the seeds may absorb moisture while being transferred from storage container to shipment container. Seeds of recalcitrant species, on the other hand, are best left unsealed, since the effect of some loss of moisture is less harmful than that of the overheating which can occur as a result of rapid respiration in sealed bags at ambient temperatures. They should be well-mixed with pulverized sphagnum moss, ground peat, coconut fibre or sawdust, that has been moistened and squeezed dry. A mixture of equal weights of dry packing and water will give adequate moisture content to these materials. In the case of international transit, however, an inert non-organic substance such as moist vermiculite is likely to be more acceptable to quarantine authorities.

Sealed moisture-proof containers should always be used for long journeys, *e.g.* form one country to another, of orthodox species of short longevity, provided that the initial MC is correct. But if orthodox seeds are being forwarded soon after collection and without having been dried to the appropriate MC for storage, it is preferable to ship in bags permeable to air rather than to seal with excessively high MC. A number of species with resistant seedcoats or pericarps, such as Tectona and many leguminous species, are able to withstand prolonged periods in ambient conditions; cotton or paper bags or hessian sacks are perfectly suitable for these species.

Large, moist seeds can be sealed individually with paraffin wax or latex. In the method described by Baldwin (1955), paraffin wax is heated to 71°–77° C and seeds or nuts dipped for a few seconds in a screen-type container, which should be shaken vigorously during the immersion. The waxed seeds should be packed in soft material so that the wax is not scraped off during transit. At the time of sowing the wax must be partly scraped off to permit the entry of water.

Protection against high or rapidly fluctuating temperatures is more difficult, but care should be taken to avoid placing the seeds close to local hotspots such as radiators and hot pipes. For very sensitive seeds, temperature effects can be mitigated by the use of insulating material in the packaging.

Sub-zero temperatures do not usually affect dry seeds but may cause damage to recalcitrant seeds which must be kept moist. Premature germination is another risk which affects moist seeds. During storage, germination can be restricted by the use of low temperatures just above freezing, but the higher temperatures encountered during transit may induce germination in a substantial number of seeds. Seeds which are prone to germinate when held in moist packing may be treated with an inhibitor such as maleic hydrazide. No matter what type of seeds is being despatched, it is necessary to take precautions against mechanical damage to seeds and against losses due to damage to the containers in transit.

Double wrapping is often advisable, for example a sealed polythene bag should be placed inside a stout canvas bag. Stout drum cartons with sealed polythene or aluminium-foil containers inside provide an especially effective combination for seeds which need to be kept dry. If the inner bag is labelled, this is also an insurance against accidental defacement of the outer label. Clear labelling is essential and the consignee should be advised of despatch by means of an appropriate seed consignment note or seed issue form.

Stein et al. (1974) have provided a useful check list of helpful practices in seed shipment, reproduced hereunder:

1. Double wrap the seed. Enclose the seed container in a sturdy, preferably rigid, outer container.
2. Small or moderate size containers generally withstand shipment better than large containers.

3. Fill containers completely to minimize air content and jostling of seeds during shipment.
4. All packages should bear a good identifying label on the innermost covering and another one within the container.
5. For long distances, shipment of sensitive seeds by air is desirable.
6. Seed packages should permit ready opening and reclosing if destined for export to a country requiring fumigation. In addition a copy of the phytosanitary certificate should be readily available to quarantine authorities *e.g.* by sealing it in an envelope which is firmly attached to the outside of the package.

Seed storage facilities at nursery sites or district forest stations are inferior to those at the central seed store. Shipments should therefore be timed so that seeds can be sown with the minimum delay after receipt.

8

Disease Management in Seed Crops

DISEASE MANAGEMENT

Black rot management begins with the identification of potential disease sources and utilising an Integrated Pest Management (IPM) strategy including host resistance, planting disease free seed, avoiding spreading the disease and proper sanitation.

Sanitation is the main method that reduces, excludes or eliminates the initial sources of disease. General sanitation practices include crop rotation, disinfecting seed, rouging diseased plants, elimination of refuse piles and eradication of alternative hosts.

SEED TREATMENT

Seedborne inoculum significantly contributes to the spread of black rot bacteria. Growers should only plant tested certified seed < 1 infected seed in 30,000 or 0.003% contamination.

When the infection level of seed is not known or disease-free seed is not available, seed should be treated to eliminate the bacteria. Growers who purchase transplants should request proof the seedlings were grown from disease-free or treated seed. During transplanting, diseased seedlings should not be planted in the field.

Seed treatments do not always eliminate 100% of the bacteria on or in the seed, and may adversely affect seed germination and vigour. Soaking seeds in hot water at 50 °C for 25–30 min. is the most effective treatment for seedborne blackrot control. Weak seed, seed stored for several years and seed of certain crucifer crops; such as, cauliflower, kohlrabi, kale, rutabaga and summer turnip, may be damaged by hot water treatment; soak for 15 min. at 50 °C only.

The effect of the hot water seed treatments on every variety of each individual crucifer crop has not been investigated. Growers are encouraged to treat a small portion of seed and plant in pots to determine the effect of the seed treatment on germination and vigour, prior to treating the entire seed lot.

AVOID DISEASE SPREAD

Use new seed trays each year to avoid contaminating this year's crop with residual black rot bacteria from the previous year. If purchasing new trays each year is not economically feasible, used trays can be sterilised with steam, boiling water or chemical disinfectants to eliminate potential contamination. Destroy infected seed trays immediately to prevent disease spread to other seedling trays.

Avoid soaking crates or bundles of transplant seedlings in tubs of water before transplanting. The black rot bacteria can spread from diseased to healthy seedlings by infecting leaf scars and wounds on roots when soaked in water.

Black rot bacteria can contaminate the surface of clothing, equipment, tools and water sources. Reducing seeding rates and densities to promote good air circulation, facilitating the quick drying of plants, timing irrigation when plants will dry quickly and restricting field activities until later in the day when fields are dry will help reduce disease spread. Working in diseased fields last will also avoid disease spread from infected to non-infected fields. Wash and disinfect equipment before moving from one field to another.

FIELD SELECTION

Field selection is very important due to the distance the pathogen can spread. Whenever possible, select fields as far away from fields grown to crucifer crops the previous year. Select fields that are well drained and will not receive run-off water from areas or fields where crucifers have been grown previously.

Well drained, light soils are best for crucifer production because they can be worked early in the season and facilitate earlier planting of transplants. Planting early can help avoid disease because environmental conditions are usually not conducive for the development and spread of black rot bacteria.

SEED INFECTION MECHANISMS

The area of science that studies the relationship between pathogens and seeds is Seed Pathology. It does not only identify the pathogens, it also includes the role of the seed as source of inoculum, the survival of the pathogen and the actions taken to control the pathogens associated to it. It uses the knowledge of General Pathology, Microbiology and Seed Analysis. To obtain a perspective of seed borne diseases, seed borne microorganisms can be considered fewer than four classes.

The first consists of pathogens for which the seed is the main source of inoculum; when seed infection is controlled, the disease is controlled. An example would be lettuce mosaic virus. For these pathogens, the importance of seed-borne inoculum has long been recognized, and control practices have been developed. The second class consists of important pathogens in which the seed borne phase of the disease is of minor significance as a source of inoculum.

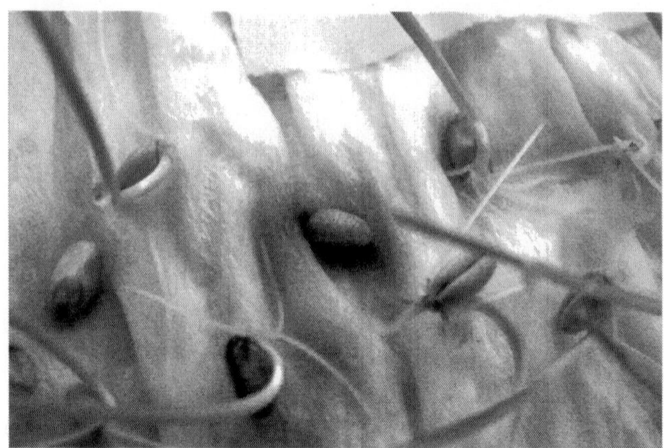

Fig. Infection of the Seed

Examples are those in which the crop residues in the field were the major source of inoculum. The third and largest group of seed borne microorganisms consists of those that have never been shown to cause disease as a result of their presence on seeds. The fourth class is a group of microorganisms that can infect the seed either in the field or in storage and reduce yield and seed quality. Examples of field fungi are *Diplodia, Fusarium, Cladosporium,* etc.

The storage fungi *Aspergillus* and *Penicillium* can invade most types of seeds under high-moisture storage conditions. The process of seed infection is influenced by the conditions under which the crop grows.

Between the facts that influence in the process of infection are: the host and its genotype, the pathogen and its genotype end environmental facts. An infected seed will not always an infected seed will be the cause of the infection of the infection in the plant which originates, so, it is to say that the infection caused from an infected seed is an exception, not something to happen generally.

There are two situations, Systemic infection of the seed and Contamination or Infestation of the seed.

Systemic Infection of the Seed

The establishment of a pathogen in any part of the seed is refereed as seed infection. It can be systemic, by the vascular system or plasmodesmata or directly by natural or artificial wounds. The same pathogen can infect the seed using one or more of these mechanisms.

For example: *Xanthomonas campestris* pv. *phaseoli* can infect seed through the vascular system, by natural openings (From the pod suture goes to the funiculus, then to the raphe and tegument, or it can also happen thought the micropyle).

Systemic Infection through Flowers, Fruits or Funiculus

Most of the systemic seed-borne bacteria and fungus reach and infect the embryo through the flower or from the peduncule of the fruit, via funiculus. Viruses go to the embryo from the systemically infected mother plant and the infected or contaminated pollen. They rarely reach the embryo during the formation of the seed or the embryo itself.

Examples of some infections that occur through the vascular system are: Some forma specialis of Fusarium oxysporum in pumpkin, pea and tomato; Plasmopara halstedii in sunflower; Septoria glycines in soy; Verticilium dahliae in spinach and sugar beet; Certain pathovares of Xanthomonas campestris in bean, cabbage, rice and sweet pepper; and Pseudomonas syringae pv. Lachymans in cucumber.

Penetration through the Stigma

During the infection, some pathogens follow the same path as the pollen grains do. Spores of some fungus reach the stigma and germinate, producing an hypha that reaches the ovary through the style, where they can stay as a dormant mycelium until seed germination.

For example: Ustilago nuda and U. triticci in barley and wheat, and Alternaria alternata in sweet pepper. Viruses can infect through infected pollen, the male gamete carries the virus, when joining the ovule it generates an infected embryo. If both, the male and female gamete are infected they can even produce an infected endosperm.

Penetration through the Wall of the Ovary

Some fungi, like *Ustilago nuda* and *U. triticci* penetrate through the wall of the ovary as a result of the germination of the Teliospores on the stigma or the wall of the ovary.

The pro-mycelium goes through the wall and other tissues until it reaches the embryo. In some other cases, penetration occurs through breakages on the testa, establishing itself in the endopleura or the endosperm. In fleshy fruits, like cucumber, melon, eggplant, tomato, sweet pepper and others, contamination can occur directly through the funiculus or in the tegument, during the process of seed formation.

Examples of this are Colletotrichum lagenarium in watermelon and Rhizoctonia solani, when it invades fleshy fruits, it is capable of infecting from the placenta and penetrate to the developing ovule or seeds that are still in its formation process and have not lignified its cover.

Penetration through Wounds and Natural Openings

Natural openings like the hilum and the micropyle or wounds generated during the thresh are spots where pathogens like *Xanthomonas campestris* pv. *phaseolicola* in bean and *Pseudomonas syringae* pv. *lachrymans* in cucumber.

SEED CONTAMINATION OR INFESTATION

Contamination or infestation refers to the passive relationship of a pathogens and seeds. The pathogen itself or parts of it can stick or can get mixed with the seeds during any of the processes during seed recollection: harvesting, extraction, thresh, selection and packing.

Pathogens that Stick to the Surface of the Seed

Pathogens that stick to seeds during harvest or postharvest do so by their spores (Clamidospores, Oospores, Teliospores, Uredospores), bacterial cells and in some cases, virions. Spores of the following fungi can be carried on seed coat surfaces: Alternaria brassicae and A. brassicicola in crucifers; A. longipes in tobacco; A. radicina in carrot; Ascochyta pinodella in pea; Drechslera sorokiniana and D. oryzae in rice; D. avenae in oats; Fusarium oxysporum f. sp. callistephi in China aster; Sclerotia of Rhizoctonia solani in eggplant, pepper, and tomato; Tilletia caries, T. foetida, T. contraversa, and Urocystis agropyri in wheat.

A number of bacteria, such as the following, contaminate seed surfaces: Corynebacterium flaccumfacjens pv. flaccumfacjens in bean, Pseudomonas syringae pv. phaseolicola in bean, P. syringae pv. lachrymans in cucumber, P. syringae pv. tomato in tomato, C. rnichiganense pv. michiganense in pepper, *Xanthomonas campestris* pv. *campestris* in cabbage Also some viruses as Tobacco mosaic virus, Tomato mosaic virus, Pepper Mosaic virus.

ACCOMPANYING CONTAMINATION

This type of contamination refers to physical mixing of the seed with pathogen's propagation organs like the sclerotium, nematode's galls, contaminated plant parts or soil particles containing pathogens.

Structures of the Pathogens

Some fungi produce resistance structures named sclerotiums that are made of compacted mycelium. Under certain conditions of temperature and humidity it can germinate, it is common that this occurs together with the seed, producing its infection, by direct penetration or by the spores generated by the fructification organs (Apothecium or Perithecium). These spores are carried by the wind and taken to susceptible tissues of the plant, generally, some parts of the flower. Sclerotiums have many shapes and during thresh can be easily included with the rest of the seed.

Some examples of this are Sclerotinia sclerotiorum in horticultural, grain and flower crops, Sclerotinia cepivorum in garlic and Claviceps purpurea in barley.

Mix with infected Plant Parts

Infected plant's parts and residues can carry fructifications or spores of fungi and bacteria. This situation is particularly common in cereal, forage and oily crop's seeds. Examples of this are Septoria nodorum in wheat, Puccinia malvacearum in hollyhock (malvarrosa), Colletotrichum trifolii in clover, Erwinia carotovora pv. carotovora in tobacco, Pseudomonas syringae pv. phaseolicola in bean and Sclerotinia sclerotiorum in sunflower, soy and peanut.

Soil

Seeds can be mixed with contaminated soil and, carry micro-sclerotiums of *Macrophomina phaseolina* in dirty seeds of kidney bean, Verticillium oxysporum f.sp. phaseoli in cotton seeds, Plasmodiophora brassicae in turnip and Fusarium oxysporum f. sp. phaseoli in bean.

SEED TRANSMITION

We saw how the seed gets infected or infested. Now we will see how these seeds can produce an ill plant. This is to say, how the inoculum goes from the seed to the plant. We call seed-borne pathogens only those that can produce an infection. Because they have to be distinguished from those that can associate to the seed but do not produce an infection, these ones are called pathogens not transmitted by seed.

The virus that causes curly top in sugar beet can be present in the perisperm but does not cause an infection in the plant. In general terms, infection can be classified into systemic and non systemic. It is systemic when the pathogen introduces itself to the plant when the seed germinates, and develops with it. Non systemic infection occurs when there is a localized infection caused by the pathogen in the seedling at the stage of pre or post emergency, in this situation there are no systemic symptoms.

Systemic Transmition

This type of infection can be produced by pathogens that are carried with the seed in various parts, like the embryo, endosperm or episperm, or by contamination of the outer portion of this one.

Infection of the Embryo

When the seed germinates, if the embryo is infected, the pathogen initiates its growth together with the plant. Symptoms can show up during different stages of development. In the case of *Ustilago triticci*, the mycelium grows together with the plant and expresses symptoms only at the flowering stage, which are that all the tissues of the spike, exept the rachis, are replaced by spores.

Some pathovares of *Xanthomonas campestris* that infect cabbage, bean or sweet pepper move between the cells of the host until they reach the vascular system and produce symptoms in leaves or stems. Most of seed-borne viruses persist inside the embryo. Its multiplication and movement accompanies the plant during its development, and there can be symptoms at any stage, from the formation of first leaves until flowering or fructification.

Non Embryo Infection

Infection of the episperm (testa and endopleura) occasionally conducts to a systemic infection. Some bacteria, like Corynebacterium michiganense pv. michiganense in tomato and Xanthomonas campestris pv. campestris in cabbage, penetrate through stomata of cotyledons, and from there, reach the vascular system, initiating the systemic infection.

Episperm Contamination

In some few cases this type of contamination conducts to a systemic infection. Most of these exceptions are fungi that are highly specialized in their pathogenesis and produce in cereals the so called smuts, rusts or mildews.

In these cases, generally, spores are carried outside the seed, they germinate, penetrate the coleoptile, and start a systemic infection. This can occur directly o by a more complex system of haploid hypha fusion in genera like Tilletia, Ustilago, Uromyces, Sclerospora, Pernospora and Puccinia.

Non Systemic Transmition

Non systemic infection is very common, and in the same way as systemic infection, it can come from an infection, outside contamination or by pathogens mixed with the seeds.

Infection of the Embryo

This case is restricted to some pathogenic fungi that maintain itself in the embryo or the episperm as hypha inside the seed. Primary infection starts as injuries in the cotyledons or primary leaves, stems or petiole. Fungi fructifications (Pycnidium, Acervulus) can develop on these organs, under certain favorable conditions (temperature and humidity).

These fructifications produce spores that, with the action of water and wind, disperse the disease to other parts of the plant and other plants. This occurs in *Ascochyta pisi* in pea, *Colletotrichum lindemuthianum* in bean and *C. truncatum* in soy.

Infection of the Episperm

Generally, seeds in which the episperm is infected, do not geminate, or germinate and contaminate the soil. Rarely produce a systemic infection, but can infect the seedling from the outside. Over the injuries, new inoculum is produced; this one infects other parts of the plant or other plants. *Septoria nodorum* in wheat under high humidity conditions forms Picnidium in the coleoptile, whose spores disseminate the disease to other plants.

Contamination of the Episperm

Contamination outside the seed's testa can produce healthy seedlings, but the inoculum infects the soil and from there it can cause infections at more developed stages of the plant. Wheat and rice grains that are contaminated with Teliospores of Neovossia indica in wheat and N. torrida in rice, produce Esporidisporas in the soil, that can be carried by the wind and infect flowers, originating the smuts.

Accompanying Contamination

The mixture of seeds with sclerotiums of the fungi or contaminated soil particles produces non systemic infections, at any stage of the plant growth and development. It is common for *Sclerotinia sclerotiorum* to produce mycelium from the sclerotiums. These mycelium can directly infect the plant or produce fructifications that produce spores that can be carried by the wind and infect flowers of different species, like sunflower, soy, peanut, etc.

Prevention of Seedborne Diseases

Prevention of seed-borne diseases is based in two main facts. On one hand there is the usage of material free from pathogens and contaminants that could originate a disease and on the other there are the treatments to eliminate the inoculum in the seed or propagule. When using material free from pathogens it is important to have adequate pathogen detection methods, this is to say, sensible and specific. No production of health quality seed can be done without valid methods to detect the possibility of contamination. Some treatments can be necessary to eliminate the pathogens from the seeds, or at leafs, maintain an economically viable balance.

Health Testing of the Seed

One of the most relevant aspects about controlled health quality seed is refereed to the detection methods used on every case. ISTA (International Seed Testing Association) appeared in 1924, by was only interested in seed's gene purity and germination aspects. But detection methods were applied.

According to the criterion of each laboratory, so their results were not comparative. So, in 1957, PDC established a comparative health testing programme and standardize applied methods. The complexity for comparison of these methods made a considerable step back on the publication of working sheets (protocols). ISTA approved methods became Rules. By 1999 only 64 protocols were being compared, and only 14 of them were accepted as Rules.

As the solutions proposed by ISTA did not satisfy the needs of international seed industry, this industry, in 1994, organized The International Seed Health Initiative for Vegetables (ISHI-Veg) chartered by the vegetable seed industries in The Netherlands and France.

The Initiative was soon joined by the seed companies in the United States, Israel and Japan. This group represents the production of over 75% of the worlds vegetable seed supply. ISHI published many protocols, organized by crop. For horticultural crops, the ISHI-VEG, the Seed Health Testing Methods.

ISHI-Veg Status

- An ISHI Validated Reference Method is a method that has been through the ISTA/ISHI comparative testing process and is being published as ISTA wn ISHI Reference Method is a method that has been through ISHI comparative testing and is under review by the ISTA/ISHI reviewers for publication as ISTA Working Sheet.
- An ISHI Accepted Method is a method that is commonly used by the seed industry for determining seed health. The method has been published in a journal, as a working sheet and/or is publicly available for comparative testing and commonly in use by the seed industry. The method is in comparative testing or aspects of the test are being tested for inclusion or deletion from the method.
- An ISHI Reviewed Method is a method that is publicly available or published, documented and prioritized by the ITG, but not subjected to a comparative test at this time. The method usually is accepted by other agencies or groups.

In 1995, during the second ISTA-PDC symposium it was decided that a Joint ISTA/ISHI Guidelines for Comparative Testing of Methods for Detection of Seed Borne Pathogens would be edited.

Through this combined effort many laboratories, organized under this alignments, the ISTA developed the Manual of Seed Health Testing Methods that has two sections:

- Validated Test Methods
- Peer Reviewed Methods.

The transition process from working protocol to ruled method is long and laborious, so working protocols can be considered as adequate methods for seed analysis. Nowadays, it is common that technological advances may become obsolete even before they are published.

The protocol to analyse *Xanthomonas campesrtis* pv.*campestris*, which causes black rot in crucifers, was published in 1982, there, it was recommended one minute to extract the bacteria from the seeds to starch culture medium. 2 years later, this time was raised to 2 hours, and other specific culture mediums. In 1987 the protocol was replaced, again in 1996 and at present time again it is under revision.

Minimum requirements to Produce Certified Seed

The Organization for Economic Cooperation and Development (OECD) establishes steps to be taken in order to produce basic and certified seed. Between these steps it is emphasized that generations from the mother material to the basic material are strictly limited in number. The number of certified seed generations, after the basic seed, varies from crop to crop and technical conditions. In both, sexually and vegetative reproduced species, the schemes include similar steps. It generally starts with selection of a plant with genetic and health advantages. If health condition was not satisfying, certain actions should be taken in order to release it from considered pathogens. This Mother Plant is maintained under strict special security conditions in order to maintain healthiness (isolated glasshouses or chambers).

In the Certification Schemes two particularities abut seed borne pathogens should be considered, those transmitted exclusively with the seed and those that can also be transmitted with contaminated soil and plant rests. Diseases transmitted exclusively with the seed can be controlled by using clean healthy seed.

This seed can be obtained by Seed Certification Programmes or by treatments that are really effective, like *in vitro* culture, thermotherapy, chemotherapy or a combined action of these, for systemic pathogens.

Tolerance levels in certified seed depend on the multiplication speed and dispersion of the inoculum of the pathogen of the considered disease. If a small amount of inoculum may be epidemic, like Anthracnose and the bean bacterial blight, if levels above zero chemical control may not be enough to stop the disease.

But pathogens whose inoculum multiplies slowly generation after generation, the disease may be evident only several generations later, so, if contamination levels are above zero may be accepted in adequate control methods are applied to the seed. This is what happens in wheat, with the Loose Smut, continuous treatments diminish inoculum levels, and make it perfectly tolerable.

New Techniques for Seed Health Analysis

Pathogen detection methods have suffered great changes since 1957. Back then it was mainly targeted to fungi, and based exclusively on incubation and identification, or the cultivation of these seeds and counting the percentage of unhealthy seedlings. These analyses were very laborious and required a high infection level in order to be detected.

Nowadays, technological advances have allowed detecting even low infection levels, evaluating 1,000 to 10,000 seeds at once. Specificity of culture mediums, usage of antibiotics and other products has allowed isolating the pathogen from the seed.

Together with this, the use of reactives as mono or polyclonal antiserums and molecular markers is also available. The Lettuce Mosaic Virus (LMV), back in 1950 the symptoms were observed over many thousands of seedlings from the seed that were going to be analyzed.

Later, technical advances determined that 30,000 seedlings were needed, and even later, in 1983, an alternative to the *Chenopodium quinoa* test was discovered, the ELISA test.

Watermelon Fruit Blotch caused by bacteria is a severe problem in glasshouse winter production. It was determined that a single contaminated seed in 9,000 of them was enough too widespread the bacteria to the rest of the plants. That is the reason why screening was done over 10,000 seedlings. Nowadays PCR technique is used.

DISEASES OF ASPARAGUS

Three diseases caused by fungi affect asparagus in Ontario. Fusarium root and crown rot is caused by a soil-borne fungus, while asparagus rust and Botrytis blight are caused by fungi whose spores are primarily carried about by wind. The life cycles and characteristics of these fungi will be discussed in this Factsheet to develop an understanding of how they may be recognized and controlled.

FUSARIUM ROOT AND CROWN ROT

Description

The major disease problem of asparagus is caused by two species of fungi called *Fusarium*. *Fusariummoniliforme* causes decay of storage roots, stems and crowns. This fungal species is present in all agricultural soils and infects corn, grasses, and other monocotyledonous plants as well as asparagus. *Fusarium oxysporum* f. sp. *asparagi* causes root rot and seedling blight, and may also plug the water conducting vessels, causing wilting of spears and fern.

These species of soil-dwelling fungi are very prolific, long-lived, and capable of growing as saprophytes, feeding on decaying asparagus residues and soil organic matter. They colonize old roots and crowns, invading directly through root tips or through wounds from implements, cutting tools, or insect feeding. Asparagus plants which are under stress are more susceptible to infection than those which are growing vigorously.

Affected spears may shrivel and rot in spring before or after emergence. Infected crowns have hollow, rotted feeder and storage roots. When crown and stem tissue is sliced open, a reddish-brown discoloration is visible. Symptoms on fern includes stunting, yellow to brown discoloration of one or more stalks per crown, and fewer stalks per crown. Affected crowns decline in vigour and die. Such damage is scattered throughout the field and increases until the stand is too sparse to harvest economically.

Fig. Yellow Discoloration of Asparagus Fern Infected by *Fusarium*.

Fig. Cross-section of Asparagus Crown Infected by *Fusarium*.

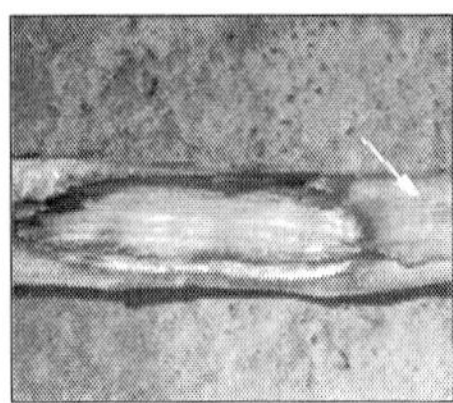

Fig. Cross-section of Asparagus Spear Infected by *Fusarium*.

CONTROL

Because *Fusarium* species are present in most agricultural soils, the diseases are almost impossible to avoid. Control is therefore directed at minimizing infection early in the life of crowns, and at maintaining a vigorous, long-lived asparagus stand by careful management.

Suggested management practices include:

1. Avoid replanting in land which previously grew asparagus. *Fusarium* builds up to extremely high levels during the long period of asparagus culture in the field, and survives many years in that soil even after the crop is removed.
2. Soil fumigation with a suitable material will reduce infection in crown nurseries or direct-seeded asparagus.
3. Use treated seed. Untreated seed should be disinfested with 1 part sodium hypochlorite bleach in 5 parts water for two hours. Rinse seed in fresh water and dry before planting.
4. Use only vigorous, one year old crowns and encourage proper handling procedures for transplanting. Weak crowns are highly susceptible to infection.
5. Minimize stress in young and established plantings. Stresses which weaken asparagus crowns and thus promote *Fusarium* diseases are:
   - Overextending the harvest season, or picking before crowns are well established;
   - Insect and disease damage;
   - Weed competition;
   - Mechanical wounding from tillage equipment, cutting knives, etc.;

- Poor soil drainage;
- Acidic soil pH;
- Drought;
- Injury from misapplication of pesticides, fertilizers, and soil ammendments;
- Low fertility; and
- Soil compaction.

6. Use resistant varieties. At present, no varieties recommended in Ontario are resistant. However, lines and varieties being developed in the University of Guelph breeding programme are being screened for *Fusarium* resistance.

ASPARAGUS RUST

Description and Life Cycle

Puccinia asparagi, which causes asparagus rust, has a complicated life cycle consisting of several stages, all of which occur on asparagus. Some members of the onion family, such as cooking onions and chives, are also susceptible. There are no alternate hosts such as is common with other *Puccinia* rusts on wheat and oats. The fungus overwinters as teliospores on asparagus debris. Teliospores germinate in spring, producing small, basidiospores which are blown onto emerging spears and cause infection. Later, from April until July, small upraised, light-green, oval lesions (patches) called aecia occur on the lower portion of the infected fern stalks. As the aecial lesions turn creamy orange in colour, aeciospores are released and re-infect asparagus fern during several hours of continuous leaf wetness. Twelve to fourteen days after re-infection, upraised tan blisters called uredia appear on asparagus stalks and foliage. Uredia break open to expose masses of rusty-coloured spores called uredia spores, after which the disease is named. Urediospores repeatedly re-infect asparagus from June until September. Warm weather with heavy dew, fog, or light rainfall enhances rust development. Late in summer, telia develop, producing black teliospores, completing the yearly life cycle.

DAMAGE

The fungus develops in the tissue of asparagus fern and drains the plant of vital nutrients. The foliage then dries out and falls prematurely, further reducing the production and storage of food for the following crop. Successive years of infection by rust will weaken asparagus crowns.

CONTROL

1. Cultural control involves breaking the continuity of the rust life cycle. Inoculum from overwintered teliospores in fern residue may be spread from wild or neighbouring asparagus, uncut spears, young or seedling plantations, or volunteer seedlings in the asparagus crop. It is important to:

- Destroy wild and volunteer asparagus;
- Clean cut all fields during harvest; and
- Locate nurseries and young plantings remotely from established fields if possible.

2. Use resistant asparagus varieties. Viking varieties have some degree of tolerance to rust, but will become infected under conditions of favourable environment and abundant inoculum. Breeding and variety trials at the University of Guelph are including rust resistance as an important criterion in the development of new varieties.
3. Control with fungicides is presently limited to the chemical zineb. Spraying should begin when the aecial or initial uredial stages are evident on the fern - in May on young plantings, in late June in harvested plantings. Because zineb is a protectant fungicide, adequate coverage *before* rust appears and throughout the season is essential to achieve control.

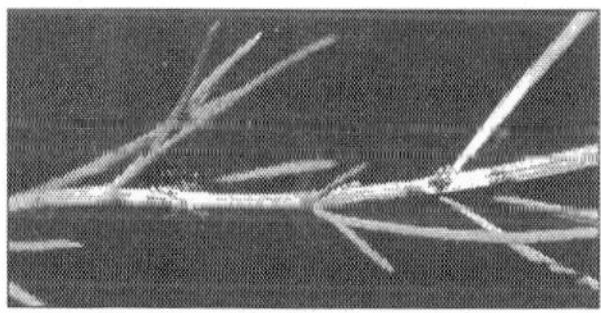

Fig. *Botrytis* Lesions on Asparagus Fern Branches. Note the Tufts of Fruiting Bodies Produced by the Fungus in these Lesions.

Fig. Extensive Blighting of the Lower fern Canopy Caused by *Botrytis*.

Fig. Aecial Lesion of Asparagus Rust. Note Oval to Elliptical Shaped, Upraised Lesion.

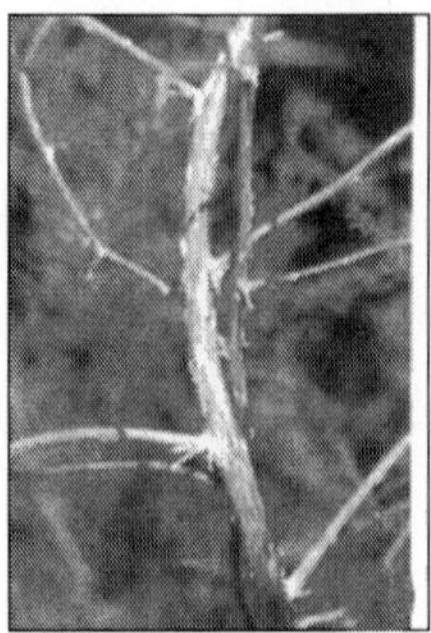

Fig. Uredial Lesions of Asparagus Rust. Note Tan Blisters which have Opened, Exposing the Rusty Coloured Spore Mass.

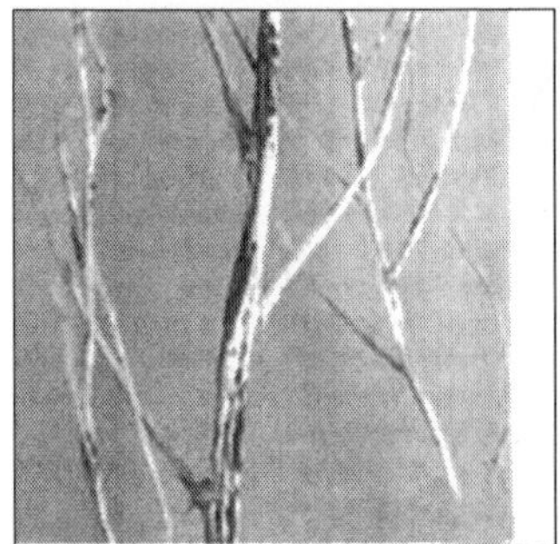

Fig. Telial Lesions of Asparagus Rust. Note Black Teliospores on Overwintering Asparagus Fern.

- Botrytis Blight
- Description and Life Cycle

This disease, caused by the fungus *Botrytis cinerea,* occurs during summer, causing browning of the lower fern canopy (Figure). *Botrytis* progresses most rapidly during hot, moist weather when the fern does not dry adequately. This fungus attacks many other crops, such as strawberries and other small fruit, potatoes, beans, and other vegetables, and many ornamentals. The disease begins on senescing (dying) flowers or injured fern. Botrytis spores are spread by wind and rain within the dense fern canopy. Individual lesions are tan with dark brown borders, often surrounded by a fellow halo (Figure). When wet weather persists, newly-emerged spears may be completely blighted, turning brown or black, and often covered with grey, fuzzy, spore-bearing fungal growth.

Control

Zineb will also give some control of *Botrytis* when used regularly for rust control.

Viruses

Though several viruses have been identified in asparagus in Europe and Western USA, they have not yet been reported in Ontario.

Pesticide Application

As fern growth becomes dense, more water and higher rates of pesticides must be used to obtain adequate spray coverage. Good coverage is essential, especially for protective fungicides like zineb. Drop nozzles are recommended. Up-to-date pesticide recommendations are available for commercial growers in OMAFRA Publication 363, *Vegetable Production Recommendations*. Homeowners are referred to OMAFRA Publication 64, *Insect and Disease Control in the Home Garden.*

In conclusion, as a perennial crop, asparagus is a long-term investment. It will pay to monitor pests regularly throughout the harvest and growing seasons. Only then can problems be detected and controlled before they cause long-term economic damage.

9

Seed Storage in Fruit and Vegetable Crops

PROCESSED OF FRUIT AND VEGETABLES

Practically any fruit and vegetable can be processed, but some important factors which determine whether it is worthwhile are:

- The demand for a particular fruit or vegetable in the processed form;
- The quality of the raw material, *i.e.* whether it can withstand processing;
- Regular supplies of the raw material.

For example, a particular variety of fruit which may be excellent to eat fresh is not necessarily good for processing. Processing requires frequent handling, high temperature and pressure.

Many of the ordinary table varieties of tomatoes, for instance, are not suitable for making paste or other processed products. A particular mango or pineapple may be very tasty eaten fresh, but when it goes to the processing centre it may fail to stand up to the processing requirements due to variations in its quality, size, maturity, variety and so on.

Even when a variety can be processed, it is not suitable unless large and regular supplies are made available. An important processing centre or a factory cannot be planned just to rely on seasonal gluts; although it can take care of the gluts it will not run economically unless regular supplies are guaranteed. To operate a fruit and vegetable processing centre efficiently it is of utmost importance to pre-organise growth, collection and transport of suitable raw material, either on the nucleus farm basis or using outgrowers.

PROCESSING PLANNING

The secret of a well planned fruit and vegetable processing centre is that it must be designed to operate for as many months of the year as possible. This means the facilities, the buildings, the material handling and the equipment itself must be inter-linked and coordinated properly to allow as many products as possible to be handled at the same time, and yet the equipment must be versatile enough to be able to handle many products without major alterations.

A typical processing centre or factory should process four or five types of fruits harvested at different times of the year and two or three vegetables. This processing unit must also be capable of handling dried/dehydrated finished products, juices, pickles, tomato juice, ketchup and paste, jams, jellies and marmalades, semi-processed fruit products. Advanced planning is necessary to process a large range of products in varied weather and temperature conditions, each requiring a special set of manufacturing and packaging formulae. The end result of the efforts should be a well-managed processing unit with lower initial investment. A unit which is sensibly laid out and where one requirement co-relates to another, with a sound costing analysis, leads to an integrated operation. Instead of over-sophisticated machinery, a sensible simple processing unit may be required when planned production is not very large and is geared mainly to meet the demand of the domestic market.

Location

The basic objective is to choose the location which minimises the average production cost, including transport and handling. It is an advantage, all other things being equal, to locate a processing unit near the fresh raw material supply. It is a necessity for proper handling of the perishable raw materials, it allows the processing unit to allow the product to reach its best stage of maturation and lessens injury from handling and deterioration from changes during long transportation after harvesting.

An adequate supply of good water, availability of manpower, proximity to rail or road transport facilities and adequate markets are other important requirements.

Small-Scale Processing systems

- Small-Scale Processing. This is done by small-scale farmers for personal subsistence or for sale in nearby markets. In this system, processing requires little investment: however, it is time consuming and tedious. Until recently, small-scale processing satisfied the needs of rural and urban populations. However, with the rising rates of population and urbanisation growth and their more diversified food demands, there is need for more processed and diversified types of food.
- Intermediate-Scale Processing. In this scale of processing, a group of small-scale processors pool their resources. This can also be done by individuals. Processing is based on the technology used by small-scale processors with differences in the type and capacity of equipment used. The raw materials are usually grown by the processors themselves or are purchased on contract from other farmers. These operations are usually located on the production site of in order to assure raw materials availability and reduce cost of

transport. This system of processing can provide quantities of processed products to urban areas.

- Large-Scale Processing. Processing in this system is highly mechanised and requires a substantial supply of raw materials for economical operation. This system requires a large capital investment and high technical and managerial skills. Because of the high demand for foods in recent years many large-scale factories were established in developing countries. Some succeeded, but the majority failed, especially in West Africa. Most of the failures were related to high labour inputs and relatively high cost, lack of managerial skills, high cost and supply instability of raw materials and changing governmental policies. Perhaps the most important reason for failure was lack of adequate quantity and regularity of raw material supply to factories. Despite the failure of these commercial operations, they should be able to succeed with better planning and management, along with the undertaking of more in-depth feasibility studies.

It can be concluded that all three types of processing systems have a place in developing countries to complement crop production to meet food demand. Historically, however, small and intermediate scale processing proved to be more successful than large-scale processing in developing countries.

PROCESS OF SEED GERMINATION

Seed germination is a process by which a seed embryo develops into a seedling. It involves the reactivation of the metabolic pathways that lead to growth and the emergence of the radicle or seed root and plumule or shoot. The emergence of the seedling above the soil surface is the next phase of the plants growth and is called seedling establi-shment.

Three fundamental conditions must exist before germination can occur:

1. The embryo must be alive, called seed viability.
2. Any dormancy requirements that prevent germination must be overcome.
3. The proper environmental conditions must exist for germination.

Seed viability is the ability of the embryo to germinate and is affected by a number of different conditions. Some plants do not produce seeds that have functional complete embryos or the seed may have no embryo at all, often called empty seeds. Predators and pathogens can damage or kill the seed while it is still in the fruit or after it is dispersed. Environmental conditions like flooding or heat can kill the seed before or during germination. The age of the seed affects its health and germination ability: since the seed has a living embryo, over time cells die and cannot be replaced. Some seeds can live for a long time before germination, while others can only survive for a short period after dispersal before they die. Seed vigour is a measure of the quality of seed, and involves the viability of the seed, the germination percentage, germination rate and the strength of the seedlings produced. The germination percentage is simply the proportion of seeds that germinate from all seeds subject to the right conditions for growth. The germination rate is the length of time it takes for the seeds to germinate. Germination percentages and rates are affected by seed viability, dormancy and environmental effects that impact on the seed and seedling. In agriculture and horticulture quality seeds have high viability, measured by germination percentage plus the rate of germination.

This is given as a per cent of germination over a certain amount of time, 90% germination in 20 days, for example. 'Dormancy' is covered above; many plants produce seeds with varying degrees of dormancy, and different seeds from the same fruit can have different degrees of dormancy. It's possible to have seeds with no dormancy if they are dispersed right away and do not dry (if the seeds dry they go into physiological dormancy). There is great variation amongst plants and a dormant seed is still a viable seed even though the germination rate might be very low. Environmental conditions effecting seed germination include; water, oxygen, temperature and light. Three distinct phases of seed germination occur: water imbibition; lag phase; and radicle emergence. In order for the seed coat to split, the embryo must imbibe (soak up water), which causes it to swell, splitting the seed coat. However, the nature of the seed coat determines how rapidly water can penetrate and subsequently initiate germination.

The rate of imbibition is dependent on the permeability of the seed coat, amount of water in the environment and the area of contact the seed has to the source of water. For some seeds, imbibing too much water too quickly can kill the seed. For some seeds, once water is imbibed the germination process cannot be stopped, and drying then becomes fatal. Other seeds can imbibe and lose water a few times without causing ill effects, but drying can cause secondary dormancy.

Inducing Germination

A number of different strategies are used by gardeners and horticulturists to break seed dormancy. Scarification which allows water and gases to penetrate into the seed, include methods that physically break the hard seed coats or soften them by chemicals.

Means of scarification include soaking in hot water or poking holes in the seed with a pin or rubbing them on sandpaper or cracking with a press or hammer. Soaking the seeds in solvents or acids is also effective for many seeds. Sometimes fruits are harvested while the seeds are still immature and the seed coat is not fully developed and sown right away before the seed coat become impermeable. Under natural conditions seed coats are worn down by rodents chewing on the seed, the seeds rubbing against rocks (seeds are moved by the wind or water currents), by undergoing freezing and thawing of surface water, or passing through an animal's digestive tract. In the latter case, the seed coat protects the seed from digestion, while often weakening the seed coat such that the embryo is ready to sprout when it gets deposited (along with a bit of fertilizer) far from the parent plant.

Microorganisms are often effective in breaking down hard seed coats and are sometimes used by people as a treatment, the seeds are stored in a moist warm sandy medium for several months under non–sterile conditions. Stratification also called moist–chilling is a method to break down physiological dormancy and involves the addition of moisture to the seeds so they imbibe water and then the seeds are subject to a period of moist chilling to after–ripen the embryo.

Sowing outside in late summer and fall and allowing to overwinter outside under cool conditions is an effective way to stratify seeds, some seeds respond more favourably to periods of oscillating temperatures which are part of the natural environment. Leaching or the soaking in water removes chemical inhibitors in some seeds that prevent germination. Rain and melting snow naturally accomplish this task. For seeds planted in gardens, running water is best—if soaked in a container, 12 to 24 hours of soaking is sufficient. Soaking longer, especially in stagnant water that is not changed, can result in oxygen starvation and seed death. Seeds with hard seed coats can be soaked in hot water to break open the impermeable cell layers that prevent water intake.

Other methods used to assist in the germination of seeds that have dormancy include prechilling, predrying, daily alternation of temperature, light exposure, potassium nitrate, the use of plant growth regulators like gibberellins, cytokinins, ethylene, thiourea, sodium hypochlorite plus others. Some seeds germinate best after a fire, for some seeds fire cracks hard seed coats while in other seeds chemical dormancy is broken in reaction to the presence of smoke, liquid smoke is often used by gardeners to assist in the germination of these species.

DEVELOPMENT OF GERMINATION

When the pod of the bean is developing, the embryo in the seed is being fed by the parent and visibly grows until ripeness is attained. The young plant then assumes a dormant or resting state within the seed without showing any signs of life. Under certain conditions, however, the plantlet begins to wake up, and soon escapes from its protective coat to lead a separate and independent life.

This awakening from a resting condition to a state of active growth is called germination and is dependent upon an adequate supply of:

- Water
- Heat
- Air or oxygen.

It is also essential, of course, that the plantlet in the seed must be alive. The exact nature of the dormant state of seeds is not understood, but in old seeds and those which are gathered in an immature condition or badly stored the embryo is often weakened or actually dead; in the latter case no germination is possible.

The exact length of time which seeds may be kept before death of the embryo takes place has never been satisfactorily determined; it varies with the species of the seed, its ripeness and composition, and also with the method of storage. In the case of most farm and garden seeds kept in the ordinary way, few of them are found capable of growth after ten years, and a large number die in two or three years.

For present purposes it will suffice merely to mention that age is a determining factor in the germination of seeds. Water is necessary is well known, as beans may be kept indefinitely in a sack or drawer at various temperatures and with access to air without germination taking place.

When placed in moist ground, or between damp blotting-paper, they absorb water very readily. This is most easily observed when beans are soaked for twelve hours in a dish containing water. The water is transmitted through all parts of the coat, but much more quickly and easily through the micropyle and the line of softer material which runs the whole length of the centre of the hilum.

It is rapidly brought into contact with the part of the embryo which grows first, namely, the radicle. The soft spongy thicker part of the inside of the testa lying beneath the hilum stores up a considerable amount of water for the benefit of the developing plant, and the whole of the embryo and the seed-coat absorb water and become softer and larger in consequence; it is only after this swelling has happened that a bean begins to show any signs of germination.

The need of an adequate temperature for germination is a matter of common knowledge among those accustomed to sow seeds. If soaked beans are placed in the ground in midwinter they show little or no signs of waking from their dormant condition, yet when placed under a glass on damp blotting-paper indoors, the radicle makes its exit from the seed in a few days.

Seeds differ in the temperature which is necessary to induce them to germinate, the embryos in some commence to extend their radicles and push their way through the seed coat even if just kept above freezing point; others require a temperature of 9° or 10°C to start growth.

If attempts are made to grow beans at 45°C it will be found to be too hot, and they make little or no progress. Between this high temperature at which growth appears impossible, and the freezing point where the development of the embryo of the bean is also suspended, there is a temperature at which the embryo makes the most rapid progress, and emerges from the seed-coat in the shortest possible time; this most favourable temperature is about 28°C, both above it and below it the germination of the bean is retarded. The supply of fresh air is also an essential condition for growth of the young plant from the bean seed, but the evidence for its need is not so manifest or so generally recognised as the necessity for moisture and warmth.

It will be found, however, when beans are placed in a flask or bottle containing carbon dioxide or hydrogen gas they refuse to germinate, even when they are supplied with a proper amount of water. The peculiar extension or growth of the parts of the interior of the bean seed, and the fact that a suitable supply of water, air and heat is necessary for the manifestation of these changes, suggests to us that we are dealing with a living structure.

This becomes all the more«apparent when we observe that the oxygen of the air is absorbed, and in its place carbon dioxide is given off into the surrounding air, for this is what happens in the breathing of a living animal. Carbon dioxide is produced when beans germinate.

Place twenty soaked beans in a wide-mouthed bottle, and cork them up after showing that a match burns freely in the bottle. Leave them in a warm place for twenty-four hours, and try if a match will now burn in the bottle. The carbon dioxide gas can be poured out into a beaker containing lime water; on shaking, its presence is proved by the lime water becoming 1 milky ' owing to the precipitation of carbonate of lime. The particular use of the water, heat and air to the plant we cannot at present discuss.

Without water the embryo would have little chance of becoming free from the tough and hard seed-coat surrounding it; water softens the latter, and makes it more easily torn by the extending radicle and plumule. In the early stages of the life of the bean plant, from the commencement of germination up to the time when the first green leaves are unfolded, the development and building up of the elongating rootlet and shoot depend upon the thick cotyledons.

At first the latter are thick and fleshy, but as the radicle and plumule grow the cotyledons become softer and thinner, ultimately shrivelling considerably. The cotyledons are leaves, the interior of which is packed with food for the rest of the growing embryo, and a large amount of the water absorbed by the seed is used for the purpose of dissolving the nutrient material in them, and carrying it from them to the various parts of the root and shoot of the young plant where growth is going on the cotyledons are esential to the development of the root and shoot of the embryo by cutting them off as soon as the two latter parts have emerged from the seed-coat. Try separating one cotyledon and then two at various stages of development, and see if the axis (root and shoot) can be made to develop without them.

The growth should be allowed to continue some time in order to obtain well-marked effects. Not only do the changes observed innhe embryo of a germinating bean point to the conclusion that it is a living structure and like an animal dependent on a proper supply of water, heat and air for the manifestation of its life, but the parts of a young bean plant after emerging from the seed soon give evidence of the possession of peculiarities which are associated with life. When put in the ground, the radicle, in coming out of the seed, turns straight downwards and continues to grow in this direction. This is the case no matter in what position the seed is placed. If, after germination has commenced, it is taken and replanted with the primary root pointing to the surface of the soil, the tip of the root soon begins to curve downwards again, and will maintain this course until again disturbed.

The plumule behaves in exactly the opposite manner; after emerging from the seed-coat its bent tip grows upwards and away from the root; if the seed is reversed and replanted the plumule begins to curve in such a manner that its tip is driven upwards towards the surface of the ground. That these peculiarities are somehow connected with life is clear, as dead embryos show no such behaviour. Sow soaked beans in a flower pot or box filled with ordinary garden soil placing them in various positions in it, some laid on the flat side, some with the hilum directed upwards, and others with the hilum downwards. Allow them to grow in a warm place: take them up as soon as signs of germination are noticed, and observe the direction the root and shoot have taken. The peculiar tendency for the root always to go downwards and stem upwards can be investigated by sowing beans in ordinary garden soil and afterwards reversing them.

To avoid error all should be taken up, and then placed again in the soil in various positions—some as they were, a few with their roots and stems reversed, and others laid in a horizontal position. They may be re-examined at the end of a week. When the roots have extended about half an inch take two seeds and suspend them by means of thread side by side in a bottle with their roots downwards and stem upwards. The bottle should contain a little water to keep the air damp.

When the roots have grown about two inches reverse one of the seeds so that its root points upwards tad stem downwards. Notice that the tip of the root of the reversed seed in about twelve hours begins to turn downwards, while the plumule more slowly bends in such a way as to assume the position it had before it was reversed. The bottle should be placed in a dark box or cupboard to avoid the influence of light on the plant, and fresh air should be blown into the bottle twice a day. Although seeds vary almost indefinitely in regard to size and shape they are similar to the bean in so far as they all contain a young plant packed away within the seed-coats. In this essential feature all seeds agree with few exceptions, and it is on account of the existence of a young plant within them that they are of use in the raising of crops or plants. The manner, however, in which the embryo is arranged, and the relative size and appearance of its various parts, differ considerably in seeds; moreover, the growth during and after germination is not the same in all cases. A few of the more important and common variations in these respects must be noticed.

TROPICAL AND SUBTROPICAL FRUIT CROPS

Bacterial citrus canker and now citrus greening have ravaged citrus plantings throughout Miami-Dade and Broward Counties and much of south Florida. Infected trees may severely defoliate and management options are very limited; sanitation procedures and following quarantine regulations are very important.

Because of the highly infectious nature of these citrus diseases fruit crops other than citrus should be considered for planting in the home landscape. For some residents, that could mean the opportunity to grow some exciting tropical and subtropical fruit crop alternatives. The south Florida climate offers the opportunity to successfully produce a surprising variety of tropical and subtropical backyard fruits.

Many of them have strange names, odd shapes and exotic flavours, particularly for those of us who have come from up north. Yet many of these fruits are tempting, nutritious, and luscious old favourites enjoyed by the peoples of the tropics around the globe. Examples include, avocado, mango, guava, and papaya or perhaps the even more exotic carambola, jakfruit and jaboticaba. Although many of these species can attain the size of large trees, routine pruning can keep them to a manageable size for most backyard situations.

The University of Florida, Institute of Food and Agricultural Sciences has a wealth of information on how to grow and enjoy these and many other tropical and subtropical fruits right here in south Florida. Learn which varieties are adapted to our area, recommended cultural practices, pitfalls, and preparation techniques. An overview of some of the fruits we think local residents might want to grow and several (guava and sapodilla) that the University of Florida, IFAS does not recommend because of their invasive status.

Atemoya (Custard Apple)

The atemoya is a hybrid between the sugar apple and cherimoya. It produces a fruit very similar to the sugar apple, also called sweetsop or anon. It is a small to medium sized, open, deciduous tree with a rounded canopy rarely exceeding 20 feet in height and width.

Because of their relatively small stature, atemoya trees are suited to small yards. A relative of the atemoya, sugar apple has similar characteristics. The fruit is heart-shaped, round, ovate or conical, from 2 to 4 inches in diameter. The pulp is white or creamy white, with a custard-like consistency and a sweet, pleasant flavour.

When ripe, atemoyas become light green or yellow-green in colour. The pulp of atemoya is consumed principally as a dessert fruit and has an excellent flavour. It may be eaten fresh but it may also be used to make a tasty ice cream or milkshake. Season: late Aug.-Oct., sometimes Dec.-Jan.

Avocado

Perhaps one of the best known subtropical fruits, avocado trees are typically medium to large in size (40-60 feet), easy to grow and can produce large quantities of fruit. Limbs are vulnerable to wind damage and can even break under the weight of a heavy fruit load. The fruit is a very large berry consisting of a single large seed surrounded by a buttery pulp.

Mature fruit are generally green, although some cultivars may be black, red or purple when soft (ripe). Avocados do not ripen until they are picked or fall to the ground. Mature fruit size varies considerably depending upon cultivar and growing conditions. Avocados are highly nutritious. Fruit are popular eaten fresh, in salads or used to make guacamole and other dishes. Planting in a well-drained site is a must, as avocados do not tolerate flooding. Productivity, season of maturity, cold tolerance, and disease tolerance vary greatly depending upon the variety under consideration. Season: late May to March.

Banana

A true tropical favourite, bananas are perhaps the best known of the tropical fruits. There are many different cultivars available with a wide variation in fruit type and quality. Bananas are not cold tolerant and some varieties are better adapted to south Florida than others. Growth is extremely rapid during the very warm, wet summer months. Because they reproduce from underground rhizomes and not seed, a single plant can quickly spread out producing multiple trunks in a matter of a few months.

Fruits develop in clusters on the end of flower stalks usually within 1-2 years. Shortly after bearing, the stem dies. Bananas like full sun and moist but well drained soil. They will tolerate partial shade, but best growth and fruit production is in full sun. Bananas may be eaten fresh, fried, baked, and added as a component to deserts and drinks. Season: year round.

Caimito (Star Apple)

Native to Central America and the West Indies, this evergreen tree is highly ornamental and produces a sweet purple or green fruit. Trees may reach 25 to 100 ft. in height and have a dense, broad crown. The attractive leaves are glossy and green on the upper surface and golden brown on the under surface.

Trees produce either light green or purplish coloured fruit. The fruit is round, 2-4 inches in diameter, and may have a dark-purple or white flesh. The flesh is milky, sweet, and gelatinous and surrounds up to 10 seeds. Plant trees in a well drained soil and sunny location. Season: Jan.-June.

Carambola (Star Fruit)

Originally from southeast Asia, star fruit is becoming increasingly popular and available in the U.S. Locally, commercial production occurs in the southern half of south Florida. Trees are generally small to medium in height (35 feet max.) and spreading with the majority of fruit borne on branches in the mid-canopy region. The fruit is a large fleshy berry, 2-6 inches in length, yellow skinned with a waxy surface and star-shaped cross section.

Flesh is light to dark yellow, translucent, crisp and very juicy. Better cultivars have a nice, slightly sweet to subacid flavour. Eat fresh, cut up in fruit salads, or as an iced juice drink. The fruit may also be canned, preserved, and dried. Carambola trees also have ornamental value with their dark green foliage and attractive flowers and fruit. Plant in a well-drained soil and sunny location, which has protection from wind. On alkaline soils (those with high pH) watch for signs of minor element deficiencies, particularly zinc, iron, and manganese and treat accordingly. Season: July-September, Nov.-Feb.

Canistel (Egg Fruit)

Native to southern Mexico, Belize, Guatemala, and El Salvador, this evergreen tree is ornamental and produces a sweet yellowish-orange top-shaped fruit. Trees may reach 25 ft. or so in height with leaves mostly grouped at the ends of branches. The attractive lanceolate-shaped leaves are glossy and green. Trees produce 3-5 inch long by 2-3 inch diameter, conical-shaped fruit. The maturing fruit changes from green to light green to golden-yellow or pale orange-yellow at maturity. After picking allow fruit to become soft before eating. Soft fruit will have a yellow-orange flesh, which is smooth with a sweet to musky flavour. Eggfruit is eaten fresh and used in making milk shakes. Plant trees in a well-drained soil and sunny location. Season: Dec.-May.

Jaboticaba

Native to Brazil, jaboticaba is an unusual tree that produces a purple, grape-like berry directly upon the trunk and larger branches either singly or in clusters. Under the skin is a whitish pulp with 1 to 4 seeds. The fruit has a pleasant flavour and taste a bit like grape. They can be eaten fresh or made into jam, jellies or wine. Flowering and fruiting occur periodically throughout the year so multiple crops are produced. Other than its fruit, jaboticaba is known for its beautiful multicolored bark, which gives it value in the landscape. The tree is small, slow growing, and bushy, and seldom exceeds 20 feet in Florida. Flowers are small and white, interesting but rather inconspicuous and borne right on the trunk and larger branches.

Jaboticaba is relatively hardy but will not tolerate drought. Plant in full sun in a moist but fairly well drained soil. The tree prefers a slightly acid soil, so some special attention will be required to provide the proper nutrition on our alkaline soils. If provided with an ideal growing environment, jaboticaba can be a relatively low maintenance tree requiring little pruning or spraying for pests. Season: variable.

Jackfruit

Native to India and Malaysia, this attractive large tree has glossy, dark green leaves and produces a very large, oval shaped, rather unusual looking segmented (spiked) fruit. Typically, fruit are produced in clusters of 2 or more at various height intervals along the trunk. The skin of the fruit must be peeled to reveal the succulent, yellow to orange coloured pulp. Because of sticky latex in the peel, coat your hands with vegetable oil prior to peeling a jackfruit. This will make clean up a lot easier. The flavour is sweet, not unlike that of banana or pineapple, but with a strong, fruity aroma and taste. Fruit may be used fresh, fried green, pickled or roasted (seeds). Jackfruit must be planted in flood free, well-drained soils. In south Florida, jakfruit trees have few serious pest or disease problems. Season: spring-fall (some all year round).

Longan

Indigenous to Mayanmar (Burma), southern China, southwest India, Sri Lanka, and the Indochinese peninsula, longan is large tree (30-40 ft.). Longan is an excellent fresh fruit with a pleasant, unique, sweet flavour. Fruit are relative small (about the size of a typical strawberry), round to oval, and borne in loose clusters. When fruit are ripe, the leathery skin develops an attractive golden brown colour. Longan trees are attractive having a dense, round to upright, symmetrical canopy of dark green foliage. They may grow as high as 40 feet. Plant in sunny, well-drained sites. Trees may begin to bear within 3-5 years of transplanting. However, unreliable bearing is a major constraint. Season: July-Aug.

Lychee (Litchi)

Lychee trees are native to southern China and southeast Asia and may grow to 40 or more feet in height. Lychee is an excellent fresh fruit with a pleasant, sweet flavour. Fruit are a relatively small (about the size of a typical strawberry), round to oval, and borne in loose clusters. When fruit are ripe, the leathery skin develops an attractive pinkish red colour. Lychee trees are attractive having a dense, rounded, symmetrical canopy of dark green foliage. They may grow as high as 40 feet. Plant in sunny, well drained sites, preferably where there is some protection from wind. Trees may begin to bear within 3-5 years of transplanting. However, unreliable bearing is a major constraint. Season: June, early July.

Mamey Sapote

Native to Mexico and the Central American lowlands, mamey sapote trees are large and erect reaching a height of about 40 feet in Florida. Mamey sapote makes an excellent specimen tree in the home landscape with handsome glossy foliage. The fruit has a brown, scurfy peel and the flesh of mature fruit is salmon pink to reddish brown in colour and has a unique, sweet flavour. The tree grows well in a variety of soils, but requires good drainage. Mamey sapote can be eaten fresh, and is also excellent for use in ice cream, sherbets, jellies, and preserves. Milkshakes may also be prepared from the flesh. It requires little care and yields a useful, good tasting fruit. Occasionally, an insect known as the Cuban May beetle may cause some defoliation. Season: Jan.-Sept. (some all year).

Mango

Mango trees are native to southeast Asia and India. Hardly a new comer to Florida, mango trees have been planted and enjoyed for over 100 years in Florida. Mango trees are medium to large trees (up to 100 ft.) and there is a wide variety of cultivars available with varying shapes, colours and maturities. Most mangoes are ripe when the fruit softens slightly and takes on a yellow to orange or red colour and carries a subtle sweet fragrance. Mangoes are a good source of vitamins A and C. They can be eaten fresh or pureed. Perhaps more than any other tropical fruit, mangoes are readily available at many local nurseries and garden centres. Trees are tough and relatively easy to grow. Most varieties of grafted trees will bear in as little as 3-5 years after transplanting. Season: May-Oct.

Papaya

Indigenous to southern Mexico, Central and South America, papaya plants are relatively short lived (1-3 years) and are easily propagated from seed. Papayas are relatively easy to grow so long as they are sited in full sun and have excellent drainage. Common throughout the tropics, papayas are small to large fruits born on the stem of upright semi-herbaceous trunks. Fruit are sweet, have orange to reddish-salmon coloured flesh and contain numerous small black seeds in the interior cavity. Papaya fruit is typically peeled, sliced and consumed fresh. Papaya ring spot virus can be a problem causing stunting and fruit loss; at present there is no control for this disease. Papaya fruit fly is another problem but can be overcome by placing a 3-5 pound paper bag over developing fruit. Amend the soil with plenty of organic matter and fertilize often. Season: year round.

Passion Fruit

Native to South America, passion fruit is a vigourous vine that produces purple, yellow, or reddish coloured fruit containing seeds surrounded by an orange, sweet, watery pulp. The juice is very aromatic and is commonly used to make juice or punch. Plant vines next to a fence or along a trellis in a well drained soil area with full sun. Vines begin to bear within 3-6 months of planting. Season June-Dec.

Sugar Apple (Annon, Sweetsop)

Native to tropical America, the sugar apple, is also called sweetsop or anon, has been widely planted in home gardens of south Florida because of its high quality fruit. It is a small, open, deciduous tree with a rounded canopy, rarely exceeding 20 feet in height and width. Because of their small stature, sugar apple trees are suited to small yards.

The fruit is heart-shaped, round, ovate or conical, from 2 to 4 inches in diameter. The pulp is white or creamy white, with a custard-like consistency and a sweet, pleasant flavour. When ripe, sugar apples become light green or yellow-green in colour. The sugar apple is consumed principally as a dessert fruit. A relative of the sugar apple called atemoya has similar characteristics to sugar apple. The pulp of sugar apple has an excellent flavour and is usually eaten fresh but it may also be used to make tasty ice cream or milkshakes. Season: July-Sept., Nov.-Jan.

Tamarind

Indigenous to tropical Africa, tamarind is a sturdy, slow growing but large (up to 80 ft.) leguminous tree with rough bark, twisting spreading branches and finely pinnate, evergreen foliage. Flowers are pale yellow with red veins and born in small racemes in early summer. The fruit are brown pods measuring 3-8 inches in length containing large seeds embedded in a sticky brown pulp, which is edible.

The pulp may have a sweet to sour molasses-like flavour, can be extracted and used to flavour cooking sauces. It is a primary ingredient in certain brands of steak sauce. Because of the many fissures found on its rough textured bark, tamarind trees provide an excellent substrate for growing orchids. Season: April-June.

CARE AND MANAGEMENT OF ORCHARDS

TRAINING AND PRUNING

Modified central leader system of training has been recommended for both apple and pear on seedling rootstocks. The proportional heading back and thinning out system of pruning is adopted after the juvenile phase of plant growth. Spur pruning encourages vegetative growth and helps in new spur development in old plantations.

For high density planting on semi-dwarfing and dwarfing rootstocks spindle bush, dwarf pyramids and cordon system of training are suggested. Spindle bush on M7 and modified central leader on MM106 rootstocks are successful.

Peach is generally trained to open centre system, while in sweet cherry modified central leader system is followed. Pruning is done in such a way so that 20-50 cm new growth in young trees and 25-30 cm in older trees are secured every year. In sweet cherry about 10% of fruit bearing area should be removed annually, whereas in apricot slightly heavier pruning should be practised to keep the spur system renewed.

Manures and Fertilizers

Nutrient requirement varies from place to place. In Himachal Pradesh widespread deficiencies of N, P, K, Ca, Mn, Zn and B have been recorded. A fertilizer dose of 700:350:700 g of N, P and K for full grown bearing apple tree has been recommended.

It has been observed that VAM (Vesicular arbuscular mycorrhizea) fungi increases P uptake by apple roots. The following corrective measures for nutrient deficiencies have been suggested for apple.

In peach cv. Sharbati, 17.5 g N/tree/year and in apricot 70-100 g N/tree/year have been found to be optimum. For peach, N has been the main limiting element in H.P, while for plum, maximum fruit yield was recorded when leaf N, P and K contents were 2.89, 0.28 and 0.89 respectively.

Weeding and Mulching

In apple orchards, grass mulching (10 cm thick) with one application of post-emergence herbicide Glyphosate (0.8 kg/ha) was found to be effective. The shrubby weeds were, however, best controlled by 500 ppm Gramaxone + 100 ppm 2,4,5 -T application. Moisture conservation was maximum under grass mulching and, therefore, recommended for apple in H.P. Mulching with oak leaf was effective in U.P.

Supplementary Irrigation

Trickle irrigation at 75 per cent of field capacity results in better tree growth and higher fruit yield in apple when raised on semi-dwarfing rootstocks. For apple, most critical period of water requirement is April to August and peak water requirement is just after fruit set. 114 cm of water during whole year through 19 irrigations have been recommended. Drip irrigation saves water considerably. Field trials indicated a total irrigation requirement of 3840 liters water per tree under conventional system of irrigation; under drip system 1695 liters of water was enough. The application efficiency under drip system is about 2.27. In Santa Rosa plum, irrigation at 50% field capacity gave better growth and economic returns.

Pest and Disease Control

Pests: In apple and pear about a dozen pests are causing serious damage to the crops. Most important ones are: San Jose Scale, Woolly apple aphid, Root borer, Blossom thrips, Coddling moth and European red mite. San Jose Scale can be effectively controlled with eco-friendly miscible spray oils at 2% concentration when applied during February-March. Woolly aphids can be controlled through soil application of Phorate or Carbofuran granules during May and October/November. For coddling moth, pheromone trapping was found to be effective, and certain bio-control agents have also been identified in apple orchards for controlling certain insect pests.

European red mite is becoming a very serious pest for apple in H.P. and J&K. The pest attack causes premature leaf fall. Late dormant sprays of miscible oil provide effective control of eggs. The mite can also be controlled by sprays of Dicotol (0.05%) followed by Malathion (0.05%). In peach, leaf curl aphid can be controlled by pre-bloom sprays (at pink bud stage) of Dimethoate (0.03%) or Monocrotophos (0.04%).

Fruit fly in peach, apricot and plum can be controlled through foliar spray of baits consisting of Malathion (0.1%) + 1% sugar.Stem and shoot borers causing damage to peach, plum, apricot, almond and cherry can be controlled by inserting 0.5g of PDCB (paradichlorobenzene) into the holes and plugging them with mud.

Diseases: Apple scab caused by *Venturia inaequalis* is a serious disease causing maximum economic loss. A sound forecasting and early warning system has been developed for prediction of scab attack. Also, a judicious fungicide spray schedule has been devised. Under high disease pressure, systematic fungicides performed better, while under low disease pressure Ergosterol biosynthesis inhibiting (EBI) fungicides were as good as protectants. Ascosporic inoculum produced by over-wintered apple leaves could be substantially reduced by giving post-harvest applications of Bavistin (0.1%) and EBI chemicals Penconazole (0.5%) and Flusilazole (0.01%) as pre-harvest fungicidal sprays control scab during storage. Powdery mildew disease can be kept under check by pruning and spraying wettable sulphur (0.2-0.3%) or Karathane (0.05%) during dormancy, bud swell and petal fall stages.

Collar rot and white root rot diseases in apple occur mainly in poorly drained soils. Proper drainage of orchards and soil drenching with 0.1% Carbendazim for white root rot and with 0.3% Mancozeb or 0.3% Ridomil MZ for collar rot are effective.Virus and virus-like diseases such as mosaic, chlorotic leaf spot, rubbery wood and others have been reported. Virus cleaning through tissue culture and supply of virus free bud wood material are being pursued to contain further spread of the viral diseases in apple.

ESTABLISHMENT OF ORCHARDS

LAND PREPARATION AND PLANTING

Fertile sandy loam to clay loam soils with pH range from 5.5 to 7.5 and free from water logging conditions are suitable for establishing deciduous fruit orchards. In flat lands/valleys, square, rectangular or triangular planting may be adopted. In hill slopes, planting on contours or terraces is recommended. In shallow slopy lands, small terraces (half-moon terrace), may be made to establish the plants and large scale disturbance of surface soil need to be avoided.

Planting of deciduous fruit plants is done during winter months from the end of December to mid March. Planting is done in the centre of pits (1x1x1m cube or circle), prepared a month before planting. While refilling the pit, 50g of aldrin powder (15%) is mixed with the soil. After planting, the young plants are supported with stakes and basins are kept free of weeds. Mulching with dry grass or polyethylene is advisable. Irrigation is provided after planting.

Spacing

Spacing varies from species to species and depends also on the type of rootstocks used. For apple raised on seedlings of crab or other commercial varieties of apple, a planting distance of 6x6m or 7x7m accommodating 277-205 plants per ha is recommended for J&K hills. On clonal rootstocks like M4, M7, M26, MM106 a spacing of 4.5m x 4.5m (555 plants/ha) is suggested. For spur type varieties and standard coloured mutants of apple, high density planting on dwarfing rootstocks like M9, M4, M7 and MM107 has been found to be feasible. Fruit yield of 30-35 tons/ha has been achieved in 12 year old orchards of colour mutants of apple on MM106 under a planting density of 2222 plants/ha (3m x 1.5m) in cooler hills of H.P.

Table. Optimum Plant Spacing for Spur Types and Standard Colour Mutants of Apple

Type	Rootstock	Spacing (m×m)	No. of trees/ha
Spur Type	Seedling (Crab)	5×5	400
Spur Type	MM111, MM109	4×4	625
Standard Type	MM106, MM109	5×5	400
Spur Type	MM106, M7	3×3	1111
Standard Type	M9	2×2	2500

In pear, for trees on seedling rootstock under normal conditions, spacing of 5×5m is recommended, but on clonal rootstocks (Quince A) spacing can be reduced to 3×3m. In peach, a spacing of 4.5 to 5m and for plum, apricot and cherry a spacing of 6m are followed.

FRUIT AND VEGETABLE PROCESSING

Each new fruit and vegetable processing centre needs a good, specific preliminary study including, among other considerations, the following aspects:

- Raw material availability;
- Raw material quality in adequate varieties for the types of finished products that will be manufactured;
- Harvesting and transport practices and organisation from the field to the processing centre;
- Processing capacity related to raw material availability: quantities, seasonability, etc.
- Processing equipment size/capacity suitable for above points;
- Availability of trained operators and resources to improve their knowledge;

- Availability of workforce in the area and resources for training them in order to be able to assure adequate trained operators;
- Availability of utilities: electricity, etc.
- Position of the future processing centre as related to raw material fields and to closest transportation means; road access, railway access.
- Last but not least, market availability for finished products and for optional semi-processed products.

The decision to invest in fruit and vegetable equipment MUST be taken case by case and only after an adequate, specific preliminary study has been carried out by specialists or a specialist organisation.

OPERATE A FRUIT AND VEGETABLE PROCESSING CENTRE

Additional recommendations and "hints" to prepare, start and operate a fruit and vegetable processing centre are as follows:

- Assure a raw material temporary storage capacity/surface for 2-5 processing days. Invest in an adequate size cold room for sensible raw materials;
- Plan the equipment to operate at the start-up for at least one working shift (about 7-9 hours) per day, for 5 working days per week; when needed, a second shift could be organized;
- Plan to operate the processing centre for a maximum number of working days per year). In order to achieve this, invest in the buildings and equipment which will be able to:
  - Process as many species of fruits and vegetables as possible/as available;
  - Use as many preservation methods as possible, *e.g.* drying, dehydration, concentration, sugar preservation, etc.
- Whenever possible, "rush" the utilisation of available raw materials during crop season by additional manufacturing of semi-processed products and transform these in consumer finished products during the off-season
- Excessive automatization of processing equipment DOES NOT directly imply a good quality of finished product;
- Raw material quality is a major element with positive impact on finished product quality;
- Initial and continuous personnel training and motivation is also an important factor in the success or failure of a processing centre and in assuring a constant finished product quality;
- Keep finished product stocks at a minimum adequate level;
- Remember that the three main "outputs" of the processing centre have to be prioritized in the following order:

  Priority 1: Finished product quality conforms to specifications and standards: national and/or international, consumer special requests, etc.

Priority 2: Continuous and reliable supply of finished products to the domestic and export markets throughout the year (or at least throughout the "marketing season" of specific products);

Priority 3: Manufacturing and transport costs as low as possible, inside the stringent need to cover the first two priorities;

- When deciding on the equipment output, take into consideration all elements specified and mainly raw material availability and market demand for a specific finished product;
- Invest in simple, modular processing lines which can, with some simple on-site configuration modifications, process various types of finished products; this is mainly important for the first technological steps (preparation of raw materials, etc);
- Plan to use as much as possible of the raw materials supplied/ received to the processing centre.

  This should be facilitated by the initial design and by a good day-to-day organisation and management; all these should enable, if necessary, to make a different use of each "quality" or grade of raw materials, e. g. using them for different finished products: one quality for drying/dehydration, an other quality for juices, etc.
- Take into account the fact that the marketability of finished products will be differ in terms of types and quality for domestic and export markets.

  Be sure that an export specialized staff/organisation will help with specific export advice. To export successfully is a different job to processing fruit and vegetables.
- Avoid investing in one "big" processing line, very sophisticated in terms of automation, etc. with a high output capacity but having potential following drawbacks: being able to "generate"/produce only one finished product type from only one raw material; having too high a degree of equipment fixation work for installation and therefore very high difficulties in using the processing equipment in a modular " interchangeable " way.
- As an initial investment prefer small size processing lines, with modular equipment arrangement (*i.e.* able to be integrated in various technological configurations for processing of as many raw materials as possible and generating different finished products).
- As compared with important processing units in developed countries, it is possible to formulate as a very general rule for developing countries, that for the usual size of equipment, for a comparable environment frame, the scale/size should be approximatively 1:10 from those actually in use in developed countries.

Fruit and Vegetable Processing Unit

- Buildings
  - Covered platform for temporary storage of raw materials and washing; surface = 25 sq. m.
  - Room for wet processing (cutting, dipping, boiling, water blanching, pasteurization, etc.) = 40 sq. m.
  - Room for dry processing = 40 sq. m.
  - Storage room for processed products = 20 sq. m.

    All areas need to be on a cemented platform.

    Area (i) is simply covered and surrounded by plastic sheeting to avoid dust contamination.
- Outside drying yard

  This area is needed in order to install various dryers: simple sun drying trays; tent sun dryer; cabinet sun dryer; cabinet solar dryer with heat collector, etc.

  Ideally, this area should be cemented to avoid excessive dust generation.

  A minimum surface of 70 sq. m. is necessary.

  Access to the drying yard must be closed to non-production personnel.

  It is useful to construct some surface for shade drying from the beginning; this surface could be also be used in order to protect, if needed, drying trays in case of rain.
- Equipment and material

  Working tables (3)
  Scales: 0-50 kg, precision 1 kg
  Hand refractometre 0-90 ° Brix (2)
  Thermometres 10-100° C (5)
  Improved stoves (3)
  Stainless steel pots 51(2)
  Stainless steel pots 101(2)
  Stainless steel pots 151(3)
  Stainless steel knives, 12-15 cm blade (10)
  Stainless steel spoons, various shapes and sizes (5)
  Stainless steel household sieves (3)
  Wooden spoons (5)
  Rigid plastic funnels, large bottom (10)
  Glass jars, various sizes and screw-on caps (500)
  Aluminium pots: 251(1); 401(2); 501(3)
  Hand-operated pulp extractor
  Electrical pulp extractor
  Bottle brushes (10)
  Plastic lemon-squeezer

Plastic colanders
Stainless steel skimmer
Aluminium ladle
Sun dryers (tent type)
Sun dryers (cabinet type)
Solar dryers (cabinet type) with heat collector
Standard wood sun/solar drying trays (30)
Bottles (0.5 1)
Bottles (0.33 1)
Crown tops
Jars (0.580 1)
Jars (0.300 1)
Screw-tops for jars
Electrical heating plates
Gas-fired heating plates and/or available fuel heating plates (wood, etc.)
Hand-operated capping device (capper)
Rigid plastic drums 501(5)
Work benches (3)
Stainless steel vegetable cutter (5)

- Simple technological recommendations
  i. Ingredients
    - Sugar and potassium metabisulphite (K2S2O5) are used as preservatives;
    - Lemon or lime juice are added to the products to rectify the acidity (this improves storage stability and taste).

  ii. Hygiene measures
    - The workers should carefully wash their hands before any product processing operations;
    - The utensils and equipment will have to be properly cleaned before and after use, in order to remove dust and any possible organic particle;
    - The packaging materials, i. e. bottles and jars, have to be washed with a hand-operated appliance, hot clean water and sand. After washing, a rinse with clean water will be carried out;
    - Damaged parts of the fresh raw materials, as well as the waste, will have to be discharged and disposed of outside of the working area;
    - Before storage, the finished products will be washed and dried (this apply to jars and bottles) and properly labelled;

- The preparation and drying areas must NOT be located in the vicinity of a stock-farm.

Fruit and Vegetable Processing Centre

- Buildings
  - Covered area for temporary storage of raw materials and washing; surface = 10 sq. m.
  - Room for wet processing (cutting, dipping, boiling, water blanching, pasteurization, etc.) = 25 sq. m.
  - Room for dry processing = 25 sq. m. iv. Room for storage of processed products = 10 sq. m.

    Area (i) is simply covered; area (ii) could be similar to a simple house kitchen; areas (iii) and (iv) could be similar to house rooms.
- Outside drying yard

  This area is needed in order to install: simple sun drying trays; tent sun dryer; cabinet sun dryer.

  At best, this area have to be cemented to avoid excessive dust generation.

  A minimum surface of 50 sq. m. is necessary.
- Equipment and material

Working tables (2)
Improved stoves (3)
Stainless steel pots 51(2)
Stainless steel pots 101(2)
Stainless steel pots 151(1)
Stainless steel knives, 12-15 cm blade (10)
Stainless steel spoons, various shapes and sizes (5)
Stainless steel household sieves (3)
Wooden spoons (5)
Glass jars, various sizes and screw-on caps (200)
Aluminium pots: 251(1); 401(2);
Hand-operated pulp extractor
Bottle brushes (10)
Plastic lemon-squeezer
Stainless steel skimmer
Aluminium ladle
Sun dryer (tent type)
Sun dryer (cabinet type)
Standard wood sun drying trays (20)
Bottles (0.33 1)
Crown tops
Jars (0.300 1)

Screw-tops for jars
Wood (or other available fuel) heating plates
Hand-operated capping device (capper)
Work bench (3)
Stainless steel vegetable cutter (5)

- Simple technological recommendations
  i. Ingredients
    - Sugar and potassium metabisulphite (K2S2O5) are used as preservatives;
    - Lemon or lime juice is added to the products to rectify the acidity (this improves storage stability and taste).

  ii. Hygiene measures
    - The workers should carefully wash their hands before any product processing operations;
    - The utensils and equipment will have to be properly cleaned before and after use, in order to remove dust and any possible organic particle;
    - The packaging materials, i. e. bottles and jars, have to be washed with a hand-operated appliance, hot clean water and sand. After washing, rinsing with clean water will be carried out;
    - Damaged parts of the fresh raw materials, as well as waste, will have to be discharged and disposed outside of the working area;
    - Before storage, the finished products will be washed and dried (this apply for jars and bottles) and properly labelled;
    - The preparation and drying areas must NOT be located in the vicinity of a stock-farm.

RECENT TRENDS IN FRUIT AND VEGETABLE PROCESSING

VARIETY OF FRUIT AND VEGETABLE PRODUCTS

The number and variety of fruit and vegetable products available to the consumer has increased substantially in recent years. The fruit and vegetable industry has undoubtedly benefited from the increased recognition and emphasis on the importance of these products in a healthy diet.

Traditional processing and preservation technologies such as heating, freezing and drying together with the more recent commercial introduction of chilling continue to provide the consumer with increased choice. This has been achieved by new heating (*e.g.* UHT, microwave, ohmic) and freezing (*e.g.* cryogenic) techniques combined with new packaging materials and technologies (*e.g.* aseptic, modified atmosphere packaging).

The overall trend in new fruit and vegetable products is "added value", thus providing increased convenience to the consumer by having much greater variety of ready prepared fruit and vegetable products. These may comprise complete meals or individual components. The suitability of products and packages for microwave re-heating has been an important factor with respect to added convenience.

Fruit and Vegetable Product Trends

Heat processed products:

- Canned fruits and vegetables
  - Combination of vegetables in sauces and vegetable recipe dishes. Exotic fruits.
- Glass packed fruits and vegetables
  - "Condiverde"/"antipasti" products based on vegetables in oil.
  - High quality fruit packs.
- Retortable plastics
  - Basic vegetables or vegetable meals
  - Fruit in jelly
- Aseptic cartons
  - Ready made jelly
- Rosti meals
  - Potato based products in retort pouches
- Fruit juices
  - New combinations of juices and freshly squeezed products
- Crisps
  - Thick and crunch skin-on crisps. Kettle or pan fried chips. Lower fat crisps.

Table. Numbers of New Fruit and Vegetable Products

	1990	1991	1992 Jan June
Frozen vegetable products	66	95	21
Chilled vegetable products	76	81	78
Heat processed vegetables	51	60	38
Heat processed fruits	13	14	5
Fruit juices and drinks	73	83	46
Potato crisps	32	33	16

New product development in the fruit and vegetable sector is most important in meeting the continued challenge of providing the consumer with choice and high quality products.

A Fresh Look at Dried Fruit

New fruit varieties and advance in drying technologies are putting a fresh twist on dried fruit applications. Fruits that have been introduced to the drying process include cranberries, blueberries, cherries, apples, raspberries and strawberries - not to mention the traditional mainstays of raisins, dates, apricots, peaches, prunes and figs. Perceived as a "value-added" ingredient, dried fruit adds flavour, colour, texture and diversity with little alteration to an existing formula. The growing interest in ethnic cuisines in U.S.A. and the change to a more healthy way of eating, has also moved dried fruit considerably closer to the mainstream.

Found primarily in the baking industry, dried fruit is coming into its own in various food products, including entrees, side dishes and condiments. Compotes, chutneys, rice and grain dishes, stuffings, sauces, breads, muffins, cookies, deserts, cereals and snacks are all food categories encompassing dried fruit. Since some dried fruit is sugar infused (osmotic drying), food processors can decrease the amount of sugar in formula - this is especially the case in baked products. Processors are making adjustments in moisture content of the dried fruit so that a varied range is available for different applications. An added bonus is dried fruits' shelf stability (a shelf life of at least 12 months). Dried fruit is more widely available in different forms, including whole dried, cut, diced and powders.

CHERRY AND APRICOT OILS ARE SAFE FOR FOOD USE

The oils obtained by cold pressing the kernels of the cherry (Prunus cerasus) and the apricot (Prunus armeniaca) have been declared acceptable for food use by the UK Ministry of Agriculture, Fisheries & Food's Advisory Committee on Novel Foods and Processes (ACNFP) by June 1993. In its assessments of the safety in use of the cherry and apricot kernel oils, the ACNFP consider specifications that included data on fatty acid composition, the presence of natural antioxidants and the content of cyanide, mycotoxins and heavy metals.

The Committee says that it gave particular consideration to the possible presence in the oils of the cyanogenic glucoside amygdalin, from which cyanide is released by enzymic action when the kernels of cherry and apricot are crushed. Amygdalin was found to be absent from the cherry and apricot kernel oils. The ACNFP is satisfied that there are no food safety reasons why the use of cherry and apricot kernel should not be acceptable provided there is compliance with the specifications shown in Table given below.

Table. Specification of Purity for Cherry and Apricot Kernel Oils as Determined by UK ACNFP

	Cherry	Apricot
Contaminants limits:		
Heavy metals (total)	0.5 mg/kg	0.5 mg/kg

Aflatoxins (total)	4.0 g/kg	4.0 g/kg
Cyanide	0.15 mg/kg	0.15 mg/kg
Pesticide residues	0.01 mg/kg	0.01 mg/kg
Tocopherols:		
Alpha/delta/gamma (mg/kg)	356-886	569-899

The oils are obtained by the mechanical mincing and cold pressing of kernels extracted from cleaned cherry or apricot stones. After filtering, the oils are stored and are to be sold in a raw, unrefined state. The cherry and apricot kernel oils are high unsaturated and are expected to be used as speciality oils for salad dressings, baking and shallow frying applications.

The Use of Fruit Juices in Confectionery Products

During the last decade, the concept of fruit juices has gained immensely on consumer popularity. The majority of new non-alcoholic and alcoholic fruit drink products were a combination of syrups, fruit juices and flavours. The confectionery industry followed suit and new products incorporated fruit juices as part of their confectionery formulations and processes. Fruit juice concentrates of high solids are often used instead of normal or single-fold juices.

Juice concentrates are made of pure fruit juices. The process starts with pressing fruits and obtaining pure fruit juice; this is stabilised by heat treatment which inactivates enzymes and micro-organisms. The next processing step is concentration under vacuum up to 40-65° Brix or 4-7 fold. The concentrates are then blended for standardisation and stored.

These fruit juice concentrates are often further stabilised by the addition of sodium benzoate and potassium sorbate and are usually stored away from light and are refrigerated or frozen. Depectinised fruit juices are also used to prevent foaming in confectionery processes and are essential for use in clear beverage products. Fruit juice concentrates which are depectinised, and have added preservatives are called stabilised, clarified, fruit juice concentrates. Fruit juices are used in confectionery products in conjunction with natural and artificial flavours which provides intense flavour impact and are cost effective for a confectionery product.

The traditional concern in using fruit juice concentrates in confectionery applications has been the effect of the natural acids on the finished product, particularly the formation of invert sugar during processing. This is a logical concern since concentrates contain differing amounts and types of acids. For example: apple, cherry, strawberry and other berries contain primarily malic acid. Grapes mainly contain tartaric acid. Cranberry is high in quinic acid. Citrus fruits and pineapple contain differing amounts of citric acid. The concentrates, when used, are normally buffered to a pH of 5-7 with sodium hydroxide.

In formulating products with fruit juice concentrates, the solids of the concentrate are considered as mostly reducing sugars and a reduction in corn syrup is made to compensate for equivalent amount of reducing sugar being added in the concentrate. The exact replacement can be determined by measuring the D.E. of the concentrate to be added. In formulations when small amounts of concentrate are used (less than 1%), no adjustment is made since the reducing sugar contribution of the concentrate is not significant. Fruit juice concentrates can also be used to provide a source of natural colour, in particular red colour. Grape, raspberry, cherry, strawberry and cranberry concentrates in small amounts are very effective in colouring cream centres.

The inclusion of fruit juices in confectionery products is now left up to the imagination of the manufacturer. These products must, of course, hold up to the standards of flavour integrity, and product excellence, during the shelf-life of these products.

CITRIC ACID AND ITS USE IN FRUIT AND VEGETABLE PROCESSING

Citric acid may be considered as "Nature's acidulant". It is found in the tissues of almost all plants and animals, as well as many yeasts and moulds. Commercially citric acid is manufactured under controlled fermentation conditions that produce citric acid as a metabolic intermediate from naturally-occurring yeasts, moulds and nutrients. The recovery process of citric acid is through crystallization from aqueous solutions.

Citric acid is widely used in carbonated and still beverages, to impart a fresh-fruit "tanginess". Citric acid provides uniform acidity, and its light fruity character blends well and enhances fruit juices, resulting in improved palatability. The amount of citric acid used depends on the particular desired flavour (*e.g.*, High-acid: lemonade; Medium-acid: orange, punch, cherry; Low-acid: strawberry, black cherry, grape).

Sodium citrate is often added to beverages to mellow the tart taste of high acid concentrations. It provides a cool, distinctive smooth taste and masks any bitter aftertaste of artificial sweeteners. In addition, it serves as a buffer to stabilise the pH at the desired level. The high water solubility of citric acid (181 g/100 ml) makes it an ideal additive for fountain fruit syrups and beverages concentrates as a flavour enhancer and microbial growth inhibitor (preferably at pH < 4.6).

In processed fruits and vegetables, citric acid performs the following functions:

- It reduces heat-processing requirements by lowering pH: inhibition of microbial growth is a function of pH and heat treatment. Higher heat exposure and lower pH result in greater inhibition. Thus the use of citric acid to bring pH below 4.6 can reduce the heating requirements. In canned vegetables, citric acid usage is greatest in tomatoes, onions and pimentos. For tomato packs, the National Canners Association recommends a pH of 4.1 to 4.3. In general, 0. 1% citric acid will reduce the pH of canned tomatoes by 0.2 pH units.

- Optimise flavour: citric acid is added to canned fruits to provide for adequate tartness. Recommended usage level is generally less then 0.15%.
- Supplement antioxidant potential: citric acid is used in conjunction with antioxidants such as ascorbic and erythorbic acids, to inhibit colour and flavour deterioration caused by metal-catalysed enzymatic oxidation. Recommended usage levels are generally 0.1% to 0.3% with the antioxidant at 100 to 200 ppm.
- Inactivate undesirable enzymes: oxidative browning in most fruits and vegetables is catalysed by the naturally present polyphenol oxidase. The enzymatic activity is strongly dependent on pH. Addition of citric acid to reduce pH below 3 will result in inactivation of this enzyme and prevention of browning reactions.

MANGO AND GUAVA PROCESSING TECHNOLOGIES

MANGO PROCESSING TECHNOLOGIES

Mangoes are processed at two stages of maturity. Green fruit is used to make chutney, pickles, curries and dehydrated products. The green fruit should be freshly picked from the tree. Fruit that is bruised, damaged, or that has prematurely fallen to the ground should not be used. Ripe mangoes are processed as canned and frozen slices, purée, juices, nectar and various dried products.

Mangoes are processed into many other products for home use and by cottage industry.

The mango processing presents many problems as far as industrialization and market expansion is concerned. The trees are alternate bearing and the fruit has a short storage life; these factors make it difficult to process the crop in a continuous and regular way. The large number of varieties with their various attributes and deficiencies affects the quality and uniformity of processed products.

The lack of simple, reliable methods for determining the stage of maturity of varieties for processing also affects the quality of the finished products. Many of the processed products require peeled or peeled and sliced fruit. The lack of mechanised equipment for the peeling of ripe mangoes is a serious bottleneck for increasing the production of these products. A significant problem in developing mechanised equipment is the large number of varieties available and their different sizes and shapes. The cost of processed mango products is also too expensive for the general population in the areas where most mangoes are grown.

There is, however, a considerable export potential to developed countries but in these countries the processed mango products must compete with established processed fruits of high quality and relatively low cost.

Green Mango Processing

Pickles

The optimum stage of maturity should be determined for each variety used to make pickles. There are two classifications of pickles - salt pickles and oil pickles. They are processed from whole and sliced fruit with and without stones. Salt is used in most pickles.

The many kinds of pickles vary mainly in the proportions and kinds of spices used in their preparation. One basic recipe for the study of the preparation and storage of pickles in oil is as follows:

Mango pieces	250 g	Tumeric powder	2 to 4 g
Salt	60 g	Fenugreek seeds	2 to 4 g
Mustard powder	30 g	Bengal gram seeds	2 to 4 g
Chili powder	20 g	Gingelly oil	20 to 30 g

The ingredients are mixed together and filled into wide-mouthed bottles of 0.5 kg capacity. Three days later the contents are thoroughly mixed and refilled into the bottles. Extra oil is added to form a 1-2 cm layer over the pickles.

Chutney

The product is prepared from peeled, sliced or grated unripe or semi-ripe fruit by cooking the shredded fruit with salt over medium heat for 5 to 7 minutes, mixed and then sugar, spices and vinegar are added. Cook over moderate heat until the product resembles a thick purée, add remaining ingredients and simmer another 5 min. Cool and preserve in sterilised jars.

Spices usually include cumin seeds, ground cloves, cinnamon, chili powder, ginger and nutmeg. Other ingredients such as dried fruits, onions, garlic and nuts may be added. Drying/dehydration. immature fruit is peeled and sliced for sun-drying. The dried mango slices can be powdered to make a product called amchoo. The use of blanching, sulphuring and mechanical dehydration gives a product with better colour, nutrition, storability and fewer microbiological problems.

RIPE MANGO PROCESSING

Puree

Mangoes are processed into purée for re-manufacturing into products such as nectar, juice, squash, jam, jelly and dehydrated products. The purée can be preserved by chemical means, or frozen, or canned and stored in barrels. This allows a supply of raw materials during the remainder of the year when fresh mangoes are not available.

It also provides a more economical means of storage compared with the cost of storing the finished products, except for those which are dehydrated, and provides for more orderly processing during peak availability of fresh mangoes. Mangoes can be processed into purée from whole or peeled fruit. Because of the time and cost of peeling, this step is best avoided but with some varieties it may be necessary to avoid off-flavours which may be present in the skin. The most common way of removing the skin is hand-peeling with knives but this is time-consuming and expensive. Steam and lyepeeling have been accomplished for some varieties. Several methods have been devised to remove the pulp from the fresh ripe mangoes without hand-peeling. A simplified method is as follows: the whole mangoes were exposed to atmospheric steam for 2 to 2 1/2 min in a loosely covered chamber, then transferred to a stainless steel tank.

The steam-softened skins allowed the fruit to be pulped by a power stirrer fitted with a saw-toothed propeller blade mounted 12.7 to 15.2 cm below a regular propeller blade. The pulp is removed from the seeds by a continuous centrifuge designed for use in passion fruit extraction. The pulp material is then passed through a paddle pulper fitted with a 0.084 cm screen to remove fibre and small pieces of pulp.

Mango purée can be frozen, canned or stored in barrels for later processing. In all these cases, heating is necessary to preserve the quality of the mango purée. In one process, purée is pumped through a plate heat exchanger and heated to 90°C for 1 min and cooled to 35° C before being filled into 30 lb tins with polyethylene liners and frozen at -23.50 C. In an other process, pulp is acidified to pH 3.5, pasteurized at 90°C, and hot-filled into 6 kg high-density bulk polyethylene containers that have been previously sterilised with boiling water. The containers are then sealed and cooled in water. This makes it possible to avoid the high cost of cans.

Wooden barrels may be used to store mango pulp in the manufacture of jams and squashes. The pulp is acidified with 0.5 to 1.0% citric acid, heated to boiling, cooled, and SO2 is added at a level of 1000 to 1500 ppm in the pulp. The pulp is then filled into barrels for future use.

Slices

Mango slices can be preserved by canning or freezing, and recent studies have shown the feasibility of pasteurized-refrigerated and dehydro-canned slices. The quality of the processed product in all of these procedures will be dependent upon selection of a suitable variety along with good processing procedures. Thermal process canning of mango slices in syrup is the most widely used preservation method.

Beverages

The commercial beverages are juice, nectar and squash. Mango nectar and juice contain mango purée, sugar, water and citric acid in various proportions depending on local taste, government standards of identity, pH control, and fruit composition of the variety used. Mango squash in addition to the above may contain SO2 or sodium benzoate as a preservative. Other food grade additives such as ascorbic acid, food colouring, or thickeners may be used in mango beverages. A short description of finished products found in literature is as follows:

- Mango juice: prepared by mixing equal quantities of pulp (purée) and water together and adjusting the total soluble solids (TSS) and acidity to taste (12 to 15% TSS and 0.4 to 0.5% acidity as citric acid);
- Mango nectar containing 25% purée can be prepared using the following procedure.

Table. Brix of puree

Nectar components	15°	17°	20°
Purée	100	100	100
Sugar	45	43	40
Water	255	257	260

Commercial processing conditions may require the use of a preservative.

The pH is adjusted to approximately 3.5 by adding citric acid as a 50% solution. The time of heat processing will vary with filling temperature, can size and viscosity of the juice or nectar.

Mango squash may be prepared according to flow-sheet described below; the finished product may contain 25% juice, 45% TSS and 1.2 to 1.5% acidity and may be preserved with sulphur dioxide (350ppm) or sodium benzoate (1000 ppm) in glass bottles. Mango squash simplified flow-sheet.

Ingredients

Mango pulp	900	900
Sugar	900	1100
Citric acid	18	15
Water	900	900

Mangoes are washed, stored, peeled with stainless steel knives. The pulp is prepared by using a pulper with fine sieve (0.025-in); Sugar is mixed with water and citric acid = syrup; The pulp is added to the syrup and mixed well; The mixture is strained trough cloth; The squash is heated at 85° C and bottles are filled and closed.

For additional heat treatment bottles may need to be maintained at a product temperature of 80°C for 30 minutes if the product is to be processed without preservatives. The bottles are then left to cool in water and stored at room temperature.

Two negative points must be avoided: presence of air bubbles (which is a source of quick deterioration) and separation of squash solids (giving an undesirable appearance). The means to avoid these two phenomena are described in the fruit juices section.

A type of "squash type" beverage may also be manufactured with '/a pulp + '/a water + i/a sugar and pH adjusted to 3.7 by addition of citric acid. Using different sieve sizes affects the quality and reduces air bubbles to a certain extent but homogenisation and de-aeration of purée or squash seem to be important in order to avoid separation and air bubbles. The squash quality is evaluated on the basis of the following characteristics: pH, titrable acidity, soluble solids, ascorbic acid (by 2,6 dichlorophenol indophenol method), specific gravity.

Dried/Dehydrated

Ripe mangoes are dried in the form of pieces, powders, and flakes. Drying procedures such as sun-drying, tunnel dehydration, vacuum-drying, osmotic dehydration may be used. Packaged and stored properly, dried mango products are stable and nutritious.

One described process involves as pre-treatment dipping mango slices for 18 hr (ratio 1:1) in a solution containing 40° Brix sugar, 3000 ppm SO_2, 0.2% ascorbic acid and 1% citric acid; this method is described as producing the best dehydrated product. Drying is described using an electric cabinet through flow dryer operated at 60° C. The product showed no browning after 1 year of storage.

Drum-drying of mango purée is described as an efficient, economical process for producing dried mango powder and flakes. Its major drawback is that the severity of heat preprocessing can produce undesirable cooked flavours and aromas in the dried product. The drum-dried products are also extremely hydroscopic and the use of in-package desiccant is recommended during storage.

Canning

This preservation technology is described in various technological flow-sheets in this bulletin. Mango bar or "fruit leather" is presented in various flow-sheets.

TECHNOLOGY OF FRUIT SYRUPS PRODUCTS

Fruit syrups are products obtained by dissolving sugar in juices obtained from direct pressing of fruits. Sugar dissolving can be done at room temperature or by heating. Syrups have to contain 68% soluble substances (refractometric extract) and minimum 1 g/100 ml malic acid. Up to a maximum 10% of sugar can be replaced by corn syrup. Syrups must be manufactured from the juice of only one fruit species. Juice preparation is carried out at room temperature as described above. Sugar dissolving in fruit juice can be performed by one of the following methods:

- *Boiling in a vacuum*: The basic recipe is the same as above. Sugar and fruit juice are mixed previously in a pre-heating kettle and then transported to vacuum equipment.

Boiling is performed at 50°C and at the end the temperature is raised slowly up to 65-70°C. The syrup can be cooled directly in vacuum equipment by closing the steam inlet and by increasing the vacuum. In this boiling method it is possible to incorporate a flavour recuperation device.

- Boiling in open kettles, done by using the basic recipe:
  350-400 kg fruit juice; 650-660 kg sugar; max. 10 kg citric or tartric acid.
  The juice is brought to boiling and the sugar is dissolved; the total time has to be as short as possible in order to avoid flavour loss and a too high sugar inversion degree (optimum inversion degree is 40%). Acid is added preferably towards the end of boiling.
  During all boiling processes it is necessary to remove foam/scum. In order to avoid caramelization, the syrup has to be cooled rapidly, and this can be carried out in baths with double bottoms through which are circulated water.
  One alternative to this method is to boil syrup in closed vessels to avoid flavour losses.
- Sugar can also be dissolved at room temperature by using continuous flow percolators. These are similar to those used for salt solution preparation in vegetable canning processes. The juice goes over a sugar layer and is concentrated progressively until saturation (about 65 per cent). The syrup is then passed through a filtration section in the bottom of the percolator.
- A continuous process for syrup preparation can be carried out by dissolving components with heat while passing them through a horizontal cylinder with a screw inside.
  In the methods where sugar is dissolved by heat, it is also possible to use chemically preserved juices.
  In this case it is necessary to first perform the desulphitation of juices preserved with SO_2. This can be performed by boiling juice with optional water addition (and before any sugar addition). High quality syrups are obtained however from fresh juices.

Syrup filtration is needed in order to clarify crystals; the filtration of syrup is done in warm conditions through cloth. Filling of syrup in bottles is done in aseptic conditions as much as possible in order to avoid syrup infection with osmophile yeasts. Syrup preservation is assured by the high sugar content with respect to a low water activity [unclear]. Storage takes place in well ventilated storage rooms; avoiding sunlight at 10-15°C. The usual product range is: strawberries, cherries, wild berries, citrus fruits.

TYPES OF "FRUIT IN SYRUP" PRODUCTS

This type of product is represented by fruit (whole, halves or pieces) covered by a sugar solution and preserved by pasteurization. In these products sugar does not have a preservation effect but only a sweetening role. Technological general flow-sheet covers the following steps:

Sorting is necessary in order to choose mature fruit, whole, unblemished, undamaged and with a specific texture. Fruit of good quality but with a texture not compatible with this type of finished product are used for the production of semi-processed products, marmalades, etc. This step is done manually on a sorting belt. Washing is performed in equipment with fan and sprays. Cleaning covers removal of leaves, etc., skin removal by one of the described methods for some fruits and coring/pitting for others. Cutting is applied only to pomace fruits and is performed preferably by mechanical means. Preliminary Heat Treatment depends very much upon the fruit types and can be as different as a light blanching up to a real boiling; this step is aimed towards softening of tissues for hard fruits, elimination of waxy pruin layer (plums, etc.) or enzyme inactivation for pomace fruits.

This treatment has to be reduced to the minimum necessary in order to avoid sugar losses. Some fruit (for example apricots, black cherries, grapes, etc.) does not undergo preliminary heat treatment. Cooling is carried out in water (as cold as possible) and should not be too long to avoid soluble substances loss. Receptacle filling Fruit is introduced manually or sometimes mechanically (filling tables, etc.) in receptacles and then sugar syrup is added. Sugar syrup preparation is performed by dissolving crystal sugar in hot water (90100° C). After the sugar has been dissolved, the syrup is boiled briefly; removal of impurities and coagulated substances is then performed by foaming/scumming. An addition of about 0.3% citric acid helps syrup clarification, followed by filtration through cloth.

Preheating/Emptying of open receptacles has as its objective to eliminate air from fruit tissues and this step is largely replaced today by receptacle closing under vacuum. Preheating may be done in steam or in water in such a way that syrup temperature reaches 8090°C and is maintained for 10-15 min. Hermetic sealing is performed according to the type of receptacle and of cover/cap. Pasteurization is carried out at 100°C and may be performed in continuous installations or in normal water baths. In some countries fruits in syrup are pasteurized at 80-90°C; this contributes to a greater flavour retention. Cooling should be intensive in order to avoid discolourations and colour modifications and to reduce corrosion in metal cans. Refractometric extract is to be measured 30 days after manufacturing. In order to establish exact concentration of sugar syrup to be added in each individual case it is possible to use the following formula:

$C = 1.5 \times [(2 \times E) - F]$

C = added syrup concentration;

E = refractometric extract needed in finished products.

F = fruit refractometric extract.

Hot syrup is put into receptacles with a minimum temperature of 80°C in order to enable a corresponding vacuum after cooling. It is possible to add in the syrup 0.3-0.5% citric or tartaric acid for fruit with a low acidity. Adding ascorbic acid in proportion of 0.8% assures colour maintenance and taste improvement based on its reducing action. Avoidance of excessively soft texture in finished products may be achieved for soft fruit (strawberries, apricots, etc.) by dipping fruit in a solution of calcium chloride (5% $CaCl2$ in water). Finished product defects and production "accidents" for "Fruit in syrup" and the means of preventing them

- Change of colour in red fruit (red cherries, plums, etc.) to violet; to avoid this it is necessary to use glass jars or varnished/lacquered tinplate;
- Colour change to red for pears and quinces (pink shade) could arise in products which are over-pasteurized and insufficiently cooled;
- Change of colour of whole peaches peeled by chemical means is due to an incomplete inactivation of oxidative enzymes; to avoid this it is necessary to boil fruit after chemical peeling followed by washing with 0.5% citric or tartric acid solutions or by adding ascorbic acid in syrup;
- Softening of strawberries tissues can be prevented by a pre-treatment with warm syrup and also by addition of calcium salt.

CUTTING AND FRUIT PROCESSING TECHNOLOGY

This step is performed according to the specific requirements of the fruit processing technology.

Heat Blanching

Fruit is not usually heat blanched because of the damage from the heat and the associated sogginess and juice loss after thawing. Instead, chemicals are commonly used without heat to inactivate the oxidative enzymes or to act as antioxidants and they are combined with other treatments.

Ascorbic/Citric Acid Dip

Ascorbic acid or vitamin C minimises fruit oxidation primarily by acting as an antioxidant and itself becoming oxidised in preference to catechol-tannin compounds. Ascorbic acid is frequently used by being dissolved in water, sugar syrup or in citric acid solutions.

It has been found that increased acidity also helps retard oxidative colour changes and so ascorbic acid plus citric acid may be used together. Citric acid further reacts with (chelates) metal ions thus removing these catalysts of oxidation from the system.

Sulphur Dioxide Treatment

Sulphur dioxide may function in several ways:

- Sulphur dioxide is an enzyme poison against common oxidising enzymes;

- It also has antioxidant properties; *i.e.*, it is an oxygen acceptor (as is ascorbic acid);
- Further SO_2 minimises non enzymatic Maillard type browning by reacting with aldehyde groups of sugars so that they are no longer free to combine with amino acids;
- Sulphur dioxide also interferes with microbial growth.

In many fruit processing pre-treatments two factors must be considered:

- Sulphur dioxide must be given time to penetrate the fruit tissues;
- SO_2 must not be used in excess because it has a characteristic unpleasant taste and odour, and international food laws limit the SO_2 content of fruit products, especially of those which are consumer oriented (*e.g.* except semi-processed products oriented to further industrial utilisation).

Commonly a 0.25 per cent solution (except for semi-processed fruit products which are industry oriented and use a 6% solution) of SO_2 or its SO_2 equivalent in the form of solutions of sodium sulphite, sodium bisulphite or sodium/potassium metabisulphite are used. Fruit slices are dipped in the solution for about two to three minutes and then removed so as not to absorb too much SO_2.

Then the slices are allowed to stand for about one to two hours so that the SO_2 may penetrate throughout the tissues before processing. Sulphur dioxide is also used in fruit juice production to minimise oxidative changes where relatively low heat treatment is employed so as not to damage delicate juice flavour.

Dry sulphuring is the technological step where fruit is exposed to fumes of SO_2 from burning sulphur or from compressed gas cylinders; this treatment could be used in the preparation of fruits (and some vegetables) prior to drying/dehydration.

Sugar Syrup

Sugar syrup addition is one of the oldest methods of minimising oxidation. It was used long before the causative reactions were understood and remains today a common practice for this purpose. Sugar syrup minimises oxidation by coating the fruit and thereby preventing contact with atmospheric oxygen.

Sugar syrup also offers some protection against loss of volatile fruit esters and it contributes sweet taste to otherwise tart fruits. It is common today to dissolve ascorbic acid and citric acid in the sugar syrup for added effect or to include sugar syrup after an SO_2 treatment.

Fresh Fruit Storage

Some fruit species and specially apples and pears can be stored in fresh state during cold season in some countries' climatic conditions. Fruit for fresh storage have to be autumn or winter varieties and be harvested before they are fully mature. This fruit also has to be sound and without any bruising; control and sorting by quality are mandatory operations.

Sorting has to be carried out according to size and weight and also by appearance; fruit which is not up to standard for storage will be used for semi-processed product manufacturing which will be submitted further to industrial processing. Harvested fruit has to be transported as soon as possible to storage areas. Leaving fruit in bulk in order to generate transpiration is a bad practice as this reduces storage time and accelerates maturation processes during storage. In order to store large quantities of fruit, silos have to be built.

NATURAL ENZYME OF FRUIT AND VEGETABLES

Fruit and vegetables also contain a natural enzyme which can further hydrolyse pectin to the point where the pectin loses much of its gel forming property. This enzyme is known as pectin methyl esterase. Materials such as tomato juice or tomato paste will contain both pectin and pectin methyl esterase. If freshly prepared tomato juice or paste is allowed to stand the original viscosity gradually decreases due to the action of pectin methyl esterase on pectin gel.

This can be prevented if the tomato products are quickly heated to a temperature of about 82°C (180 F°) to deactivate the pectin methyl esterase liberated from broken cells before it has a chance to hydrolyse the pectin. Such a treatment is commonly practiced in the manufacture of tomato juice products. This is known as the "hot-break process" and yields products of high viscosity.

In contrast, where low viscosity products are desired no heat is used and enzyme activity is allowed to proceed. This is "cold-break" process. After sufficient decrease in viscosity is achieved the product can be heat treated, as in canning, to preserve it for long term storage. It is often also desirable to firm the texture of fruit and vegetables, especially when products are normally softened by processing.

In this case advantage is taken of the reaction between soluble pectic substances and calcium ions which form calcium pectates. These calcium pectates are water insoluble and when they are produced within the tissues of fruit and vegetables they increase structural rigidity. Thus, it is common commercial practice to add low levels of calcium salts to tomatoes, apples, and other vegetables and fruits prior to canning or freezing.

Sources of Colour and Colour Changes

In addition to a great range of textures, much of the interest that fruits and vegetables add to our diets is due to their delightful and variable colours. The pigments and colour precursors of fruit and vegetables occur for the most part in the cellular plastic inclusions such as the chloroplasts and other chromoplasts, and to a lesser extent dissolved in fat droplets or water within the cell protoplast and vacuoles. These pigments are classified into four major groups which include the chlorophylls, carotenoids, anthocyanins, and anthoanthins. Pigments belonging to the latter two groups also are referred to as flavonoids, and include the tannins.

The Chlorophylls

The chlorophylls are contained mainly within the chloroplasts and have a primary role in the photosynthetic production of carbohydrates from carbon dioxide and water. The bright green colour of leaves and other parts of plants is largely due to the oilsoluble chlorophylls, which in nature are bound to protein molecules in highly organised complexes.

When the plant cells are killed by ageing, processing, or cooking, the protein of these complexes is denatured and the chlorophyll may be released. Such chlorophyll is highly unstable and rapidly changes in colour to olive green and brown. This colour change is believed to be due to the conversion of chlorophyll to the compound pheophytin.

Conversion to pheophytin is favoured by acid pH but does not occur readily under alkaline conditions. For this reason peas, beans, spinach, and other green vegetables which tend to lose their bright green colours on heating can be largely protected against such colour changes by the addition of sodium bicarbonate or other alkali to the cooking or canning water. However, this practice is not looked upon favourably nor used commercially because alkaline pH also has a softening effect on cellulose and vegetable texture and also destroys vitamin C and thiamin at cooking temperatures.

The Carotenoids

Pigments belonging to this group are fat-soluble and range in colour from yellow through orange to red. They often occur along with the chlorophylls in the chloroplasts, but also are present in other chromoplasts and may occur free in fat droplets. Important carotenoids include the orange carotenes of carrot, maize, apricot, peach, citrus fruits, and squash; the red lycopene of tomato, watermelon, and apricot; the yellow-orange xanthophyll of maize, peach, paprika and squash; and the yellow-orange crocetin of the spice saffron. These and other carotenoids seldom occur singly within plant cells.

A major importance of some of the carotenoids is their relationship to vitamin A. A molecule of orange beta-carotene is converted into two molecules of colourless vitamin A in the animal body. Other carotenoids like alpha-carotene, gamma-carotene, and cryptoxanthin also are precursors of vitamin A, but because of minor differences in chemical structure one molecule of each of these yields only one molecule of vitamin A. In food processing the carotenoids are fairly resistant to heat, changes in pH, and water leaching since they are fat-soluble. However, they are very sensitive to oxidation, which results in both colour loss and destruction of vitamin A activity.

The Flavonoids

Pigments and colour precursors belonging to this class are water-soluble and commonly are present in the juices of fruit and vegetables. The flavonoids include the purple, blue, and red anthocyanins of grapes, berries, plump, eggplant, and cherry; the yellow anthoxanthins of light coloured fruit and vegetables such as apple, onion, potato, and cauliflower, and the colourless catechins and leucoanthocyanins which are food tannins and are found in apples, grapes, tea, and other plant tissues. These colourless tannin compounds are easily converted to brown pigments upon reaction with metal ions.

Properties of the anthocyanins include a shifting of colours with pH. Thus many of the anthocyanins which are violet or blue in alkaline media become red upon addition of acid. Cooking of beets with vinegar tends to shift the colour from a purplish red to a brighter red, while alkaline water can influence the colour of red fruits and vegetables towards violet and gray-blue. The anthocyanins also tend towards the violet and blue hues upon reaction with metal ions, which is one reason for lacquering the inside of metal cans when the true colour of anthocyanin-containing fruits and vegetables is to be preserved.

The water-soluble property of anthocyanins also results in easy leaching of these pigments from cut fruit and vegetables during processing and cooking. The yellow anthoxanthins also are pH sensitive tending towards a deeper yellow in alkaline media. Thus potatoes or apples become somewhat yellow when cooked in water with a pH of 8 or higher, which is common in many areas. Acidification of the water to pH 6 or lower favours a whiter colour. The colourless tannin compounds upon reaction with metal ions form a range of dark coloured complexes which may be red, brown, green, grey, or black. The various shades of these coloured complexes depend upon the particular tannin, the specific metal ion, pH, concentration of the complex, and other factors not yet fully understood.

Water-soluble tannins appear in the juices squeezed from grapes, apples, and other fruits as well as the brews from extraction of tea and coffee. The colour and clarity of tea are influenced by the hardness and pH of the brewing water. Alkaline waters that contain calcium and magnesium favour the formation of dark brown tannin complexes which precipitate when the tea is cooled.

If acid in the form of lemon juice is added to such tea its colour lightens and the precipitate tends to dissolve. Iron from equipment or from pitted tin cans has caused a number of unexpected colours to develop in products containing tannins, such as coffee, cocoa and foods flavoured with these. The tannins are also important because they have an astringency which influences flavour and contributes body to such beverages as tea, wine, apple cider, etc.

10

Seed Propagation

INTRODUCTION

- Seed Production
  - Location...........Western States (California, Idaho, Washingtion) Low humidity (less fungal infections, bacterial diseases) Mild temp (for flowering and seed set) Isolated areas (to guarantee maintenance of pure-line seeds:
  - Hybrid seed production -hybrids are a cross between two different varieties In the US -Use of male sterility genes to ensure there is no pollination from the same variety In developing countries-Hand pollination
  - Labour intensive -much labour is needed for harvesting, cleaning, and handling of the seeds
- Seed Storage
  - Optimum condition
    1. Low temperature (0-35° F)
    2. Low humidity (4-6 per cent Relitive Humidity)
  - Practical conditions
    1. 32-50° F (0-10°C) and 50-60 per cent RH
  - Longevity varies by species
    1. Anual flowers 1-2 years
    2. Lotus > 1000 years
  - Cryogenic storage (-196°C)......... Permanent
- Seed Germination
  - Viability (per cent germination)
    1. Determined by per cent germination
    2. Tetrazolium test to check tissue viability (T2T or TTC -two forms of tetrazolium)
    3. Seed priming is done for vigourous, uniform germination (seed priming- uses water imbibition to prepare seeds for germination without allowing for radicle emergence
    4. The seed are then dried again until planting)

- Breaking Seed Dormancy
  i. Scarification- methods used to overcome physical dormancy (physical dormancy- seed coat is impermeable to water and gases)
    a. Mechanical abrasive method*
    b. Hot water (170-212° F)
    c. Sulfuric acid (concentrated) 1-5 hours
  ii. Stratification method used to remove physiological dormancy (physiological dormancy- dormancy due to inhibitors within the seed)
    a Cold Stratification
  iii. Seed is stored at 32° F for 6-8 weeks with both moisture and aeration
  iv. Most popular method for woody plant seeds
- Warm Stratification
  i. Seed is stored at warm temperature (>45° F) with moisture
  ii. Required for some species (Viburnum, redbud)
- Embryo Culture (Embyo Rescue) Immature zygotic embyos are taken out of ovary and cultured *in vitro* → viable seeds cam be obtained
  i. Used in abortive interspecific crosses (potatoes)- Tuckey's medium
  ii. Embryos lacking endosperm (orchids)- Kn udson's medium

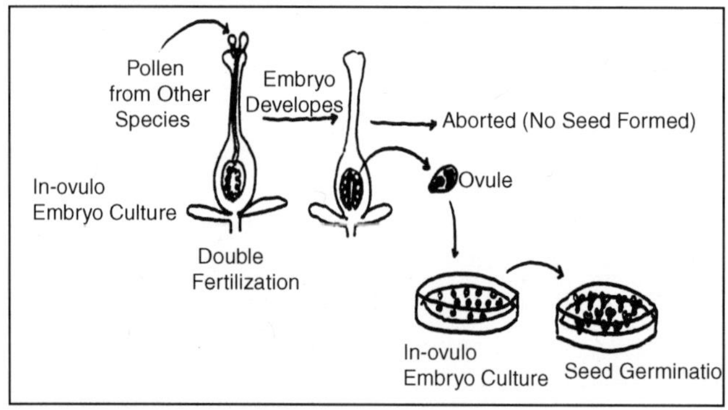

- Planting
  - Direct Seeding
    i. Use of pelletised seeds
    ii. Fluid drilling
  - Transplanting
    i. Mechanical transplanting
      a. lettuce, cabbage, cellery in California
- Germination

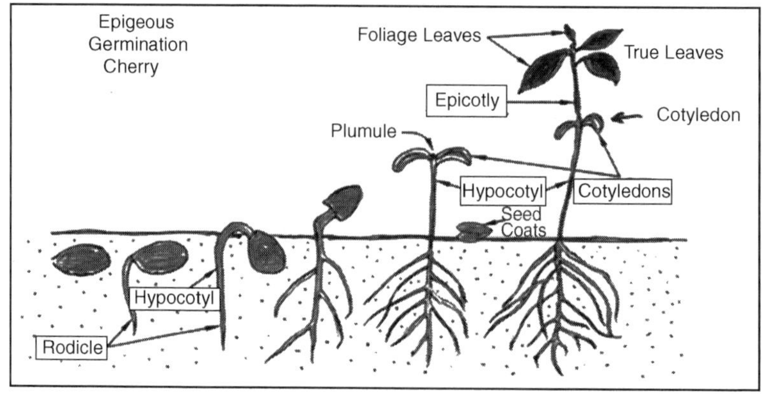

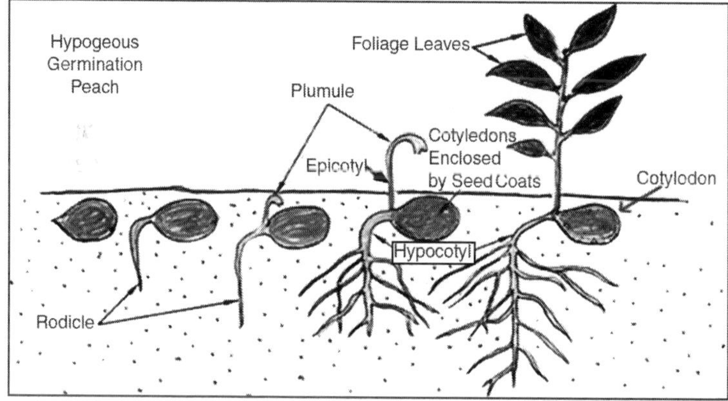

VEGETATIVE PROPAGATION METHODS (ASEXUAL)

- Apomictic seed-seed formed with out pollination
- Use of specialised vegetative structures-cuttings
- Advetitious roots and shoots-shoots and roots formed where they normally would not form
- Grafting-graft top of desired plant on a hardier rootstock.
  - Use of Apomictic Seeds
    i. Apomixis-development of seeds without fertilization-seeds are formed by megaspore mother cells and so are genetically the same as maternal parent Examples: Mango, Kentucky Bluegrass, Citrus
  - Use of Specialised Vegetative Structures
    i. Runners-stolons fo strawberry, strawberry Geranium
    ii. Bulbs Onion, tulip, Easter lily, Hyacinth
    iii. Corms- Gladiolus, crocus cormel- small corms
    iv. Rhizomes-Iris, turfgrass, lily-of-the-valley, orchid
    v. Off-shoots-pineapple (slip), Dendrobium, Cymbidium, daylily
    vi. Stem tubers-potato (eyes)

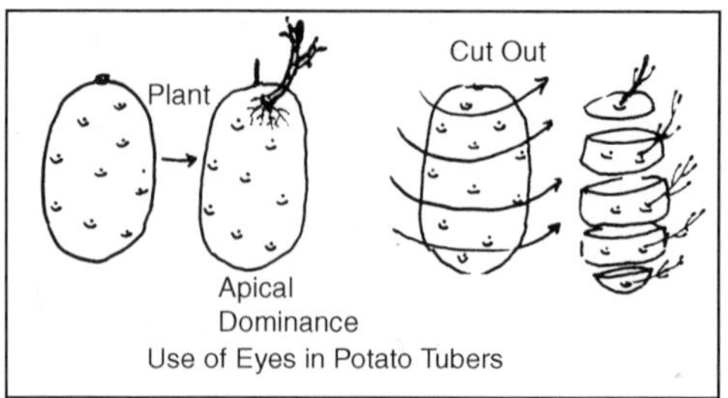

Use of Eyes in Potato Tubers

vii. Tuberous roots-sweet potato, dahlia, tuberous begonia
viii. Root suckers-Raspberry

– Induction of Adventitious Shoots and Roots
  i. Layering-Regenerated vegetative parts attached to the plant Kinds-tip, simple, trench, serpentine, mound, air

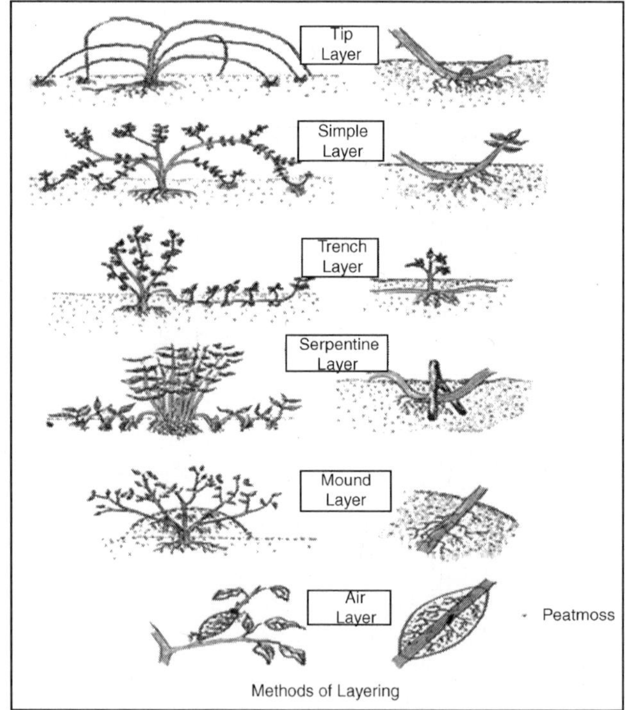

Methods of Layering

  ii. Cutting propagation- Detached stems and roots are rooted or shoot-generated -Stem cuttings- Hardwood (Rose, apple) Softwood (Peach) Semi-hardwood (most plants)

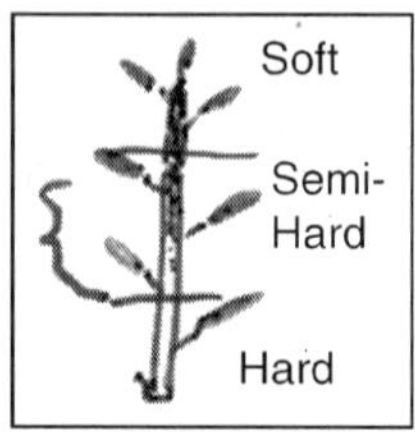

Root cuttings- Rose, sumac, lilac - root cuttings will have a problem with phenotype conversion in chimeric plants

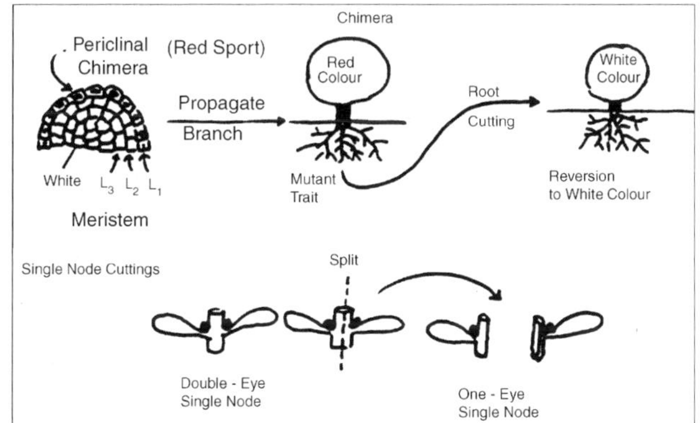

– Origins of Adventitious Shoots and Buds Adventitious buds (roots)- arising from any plant part other than root or shoot primordia (Apex, axillary bds, on nodes)- generally occurs form secondary meristematic tissues (on internodal area, leaf edges)

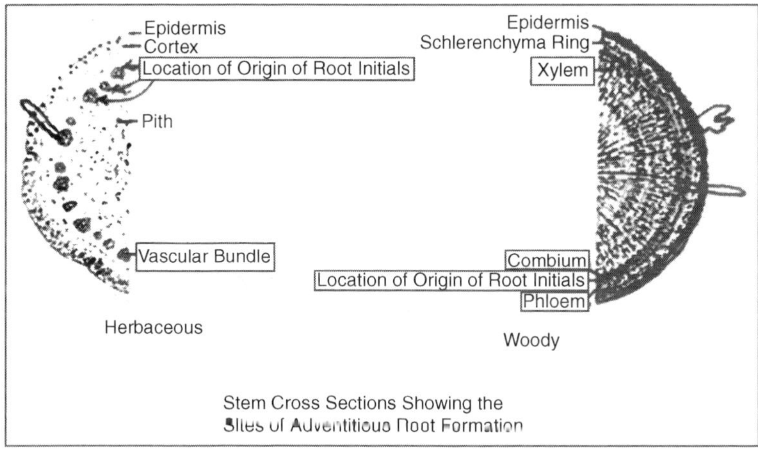

Stem Cross Sections Showing the Sites of Adventitious Root Formation

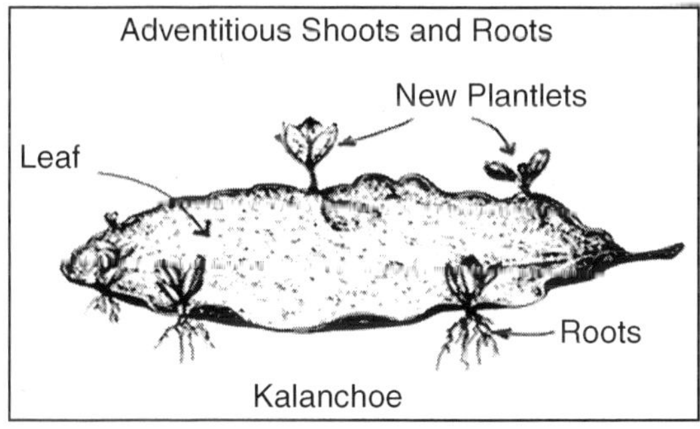

– Rooting stimulation

i. Use of rooting hormones (IBA, NAA--- Auxins)
   IBA= Indolebutyric acid
ii. Presence of light
iii. Temperature Gradient (cooler air, warmer soil)

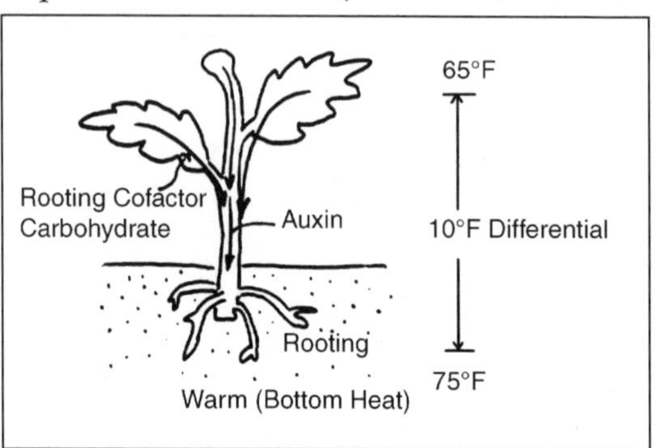

- Grafting
  i. The joining of plant parts by means of tissue regeneration
  ii. Stock (provides root) and scion (becomes stems and shoots)
  iii. Compatibility between stock and scion essential (Graft incomatibility may be shown after 10-15 years)
  iv. Alighnment of cambium layers important
- Kinds
  i. Cleft graft
  ii. Whip-and-tounge graft
  iii. Side graft
  iv. Budding (T-bud, chip bud, patch bud, flute bud)
  v. Inarching
  vi. Bridge graft
  vii. Topworking

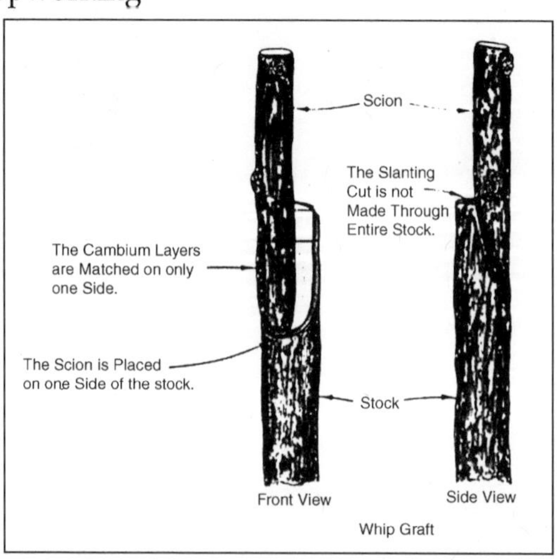

Whip Graft

MICROPROPAGATION (TISSUE CULTURE)

- *Asexual Embryogenesis (Somatic): In vitro* induction of somatic embyos which can be grown into plants.
- *Rapid Shoot Multiplication:* Apical or axillary shoot meristems are cultured shoot proliferation is fast: (One explant can produce over one million plants in one year) about 120 species are commercially micropropagated.
- *Organogenesis:* Shoot and root formation from callus tissue.

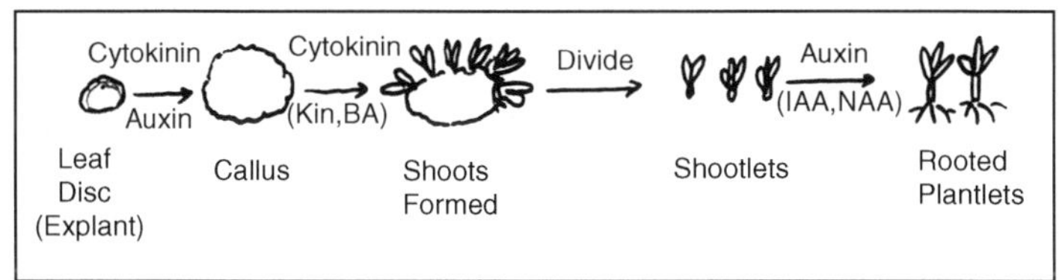

TEMPERATURE CONDITIONS ON MICROBIAL GROWTH

Impact of Time

When considering growth rates of microbial pathogens, in addition to temperature, time is a critical consideration. Food producers or manufacturers address the concept of time as it relates to microbial growth when a product's shelf life is determined. Shelf life is the time period from when the product is produced until the time it is intended to be consumed or used. Several factors are used to determine a product's shelf life, ranging from organoleptic qualities to microbiological safety. For the purpose of this report, the key consideration is the microbiological safety of the product. The Uniform Open Dating Regulation requires the shelf life of a perishable food product to be expressed in terms of a "sell by" date (NIST 2000). The "sell by" date must incorporate the shelf life of the product plus a reasonable period for consumption that consists of at least one-third of the approximate total shelf life of the perishable food product.

At Retail or foodservice, an additional period of time referred to herein as "use-period" should also be considered. As an example, fast food locations may find it operationally desirable to hold processed cheese slices at ambient temperatures for a complete shift or meal period, which may be in excess of 4 h. This practice provides operational efficiency by allowing the cheese to melt faster on a hot sandwich as well as providing a better quality sandwich. Although refrigeration may be required for safety under long-term storage conditions, for use-periods measured in hours, storage at ambient temperatures may be acceptable.

Under certain circumstances, time alone at ambient temperatures can be used to control product safety. When time alone is used as a control, the duration should be equal to or less than the lag phase of the pathogen(s) of concern in the product in question. For refrigerated food products, the shelf life or use-period required for safety may vary depending on the temperature at which the product is stored. For example, Mossel and Thomas (1988) report that the lag time for growth of *L. monocytogenes* at 10°C (50 °F) is 1.5 d, while at 1°C (34 °F) lag time is ~3.3 d. Likewise, they report that at 10°C (50 °F) the generation time for the same organism is 5-8 h, while at 1°C (34 °F), the generation time is between 62 and 131 h.

Figure shows the effect of temperature and pH on lag times of *L. monocytogenes*. The data were obtained by using the USDA Pathogen Micromodel Programme at a NaCl concentration of 2% and a_w of 0.989. It should be noted that this model was developed in broth under various salt and pH combinations, and that growth of bacteria in food systems will likely differ. According to the model results, a temperature shift from 10 (50) to 25°C (77 °F) decreases the lag time of *L. monocytogenes* from 60 to 10 h. In a similar manner, a pH increase from 4.5 to 6.5 decreases the lag time from 60 to 5 h.

In conclusion, the safety of a product during its shelf life may differ, depending upon other conditions such as temperature of storage, pH of the product, and so on. This study by Mossel and Thomas (1988), along with numerous others, illustrates that various time/temperature combinations can be used to control product safety depending on the product's intended use.

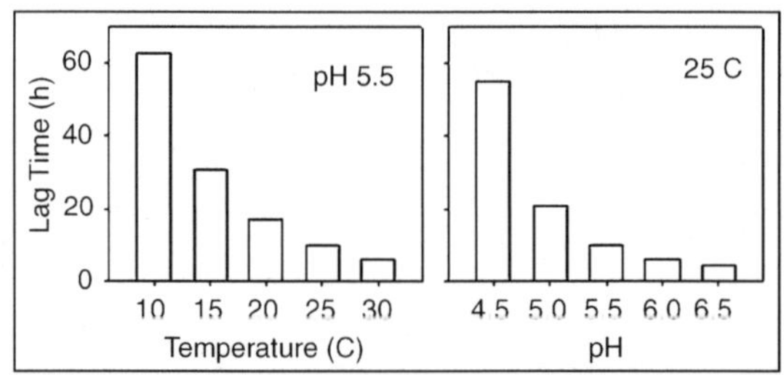

Fig. Effect of Temperature or pH on Lag Times of *Listeria Monocytogenes* from USDA PMP ver 5.1 (2% NaCl, a_w 0.989)

As stated earlier, time alone at ambient temperatures can be used to control product safety. When time alone is used as a control, the duration should be equal to or less than the lag phase of the pathogen(s) of concern in the product in question.

Impact of Temperature

All Microorganisms have a defined temperature range in which they grow, with a minimum, maximum, and optimum. An understanding of the interplay between time, temperature, and other intrinsic and extrinsic factors is crucial to selecting the proper storage conditions for a food product. Temperature has dramatic impact on both the generation time of an organism and its lag period. Over a defined temperature range, the growth rate of an organism is classically defined as an Arrhenius relationship.

The log growth rate constant is found to be proportional to the reciprocal of the absolute temperature:

- G = –μ/2.303 RT where,
- G = Log Growth Rate Constant
- μ = Temperature Characteristic (Constant for a Particular Microbe)
- R = Gas Constant
- T = Temperature (°K)

The above Relationship holds over the linear portion of the Arrhenius plot. However, when temperatures approach the maxima for a specific microorganism, the growth rate declines more rapidly than when temperatures approach the minima for that same microorganism.

A relationship that more accurately predicts growth rates of microorganisms at low temperatures follows:

- $\sqrt{r}$ = $b(T-T_0)$ where,
- r = Growth Rate
- b = Slope of the Regression Line
- T = Temperature (°K)
- T_0 = Conceptual Temperature of no Metabolic significance

At low temperatures, two factors govern the point at which growth stops:

1. Reaction rates for the individual enzymes in the organism become much slower, and,
2. Low temperatures reduce the fluidity of the cytoplasmic membrane, thus interfering with transport mechanisms (Mossel and others 1995).

At high temperatures, structural cell components become denatured and inactivation of heat-sensitive enzymes occurs. While the growth rate increases with increasing temperature, the rate tends to decline rapidly thereafter, until the temperature maximum is reached.

The Relationship between temperature and growth rate constant varies significantly across groups of microorganisms.

Four major groups of microorganisms have been described based on their temperature ranges for growth:

1. Thermophiles,
2. Mesophiles,
3. Psychrophiles, and
4. Psychrotrophs.

Tables list the temperature ranges for these four groups (ICMSF 1980) and for pathogens of concern.

The optimum temperature for growth of thermophiles is between 55 to 65°C (131 to 149 °F) with the maximum as high as 90°C (194 °F) and a minimum of around 40°C (104 °F). Mesophiles, which include virtually all human pathogens, have an optimum growth range of between 30°C (86 °F) and 45°C

(113 °F), and a minimum growth temperature ranging from 5 to 10°C (41 to 50 °F). Psychrophilic organisms have an optimum growth range of 12°C (54 °F) to 15°C (59 °F) with a maximum range of 15°C (59 °F) to 20°C (68 °F).There are very few true psychrophilic organisms of consequence to foods. Psychrotrophs such as L. monocytogenes and C. botulinum type E are capable of growing at low temperatures (minimum of - 0.4°C [31 °F] and 3.3°C [38 °F], respectively, to 5°C [41 °F]), but have a higher growth optimum range (37°C [99 °F] and 30°C [86 °F], respectively) than true psychrophiles. Psychrotrophic organisms are much more relevant to food and include spoilage bacteria, spoilage yeast and molds, as well as certain foodborne pathogens.

Growth temperature is known to regulate the expression of virulence genes in certain foodborne pathogens (Montville and Matthews 2001).

For example, the expression of proteins governed by the Yersinia enterocoliticavirulence plasmid is high at 37°C (99 °F), low at 22°C (72 °F), and not detectable at 4°C (39 °F). Growth temperature also impacts an organism's thermal sensitivity. Listeria monocytogenes, when held at 48°C (118 °F) in inoculated sausages, has an increase of 2.4-fold in its D value at 64°C (147 °F).

It Must be emphasized that the lag period and growth rate of a microorganism are influenced not only by temperature but by other intrinsic and extrinsic factors as well. For example, the growth rate of Clostridium perfringens is significantly lower at pH 5.8 versus pH 7.2 across a wide range of temperatures. Salmonellae do not grow at temperatures below 5.2°C (41 °F). The intrinsic factors of the food product, however, have been shown to impact the ability of salmonellae to grow at low temperatures.

SEED MULTIPLICATION

Governments in the region noted about 30 years or so ago that the disparity between food production and population increase in Africa south of the Sahara was ever widening with a resultant decrease in per capita food consumption. Due to population increase land availability was declining. The best approach to increasing food production was, therefore, to increase crop yields. But farmers were growing mostly their own traditional varieties, which are low yielding.

Some high yielding varieties had been released but were not available to farmers. As a result quasi-government seed companies were established in Malawi (National Seed Company of Malawi, NSCM) and Tanzania (TANSEED). After a few years of operation, these companies abandoned their self-pollinating crop seed enterprises citing poor profitability. As a result, low adoption of recommended technologies for beans still remains a major problem in the region despite the development of new high yielding varieties. Such barriers to seed dissemination are a major constraint to the determination of impact from CRSP programs. The East Africa Bean/Cowpea CRSP Project therefore embarked on development of locations and means to multiply and disseminate basic and foundation seed of improved varieties.

Earlier studies at Bunda indicated that growing beans in Malawi during the hot dry season acts as a filtration process of seedborne diseases except viruses. As a result, the programme embarked on increasing breeders' seed under irrigation on campus. In each of the past five years breeders' seed has been increased to supply basic seed. At SUA basic seed is also produced similarly.

During the past five years the Malawi Bunda CRSP multiplied basic seed of Kalima on about 5ha in each year to produce foundation seed. This seed was sold to NGOs, *e.g.*, ActionAid and Self-Help Development International, and to Ministry of Agriculture for demonstrations. In Tanzania the bean seed distribution system is significantly different from that in Malawi. Until recently, bean seed was produced exclusively on government farms with little seed being produced or reaching the farmer. As the shortcomings of the official Ministry of Agriculture multiplication and dissemination stations become apparent and as market liberalization has progressed, more decentralized seed production systems are developing.

Barriers to seed dissemination result because of a number of socio-economic constraints. Poor distribution of inputs and produce in the region results from poor infrastructure that exists. The road/bridge network is so vulnerable that vast areas are inaccessible particularly during the rainy season. Markets are not adequately established.

The quasi-government marketing organizations have better networks of markets but are poorly organized. They offer the farmers the lowest prices and generally do not have ready cash to offer the farmer when the crop is ready for sale. Entrepreneurs are coming into the system due to market liberalization. They may have ready cash and transport but lack necessary knowledge to handle seeds differently from other inputs and produce. In many instances farmers are not organized into groups or cooperative societies, which are powerful bodies that would assist them do combined efforts to break through the barriers of the marketing systems. In Malawi the credits system broke down as the country changed from dictatorship to a plural society.

It has been realized that smallscale bean farmers will:

- Buy bean seed and not just rely on their own stocks,
- Afford to buy seed of new varieties, and
- Not function efficiently in varietal diffusion.

Non-governmental organizations are increasingly filling a niche in seed dissemination activities and thereby providing temporary solutions to the barriers expressed previously and to facilitate farmers' capabilities for seed production. Their usual goals are food security and poverty alleviation. Malawi has a greater profusion of NGOs because of its burden of Mozambican refugees in the 80's and 90's. These NGOs have stayed even after most refugees were repatriated. In general, these NGOs operate two types of seed schemes; *viz.*: commercially-oriented schemes where fewer farmers purchase more seed to produce seed for sale and the food security oriented schemes where more farmers are given less seed on credit to produce seed or grain for the market or food.

The former crop is certified whereas the latter is hardly inspected but the seed is 'approved'. Approved seed is that seed which is produced either from certified or known stocks of uncertified seed. Seeds of this category do not necessarily carry a government seal but have to pass seed standards for purity and germination. This has facilitated the operations of many NGOs in seed multiplication and dissemination in Malawi.

In Malawi two organizations involved in seed distribution are described. The first is an NGO (ActionAid) that has a food security-oriented seed scheme. ActionAid developed a 'Malawi Smallholder Seed Development Project, MSSDP' after a flood disaster in 1992 when 1.1 million households benefited from 3,000 tonnes of relief seed issues.

The objectives of this Project are to:

- Develop low cost mechanisms, which would improve and sustain availability and accessibility of appropriate, improved seed to resource poor farmers in participating communities,
- Establish sustainable self-motivating community based groups, which will manage seed multiplication and distribution in the communities, and
- Train selected extension staff and participating community groups in seed production, seed quality control, on farm seed selection and storage, group dynamics and in community participatory methodologies.

In 1998 a total of 3.6 tonnes of certified bean seed was supplied and made available to 76 groups with a total membership of 1500, who in turn multiplied more than 12 tonnes of first generation approved bean seed. About 75% of this seed were Kalima and Nasaka from CRSP.

An example of the commercially oriented seed scheme is that operated by the Maize Productivity Task Force, formed in Malawi in 1995, with the aim of increasing agricultural productivity. Its Action Group Two, one of the four groups, was set up to increase availability of seeds of self-pollinating and open pollinated crop varieties, which are neglected by the large multilateral seed companies. Funded by European Union, this group embarked on seed production and marketing as a business.

The major activities of this group are:

- Technology promotion through demonstrations, radio programs and field days;
- Training of extension staff and seed growers in seed production techniques;

- Organization of farmers through establishment of farmer associations; and
- Seed production.

In 1998/99 season there were 109 farmers growing seed beans on 28ha producing 24 tonnes of seed.

The CRSP project in Tanzania released SUA90 and Rojo. These varieties have been introduced into some selected farming communities in the country. The areas where they are currently grown and marketed include Morogoro and Kilosa districts of Morogoro region, Kongwa district of Dodoma region and some areas of Tanga and Arusha regions. Seed of Rojo was recently produced at SUA, Tanzania and was purposely disseminated through two NGOs, *viz*.: LVIA (Lay Volunteers International Agency) and CCT (Christian Council of Tanzania) and through the CRSP Project.

The CRSP Project distributed the seed to Kongwa, Kisanga and Msolwa villages in addition to farmers in Maharaka and Msongozi, Muhenda and Ulaya-Mbuyuni villages of Morogoro and Kilosa districts. LVIA selected participating farmers based on their willingness to engage in seed production and on financial ability to start seed production. For SUA CRSP gender and education were major criteria. CCT selected progressive and innovative farmers in terms of knowledge on improved agricultural production. Notable amongst the results is the fact that smallholder farmers are willing to purchase foundation seed.

However, seed demand and marketing issues are clearly not addressed. Farmers were unable to sell their seeds within their villages and had to sell as cheap grain. Adequate extension is lacking. The majority of farmers are still unaware of the new seed. LVIA provided its own extension staff and farmers under their scheme were more knowledgeable. Seed promotion activities also worked better for farmers under their scheme. A general recommendation was that successful seed producers are those more financially endowed and that these are the farmers to be targeted in future.

FORMULATION FACTORS AND SEED STORAGE CONDITIONS

When evaluating a formulation, it is important to understand the range of key factors that control its microbiological stability. Intrinsic factors such as pH, aw, or preservative level may be key to preventing the growth of pathogens or to preventing spoilage that would influence the safety of the product during its intended shelf life. It is, therefore, important to test each key variable singly and/or in combination in the formulation under worst-case conditions.

For example, if the target pH is 4.8 + 0.2 and the process capability is within that tolerance range, it is important to challenge the product on the high side of that range (that is, pH 5.0). Similarly, if sorbic acid is used at a level of 0.15 + 0.05%, the product should be challenged at the low concentration of 0.10%. This is recommended to ensure that the challenge study covers the process capability range for each critical factor in the formulation. Relevant intrinsic properties such as pH, aw, and salt level should be documented for each study for future comparison and reference.

Test samples should ideally be stored in the same packaging as intended for the commercial marketplace. If the commercial product is vacuum or MAP packaged, then the samples used in the microbiological challenge study should be packaged under the same conditions using the same packaging film.

The storage temperature used in the microbiological challenge study should include the typical temperature range at which the product is to be held and distributed. A refrigerated product that may be subject to temperature abuse should be challenged under representative abuse temperatures. Products that may encounter high humidity environments should also be challenged under those conditions (Notermans and others). Some challenge studies may incorporate temperature cycling into their protocol. For example, the manufacturer may distribute a refrigerated product under well-controlled conditions for a portion of its shelf life, after which the product may be subjected to elevated temperatures immediately prior to and during use.

SAMPLE ANALYSIS

Typically, in a Microbiological challenge study, the levels of live challenge microorganisms are enumerated at each sampling point. Usually, it is desirable to have at least duplicate and, preferably, triplicate samples for analysis at each time point. In cases where higher levels of certainty are needed, a larger number of replicates should be used or the study should be replicated. The selection of enumeration media and method (for example, direct plating versus Most Probable Number) is dependent on the type of pathogens or surrogates used in the study.

If the product does not have a substantial background *microflora,* non-selective media for direct enumeration may be used. In cases where toxin-producing organisms are used (for example, *Staphylococcus aureus* or *C. botulinum*), appropriate toxin testing should be performed at each time point using the most current validated method. Toxin levels may not always be tested at each time point in the study, but should be done at frequent enough intervals throughout the desired shelf life of the product to determine if that shelf life is acceptable. Where appropriate, resucitation methods may be used to avoid erroneous result.

It is Prudent to analyse the product, including uninoculated control samples, at each or selected sampling points in the study to see how the background microflora is behaving over product shelf life. For example, if a product has a high background microflora, it may suppress the growth of the challenge inoculum. In some cases, this is useful and desirable because the product spoils before pathogens can grow. In other situations, the background microorganisms may not be universally present, leading to a potentially false sense of security. Also, under some circumstances, the background microorganisms can change the formulation parameters in the product to favour or inhibit growth of the inoculum over time (for example, molds can raise product pH; lactobacilli can decrease product pH).

It is also important to track pertinent physico-chemical parameters of the product over shelf life to see how they might change and influence the behaviour of the pathogen. Understanding how factors such as aw, moisture, salt level, pH, MAP gas concentrations, preservative levels, and other variables behave over product shelf life is key to understanding the microbiological stability of the product.

DATA INTERPRETATION

Once the microbiological challenge study is completed, the data should be analysed to see how the pathogens behaved over time.

Trend analysis and appropriate graphical plotting (that is, semi-log plots) of the data will show whether the challenge organisms died, remained stable, or increased in numbers over time.

In the case of toxin-producing pathogens, no toxin should be detected over the designated challenge period. Combining the quantitative inoculum data for each time point with data on the background microflora and the relevant physico-chemical parameters gives a powerful and broad representation of the microbiological stability of the formulation under evaluation. Based on these data, a reasonable shelf life can be established or adjustments can be made to the formulation so that it is less susceptible to pathogen growth.

When using microbiological challenge testing, as part of a process validation protocol, analysis of the data will show whether the process is capable of delivering the required level of lethality (that is, conforms with the pre-determined performance standard). Based on this information, adjustments can be made to the process, if necessary, in order to meet the lethality requirements.

The Data from microbiological challenge testing can be used in developing predictive microbiological models or in validating existing ones. Predictive models are computer-based programmes that simulate or predict how specific microorganisms will behave in a formulation under specific conditions (for example, pH, a_w, moisture, salt, and preservatives). Microbiological challenge tests are used both to generate these types of empirical models and to validate their applicability.

Overall, well-designed challenge studies can provide critical information on the microbiological safety and stability of a food formulation. They are also invaluable in validating key lethality or microbiological control points in a process. Challenge studies can be an invaluable aid in determining if a food product requires temperature control throughout its shelf life or if it can tolerate storage at room temperature for a portion or all of its shelf life.

PASS/FAIL CRITERIA

Selection of microorganisms to use in challenge testing and/or modeling depends on the knowledge gained through commercial experience and/or on epidemiological data that indicate that the food under consideration or similar foods may be hazardous due to pathogen growth. In addition, the intrinsic properties (for example, pH, water activity, and preservatives) and extrinsic properties (for example, atmosphere, temperature, and processing) should be considered. The significance of a population increase varies with the hazard characterization of each microorganism. For example, the growth of infectious pathogens should always be controlled, whereas most toxin production requires substantial growth before a hazard exists. In this case, growth of the toxigenic organism alone does not result in a health hazard, but toxin production will.

The following list identifies microorganisms that can be used in a microbiological challenge study along with the panel's recommendations and rationale for selection and assessment of tolerable growth. Toxigenic molds such as *Aspergillus, Penicillium,* and *Fusarium spp.* Challenge studies related to the need for time/temperature control for safety are not recommended because mold provides a visual clue to prevent consumption of the spoiled product.

Bacillus Cereus: The absence of toxin formation is the preferred criterion. However, since toxin measurement is difficult, a 3 log increase over inoculum levels would indicate the need for time/temperature control.

This growth limit determination is based on the following:

- Typical initial levels of B. cereus are low; therefore, 1000 CFU/g would be a conservative initial level based on the literature.
- Populations of > 106 CFU/g are needed to produce toxin at levels hazardous to health (FDA 2001).
- The emetic toxin of B. cereus is heat stable; therefore, no reduction is likely.

Other considerations:

- *Bacillus cereus* spores are relatively heat sensitive.
- Baking is likely to destroy low levels found in flour used in bread products (Kaur); however, unusual high heat resistance has been reported in pumpkin pie (Wyatt and Guy 1981).
- The potential for survival of B. cereus should be evaluated for specific products.
- Rice and potatoes have a history of association with *B. cereus* foodborne illness and therefore products containing rice or potatoes should be evaluated for time/temperature control requirements.

Campylobacter spp: No challenge testing is recommended because other organisms such as Salmonella have similar routes of contamination, are less fastidious, and are easier to culture. Furthermore, its minimal growth temperature and water activity of 32°C (90 °F) and 0.98, respectively, make Campylobacter spp. it an unlikely candidate for challenge studies.

Clostridium botulinum: The absence of toxin formation based on current methodology is the recommended requirement. Other considerations: *C. botulinum* is appropriate to consider for certain products, particularly those packaged under anaerobic and micro-aerophilic conditions such as MAP products; and those with a history of associated illness, such as products under oil and potatoes.

Clostridium perfringens: A 3 log increase is recommended based on the following facts:

- Although other products may contain surviving spores, *Clostridium perfringens* is relevant mainly to meat and poultry products, including sauces and gravies. Most products subject to the Food Code requirements will be either raw or freshly cooked.
- Vegetative cells of *C. perfringens* are easily destroyed by cooking meat and poultry products, and spore levels are typically low due to demanding sporulation requirements. An initial population of 100 CFU/g was considered to be a conservative worst case by the panel. A population of >105 CFU/g is needed to result in illness; therefore, a 3 log increase would control the hazard.

Enterohemorrhagic E. Coli: If modeling programmes are used to predict the growth of the pathogen, time/temperature holding conditions should maintain enterohemorrhagic *E. Coli* in lag phase due to the infectious nature of the microorganism. However, if laboratory challenge studies are used, the inherent variability in quantitative methods necessitates the use of a progressive increase of < 1 log as indicative that growth is controlled.

Listeria Monocytogenes: Recent risk assessments (FDA/USDA 2001) indicate that low numbers of *L. Monocytogenes* present a low risk to public health. In recognition of this, some countries such as Canada and Germany have established a tolerance for low levels of this organism that will not support growth to high levels. However, a tolerance for *L. Monocytogenes* has not been established in the United States. It is also recognized that products that support the growth of the microorganism present an increased risk. A *L. Monocytogenes* level of 100 CFU/g at the time of consumption may provide an acceptable level of consumer protection (Ross and others). However, data are insufficient to determine general worst-case initial levels. Overall, the panel concluded that a 1 log increase was an appropriate level of control for *L. monocytogenes*. This level accounts for variability in enumeration techniques and represents a view that growth of this organism to high levels represents a risk to public health that must be controlled.

Salmonella spp: Appropriately validated pathogen modeling programmes for growth can be used to verify that*Salmonella* spp. is maintained in the lag phase. Otherwise, population growth should be limited to < 1 log, following the same rationale as for enterohemorrhagic *E. Coli*.

Shigella spp: No challenge studies are recommended for Shigella spp. because it has the same potential source as *Salmonella spp*. and has more fastidious growth and survival requirements.

Staphylococcus Aureus: No detectable toxin should be formed under the time/temperature studies evaluated. As with *C. Botulinum*, current methodology should be used for toxin detection and specific toxin levels should be determined. In lieu of testing for toxin, limiting growth to <3 logs may be used. This limiting growth level is based on an initial population of 1000 CFU/g, and a minimum of 10^6 CFU/g to produce toxin.

Other Considerations: *Staphylococcus aureus* is appropriate to study in foods that receive extensive handling because of the human source of the microorganism. *Staphylococcus aureus* does not compete well with other microorganisms; therefore, it is not appropriate to consider in foods with high levels of other organisms, such as raw vegetables or properly fermented products.

Vibrio spp: Appropriately validated pathogen modeling programmes for growth can be used to verify that *Vibrio*spp. is maintained in the lag phase. Otherwise, population growth should be limited to < 1 log, following the same rationale as for *E. Coli*.

Vibrio Parahaemolyticus: Can be used as a surrogate for other *Vibrio spp*. *Vibrio parahaemolyticus* studies are only appropriate for marine foods. It should also be noted that most fish are highly perishable and therefore will be temperature controlled for spoilage reasons.

IMPROVING SEED GERMINATION

When we want to produce large amounts of plants or simply when we want to start our gardens fast and get the most out of our purchases improving seed germination becomes a large priority. One of the largest concerns of world agriculture as well as the home grower is the decrease in germination time and increase in germination percentage since both of these factors can bring great benefits. Some seeds – especially some flowers and herbs – are often quite difficult to germinate and using certain techniques to increase the rate and speed in which they sprout has been the focus of a large amount of scientific research. The use of priming to decrease germination time, especially what priming is, what types are available and which ones you can use to decrease the germination time of those very difficult seeds.

To understand the concept of priming we first need a good grasp at the general concept of seed germination. A seed is a dormant embryo which carries within it the potential for a new plant's life. The seed is alive, yet has a very slow metabolic rate due to the low mobility of substances within the embryo's cells. This low metabolic rate allows the seed to remain alive, yet survive extremely long periods of time (some seeds can survive even hundreds of years) before actually sprouting into new plants.

GERMINATION

Is the process in which we awaken the embryo – increases seed metabolism and toggles the massive reproduction that causes a new plant to grow. The main mechanism that triggers this process is simply liquid water. When water gets into the embryo and hydrates its cells, it speeds up metabolism and allows the process of cell division and growth to rapidly increase. However it is not always this simple to start this process since several impairments – both chemical and physical – can exist for successful germination.

Priming is simply a process done prior to conventional seed germination which allows the inhibiting mechanism to be broken and the metabolic speed increase to begin. There are several types of priming that can be done. A seed can be submerged in simple water (hydropriming), it can be soaked in a solution of a simple salt (halopriming) or it can be set in a non-ionic solution with high osmotic pressure (osmopriming). It is not entirely well established why one technique might work better than another but certainly some species tend to respond much more efficiently to one or another.

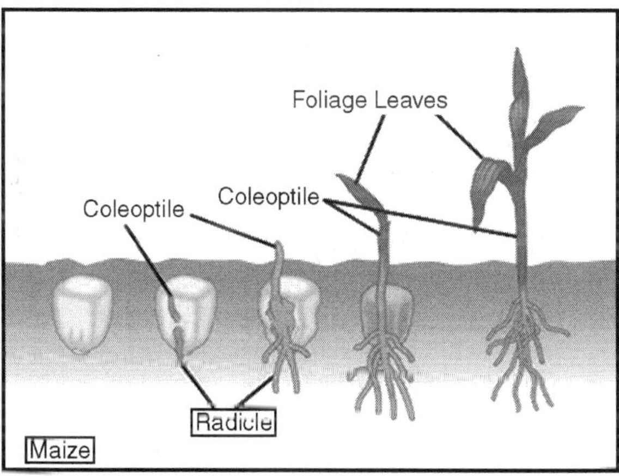

In general, priming offers the opportunity to almost always germinate seeds at much higher speeds without detrimental effects in germination percentages. For example, a two day treatment of parsley seeds with a PEG 6000 (PolyEthyleneGlycol) solution can reduce germination times substantially, from a few weeks to just a few days. Other seeds such as coriander might also benefit from similar treatments with PEG or with treatments with NaCl solutions. In general if you are looking to test priming on some difficult seeds you own you can try three small experiments to know which one works best for your particular seed variety and germination conditions. Do one experiment in which the seeds are simply soaked in water for 24 hours, another in which seeds are placed in a 200mg/L NaCl solution and another one in which the plants are submerged in a PEG 6000 20 per cent solution, then let the seeds air-dry after the treatments. After comparing the results of these experiments with a control with no priming you will be able to see which priming technique is better for you and most effectively increases your seed germination rates.

To sum it up priming your seeds is a very efficient technique to increase the speed of germination without sacrificing germination rates. This methods are not very useful for seeds such as lettuce or tomato – which germinate easily – but they are invaluable for plants such as parsley, coriander or carrots which are generally much harder to germinate. If you have some seeds that have been giving you a hard time or seem to take ages to germinate then setting up some priming experiments might be the best thing to do. Teachers may be unfamiliar with the concept of "priming" seeds. However, teachers understand the value of Early Childhood Education ... and how it can influence later classroom experiences in the higher grades! Giving a child a good start in education is important; the same can be said for seeds. Priming seeds is like giving them a "better start" to life.

Another way of explaining this process has been developed by Keith Kubik, Ph.D. a Seed Physiologist with the Harris Moran Seed Company in California. Dr. Kubik compared seed priming to building a pre-fabricated house.

Pre-fabricated houses are built Inside A Factory. The walls and roof are put together in an environment that is ideal for the workers – very important in our climate here in Canada - and building efficiencies are best. However, the house is not completed inside the factory. The pre-built walls and roof are taken to the building site where the remaining construction takes place – the electrical work, the plumbing, and the basic fixtures like countertops and inside trim.

Less time is consumed at the site, and the houses go up more quickly and more efficiently. Similarly, with seeds, the priming process starts the germination process in the laboratory. The basic chemical reactions or framework needed for the seed to germinate take place under ideal temperatures and high moisture conditions. The seed moisture is set so that the germination is ALMOST complete ... just like the prefabricated house when it is moved onto the actual home site. When you get your seeds, one of the two groups will be a control group and the other group will have been "primed" – almost ready for germination. Like the pre-fabricated house, seed germination of the primed seeds will probably take less time, because part of the germination process is already complete. You will still report on the germination process – the number of seeds planted and number of seeds germinated.

SEED PROCESSING TECHNOLOGIES

WATER ACTIVITY AND PH

The Manipulation of water activity and/or pH is the less complicated of technologies in terms of equipment, expense, and expert personnel needed. Although it may not reduce the microbial load *per se,* reducing water activity or pH may retard or impede microbial growth. (For a more extended description on how water activity and pH can be used as preservation technologies and for a list of the optimum range pH and water activity for various pathogens of concern.

When changing these characteristics of foods with the intention of safely storing a food at room temperature, those minimum pH and water activity values should be taken as guidance. At different temperatures and for different foods, these ranges may vary. For example, as the temperature moves away from optimum, a higher minimum pH is generally observed.

TECHNOLOGIES BASED ON THERMAL EFFECTS

In Addition to microbial inactivation by conventional methods of heating, microwave and ohmic and inductive heating are also considered to be heat-based processes that can inactivate microorganisms by thermal effects. Microwave and radio frequency heating refers to the use of electromagnetic waves of certain frequencies to generate heat in a material through two mechanisms-dielectric and ionic.

Ohmic heating is defined as the process of passing electric currents through foods or other materials to heat them. Ohmic heating is distinguished from other electrical heating methods by the presence of electrodes contacting the food, frequency, and waveform. Inductive heating is a process wherein electric currents are induced within the food due to oscillating electromagnetic fields generated by electric coils. No data about microbial death kinetics under inductive heating have been published.

For any of these heat-based processes, the magnitude of time/temperature history and the location of the cold points will determine the effect on microorganisms. The effectiveness of these processes also depend on water activity and pH of the product. Although the shape of the inactivation curves is expected to be similar to those in conventional heating, the intricacies of each of the technologies, however, need special attention if this technology is used for microbial inactivation.

For instance, in microwave heating a number of factors influence the location of the cold points, such as the composition, shape, and size of the food, the microwave frequency, and the applicator design. The location of the coldest-point and time/temperature history can be predicted through simulation softwares, and it is expected that food processors may be able to use them in the future. For determining the kinetics and efficiency of inactivation of microorganisms for these technologies, surrogate/indicator microorganisms could be selected from those traditionally used in thermal processing studies.

HIGH PRESSURE PROCESSING

High pressure processing (HPP), also described as high hydrostatic pressure (HHP) or ultra high pressure (UHP) processing, subjects liquid and solid foods, with or without packaging, to pressures between 100 and 800 MPa. Process temperature during pressure treatment can be specified from below 32 °F (0°C) to above 212 °F (100°C). Commercial exposure times can range from a millisecond pulse to over 20 min. Chemical and microbiological changes in the food generally will be a function of the process temperature and treatment time. The various effects of high hydrostatic pressure on microorganisms can be grouped into cell-envelope-related effects, pressure-induced cellular changes, biochemical aspects, and effects on genetic mechanisms.

HPP acts instantaneously and uniformly throughout a mass of food independent of size, shape, and food composition. Compression will uniformly increase the temperature of foods approximately 5 °F (3°C) per 100 MPa. Compression of foods may shift the pH of the food as a function of imposed pressure and must be determined for each food treatment process. Water activity and pH are among the critical process factors in the inactivation of microbes by HPP.

An increase in food temperature above room temperature, and to a lesser extent, a decrease below room temperature increases the inactivation rate of microorganisms during HPP treatment. Temperatures in the range of 113–122 °F (45–50°C) appear to increase the rate of inactivation of food pathogens and spoilage microbes. Temperatures ranging from 194–230 °F (90–110°C) in conjunction with pressures of 500-700 MPa have been used to inactivate sporeforming bacteria such as *Clostridium botulinum*. Current pressure processes include batch and semi-continuous systems, but no commercial continuous HPP systems are operating.

The Critical process factors in HPP include pressure, time at pressure, time to achieve treatment pressure, decompression time, treatment temperature (including adiabatic heating), initial product temperature, vessel temperature distribution at pressure, product pH, product composition, product water activity, packaging material integrity, and concurrent processing aids. Interestingly, because HPP acts instantaneously and uniformly through a mass of food, package size, shape, and composition are not factors in process determination.

High hydrostatic pressures can cause undesirable structural changes in structurally fragile foods such as strawberries or lettuce (for example, cell deformation and cell membrane damage). Food products that have been brought to market include raw oysters, fruit jellies and jams, fruit juices, pourable salad dressings, raw squid, rice cakes, foie gras, ham, and guacamole.

A Biphasic pressure inactivation curve is frequently encountered for both vegetative bacteria and endospores indicating the residence of a small pressure-resistant sub-population. Tailing phenomena should be investigated carefully in challenge studies. The use of pathogens rather than surrogates for highly infective pathogens may be advised.

The Elimination of spores from low-acid foods presents food-processing and food-safety challenges to the industry. It is well established that bacterial endospores are the most pressure-resistant life forms known. One of the most heat-resistant pathogens, and one of the most lethal to human beings, is *C. botulinum*, primarily types A, B, E, and F. As such, *C. botulinum* heads the list of most pressure-resistant and dangerous organisms faced by HPP. Spore suspensions of strains 17B and Cap 9B tolerated exposures of 30 min to 827 MPa and 167 °F (75°C) (Larkin and Reddy; personal communication; unreferenced). Because some types of spores of *C. botulinum* are capable of surviving even the most extreme pressures and temperatures of HPP, there is no absolute microbial indicator for sterility by HPP. Among the sporeformers of concern, Bacillus cereus has been the most studied because of its facultative anaerobic nature and very low rate of lethality.

Normally, Gram-positive vegetative bacteria are more resistant to environmental stresses, including pressure, than vegetative cells of gram-negative bacteria. Among the pathogenic non-sporeforming gram-positive bacteria, *Listeria monocytogenes* and *Staphyloccocus aureus* are the two most well-studied regarding the use of HPP processing. *Staphyloccocus aureus* appears to have a high resistance to pressure.

There appears to be a wide range of pressure sensitivity among the pathogenic gram-negative bacteria. Patterson and others (1995) have studied a clinical isolate of *Escherichia Coli* O157: H7 That possesses pressure resistance comparable to spores. Some strains of *Salmonella* spp. have demonstrated relatively high levels of pressure resistances. Given these pressure resistances and their importance in food safety, *E. coli* O157:H7 and *Salmonella* spp. are of key concern in the development of effective HPP food treatments. For vegetative bacteria, nonpathogenic *Listeria* innocua is a useful surrogate for the foodborne pathogen, L. monocytogenes. A nonpathogenic strain of Bacillus may be useful as a surrogate for HPP-resistant *E. coli* O157:H7 isolates

Commercial Implications

Current practical operating pressures for commercial HPP food treatment intensifiers and pressure vessels are in the order of 580 MPa (85,000 psi). If this pressure is specified, then the following process times may be considered as first estimates for initial process planning. It must be understood that actual process parameters must be developed from challenge test packs.

Experience with acid foods suggests that shelf-stable (commercially sterile) products, having a water activity close to 1.0, and pH values less than 4.0, can be preserved using a pressure of 580 MPa and a process hold time of 3 min. This treatment has been shown to inactivate 106 cfu/g of *E. Coli O157*: H7, *Listeria* spp., *Salmonella* spp., or *Staphylococcus* spp. in salsa and apple juice.

Acid foods with pH values between 4.0 and 4.5 can be made commercially sterile using a pressure of 580 MPa and a hold time of 15 min. Products would have an initial temperature of about 71.6 °F (22°C). Shorter hold times are possible if the product is to be refrigerated. Actual hold-time values must be determined from challenge packs and storage studies perhaps twice the length of the intended shelf life of the product.

Low-acid products can be rendered free of pathogens or pasteurized by HPP; however, satisfactory guidelines for hold times at 580 MPa for low-acid food pasteurization have not emerged. For example, the post-package pasteurization of vacuum-packed cured meat products to eliminate *Listeria* spp. represents a useful application of HPP. Ground beef can be pasteurized by HPP to eliminate *E. Coli O157:* H7, *Listeria* spp., *Salmonella* spp., or *Staphylococcus* spp. Much more work is required to develop a suggested hold time at 580 MPa due to the potential for tailing. Changes in product colour and appearance may limit the usefulness of HPP treatment pressures above 200 to 300 MPa.

PULSED ELECTRIC FIELDS

High intensity pulsed electric field (PEF) processing involves the application of pulses of high voltage (typically 20–80 kV/cm) to foods placed between two electrodes. PEF may be applied in the form of exponentially decaying, square wave, bipolar, or oscillatory pulses at ambient, sub-ambient, or slightly above ambient temperature for less than 1 s. Use of PEF can reduce energy usage compared to thermal processes, as less energy is converted into heat, which also reduces detrimental changes to the sensory and physical properties of the food.

To date, PEF has been applied mainly to extend the shelf life of foods. Application of PEF is restricted to food products that can withstand high electric fields, have low electrical conductivity, and do not contain or form bubbles. The particle size of the food in both static and flow treatment modes is a limitation. Also, due to the variations in PEF systems, a method to accurately measure treatment delivery is still needed.

Factors that affect the microbial inactivation with PEF are process factors (electric field intensity, pulse width, treatment time and temperature, and pulse waveshapes), microbial entity factors (type, concentration, and growth stage of microorganism) and media factors (pH, antimicrobials and ionic compounds, conductivity, and medium ionic strength). Many researchers have studied the effects of pulsed electric fields in microbial inactivation; however, due to the numerous critical process factors and broad experimental conditions used, definite conclusions about specific pathogen reductions cannot be made. Research that provides conclusive data on the PEF inactivation of pathogens of concern is clearly needed. Castro and others (1993) reported a 5-log reduction in bacteria, yeast, and mold counts suspended in orange juice treated with PEF. Zhang and others (1995) achieved a 9-log reduction in *E. coli* suspended in simulated milk ultrafiltrate treated with PEF by applying a converged electric field strength of 70 KV/cm for a short treatment time of 160 s. This processing condition may be adequate for commercial food pasteurization that requires 6- to 7-log reduction cycles (Zhang and others 1995).

However, numerous critical process factors exist and carefully designed studies need to be performed to better understand how these factors affect populations of pathogens of concern. Currently, there is little information on the use of surrogate microorganisms as indicators of pathogenic bacteria when PEF is used as a processing method. Selection of surrogates will require the prior identification of the microorganism of concern in a specific food and PEF system. The selection of the appropriate surrogate(s) will depend on the type of food, microflora, and process conditions (that is, electric field intensity, number of pulses, treatment time, pulse wave), and should also follow the general guidelines listed in the validation section.

IRRADIATION

Lrradiation of food refers to the process by which food is exposed to enough radiation energy to cause ionization. Ionization can lead to the death of microorganisms due to genetic damage which prevents cellular replication. For the treatment of foods, FDA has approved the use of gamma rays from decaying isotopes of cobalt-60 or cesium 137, x-rays with a maximum energy of five million electron volts (MeV), and electrons with a maximum energy of 10 MeV. An electron volt is the amount of energy acquired by an electron when accelerated by one volt in a vacuum. X-rays are produced when high energy electrons strike a thin metal film. Lethality of irradiation depends on the target (microorganism), condition of the treated item, and environmental factors. Addition or removal of salt or water, time/temperature of the treatment, or oxygen presence are factors that will influence the antimicrobial effect of irradiation.

Irradiation is considered an additive in the U.S. and as such, it needs to be approved by the FDA office of premarket approval for each new application and labeled. Two terms have been used to define the extent of pathogen reduction with irradiation. Radiation pasteurization refers to the destruction of pathogenic, non-spore-forming foodborne bacteria. In radiation pasteurization, medium dose treatments (1 to 10 kGy) reduce microbial populations, including pathogens in foods. Elimination of pathogens on meat, seafood, and poultry by medium dose irradiation has been studied. Sterilization radiation is used for radiation processes that will render the food commercially sterile or for foods that are both sterile and shelf stable. In this last case, sterilization must ensure the elimination of the most resistant pathogen, endospores of *Clostridium botulinum*.

In order to achieve this, higher doses (42–71 kGy depending on the product) than the ones currently permitted for foods (up to 10 kGy, except for spices) are needed. Only frozen meats consumed by NASA astronauts have been permitted by FDA to be sterilized through irradiation.They are, however, in the market in other countries.

Lonizing radiation is used as a means of extending the shelf life of produce (Diehl 1995; Thayer and others 1996). FDA has approved the use of ionizing radiation with a range dose 0.3–1 kGy for growth and maturation inhibition. Not much effort has been applied to the control of foodborne pathogens on fresh foods, mainly because most medium and high level doses are not appropriate for produce since they can cause sensory defects (visual, texture, and flavour) and/or accelerated senescence (Thomas 1986; Barkai-Golan 1992). Ionizing irradiation has recently been used to eliminate E.Coli O157: H7 from apple Juice, and E. Coli O157: H7 and Salmonellae from seed and sprouts. Doses in the range of <1 to 3 kGy have been shown to reduce or eliminate populations of foodborne pathogens, postharvest spoilage organisms, and other microorganisms on produce (Moy 1983; Urbain 1986; Farkas 1997). Strawberry shelf life can be extended with treatments in the range of 2 to 3 kGy (Sommer and Maxie 1966; Zegota 1988; Marcotte 1992; Diehl 1995). Research conducted since that time suggests that irradiation can be an important treatment to enhance safety of other types of produce.

FDA and the U.S. Department of Agriculture Food Safety and Inspection Service have also approved irradiation to control foodborne pathogens in raw poultry with a dose range of 1.5–3.0 kGy. Recently, the irradiation of raw refrigerated and frozen meat has been approved with maximum doses of 4.7 and 7 kGy, respectively. Radiation doses of 2.5 kGy in beef will result in 6 log reduction of *Campylobacter*, 5 log reduction of *E.Coli* O157:

H7, 3 Log reduction in Salmonellaspp., and 5 log reduction of *Staphylococcus* cells (CAST 1996). Although the potential for consumer infection by pathogens is decreased greatly and shelf life is extended by radiation pasteurization of meat and poultry, the room temperature storage of raw meat products would be highly discouraged.

Other products, such as shell eggs (up to 3 kGy), have recently been approved for irradiation for safety reasons. Shell eggs can be irradiated with the intention of significantly reducing populations of *Salmonella* spp. Reduction levels depend upon radiation dose, initial level of pathogen contamination, or other treatment-related conditions. The effect of irradiation on microbial populations suggests that it could also be used to decrease pathogensin a product with the intention of allowing microbiologically safe storage at ambient temperature for a specific time. However, the foods currently allowed to be irradiated are very limited. One could envision that in the future a food such as pumpkin pie could be irradiated to allow for safe ambient temperature storage. As with other technologies, organoleptic changes in the food would need to be considered. More important, the effectiveness of the technology will need to be validated for the specific application.

OTHER TECHNOLOGIES

Some of the technologies present greater limitations or are at a development stage that requires extensive further scientific research before they can be commercially used. For instance, high voltage arc discharge (application of discharge voltages through an electrode gap below an aqueous medium) causes electrolysis and highly reactive chemicals. Although microorganisms are inactivated, improved designs need to be developed before consideration for use in food preservation. Likewise, oscillating magnetic fields have been explored for their potential to inactivate microorganisms; however, the results are inconsistent. Data on inactivation of food microorganisms by ultrasound (energy generated by sound waves of 20,000 or more vibrations per second) are scarce, and limitations include the inclusion of particulates and other interfering substances.

Ultraviolet (UV) light is a promising technique, especially in treating water and fruit juices. A 4-log bacterial reduction was obtained for a variety of microorganisms when 400 J/m^2 was applied. Apple cider inoculated with *E. coli* O157:H7 treated in that manner achieved a 5-log reduction (Worobo 2000). Critical factors include the transmissivity, the geometric configuration of the reactor, the power, wavelength, and physical arrangement of the UV source, the product flow profile, and radiation path length.

Index

H

I

K

L

M

O

P

R

S

T

U

V

W